Year 1B

A Guide to Teaching for Mastery

Series Editor: Tony Staneff
Lead author: Josh Lury

Contents

Introduction to the author team	4
What is *Power Maths*?	5
What's different in the new edition?	6
Your *Power Maths* resources	7
The *Power Maths* teaching model	10
The *Power Maths* lesson sequence	12
Using the *Power Maths* Teacher Guide	15
Power Maths Year 6, yearly overview	16
Mindset: an introduction	21
The *Power Maths* characters	22
Mathematical language	23
The role of talk and discussion	24
Assessment strategies	25
Keeping the class together	27
Same-day intervention	28
The role of practice	29
Structures and representations	30
Variation helps visualisation	31
Practical aspects of *Power Maths*	32
Working with children below age-related expectation	34
Providing extra depth and challenge with *Power Maths*	36
Using *Power Maths* with mixed age classes	38
List of practical resources	39
Getting started with *Power Maths*	41

Unit 6 – Numbers to 20 — 44

Count to 20	46
Understand 10	50
11, 12 and 13	54
14, 15 and 16	58
17, 18 and 19	62
Understand 20	66
One more and one less	70
The number line to 20	74
Label number lines	78
Estimate on a number line	82
Compare numbers to 20	86
Order numbers to 20	90
End of unit check	94

Unit 7 – Addition and subtraction within 20 — 96

Add by counting on within 20	98
Add ones using number bonds	102
Find and make number bonds to 20	106
Doubles	110
Near doubles	114

Subtract ones using number bonds	118
Subtraction – count back	122
Subtraction – find the difference	126
Related facts – fact families	130
Missing number problems	134
Solve word and picture problems – addition and subtraction	138
End of unit check	142

Unit 8 – Numbers to 50 — 144

Count to 50	146
Numbers to 50	150
20, 30, 40 and 50	154
Count by making groups of 10s	158
Groups of 10s and 1s	162
Partition into 10s and 1s	166
One more, one less	170
End of unit check	174

Unit 9 – Introducing length and height — 176

Compare lengths and heights	178
Measure length (non-standard units of measure)	182
Measure length (using a ruler)	186
Solve word problems – length	190
End of unit check	194

Unit 10 – Introducing mass and capacity — 196

Heavier and lighter	198
Measure mass	202
Compare mass	206
Full and empty	210
Measure capacity	214
Compare capacity	218
Solve word problems – mass and capacity	222
End of unit check	226

Introduction to the author team

Power Maths arises from the work of maths mastery experts who are committed to proving that, given the right mastery mindset and approach, **everyone can do maths**. Based on robust research and best practice from around the world, *Power Maths* was developed in partnership with a group of UK teachers to make sure that it not only meets our children's wide-ranging needs but also aligns with the National Curriculum in England.

Power Maths – White Rose Maths edition

This edition of *Power Maths* has been developed and updated by:

Tony Staneff, Series Editor and Author

Vice Principal at Trinity Academy, Halifax, Tony also leads a team of mastery experts who help schools across the UK to develop teaching for mastery via nationally recognised CPD courses, problem-solving and reasoning resources, schemes of work, assessment materials and other tools.

Josh Lury, Lead Author

Josh is a specialist maths teacher, author and maths consultant with a passion for innovative and effective maths education.

The first edition of *Power Maths* was developed by a team of experienced authors, including:

- **Tony Staneff and Josh Lury**
- **Trinity Academy Halifax** (Michael Gosling CEO, Emily Fox, Kate Henshall, Rebecca Holland, Stephanie Kirk, Stephen Monaghan and Rachel Webster)
- **David Board, Belle Cottingham, Jonathan East, Tim Handley, Derek Huby, Neil Jarrett, Stephen Monaghan, Beth Smith, Tim Weal, Paul Wrangles** – skilled maths teachers and mastery experts
- **Cherri Moseley** – a maths author, former teacher and professional development provider
- **Professors Liu Jian and Zhang Dan**, Series Consultants and authors, and their team of mastery expert authors: **Wei Huinv, Huang Lihua, Zhu Dejiang, Zhu Yuhong, Hou Huiying, Yin Lili, Zhang Jing, Zhou Da and Liu Qimeng**

    Used by over 20 million children, Professor Liu Jian's textbook programme is one of the most popular in China. He and his author team are highly experienced in intelligent practice and in embedding key maths concepts using a C-P-A approach.

- **A group of 15 teachers and maths co-ordinators**

    We consulted our teacher group throughout the development of *Power Maths* to ensure we are meeting their real needs in the classroom.

What is *Power Maths*?

Created especially for UK primary schools, and aligned with the new National Curriculum, *Power Maths* is a whole-class, textbook-based mastery resource that empowers every child to understand and succeed. *Power Maths* rejects the notion that some people simply 'can't do' maths. Instead, it develops growth mindsets and encourages hard work, practice and a willingness to see mistakes as learning tools.

Best practice consistently shows that mastery of small, cumulative steps builds a solid foundation of deep mathematical understanding. *Power Maths* combines interactive teaching tools, high-quality textbooks and continuing professional development (CPD) to help you equip children with a deep and long-lasting understanding. Based on extensive evidence, and developed in partnership with practising teachers, *Power Maths* ensures that it meets the needs of children in the UK.

Power Maths and Mastery

Power Maths makes mastery practical and achievable by providing the structures, pathways, content, tools and support you need to make it happen in your classroom.

To develop mastery in maths, children must be enabled to acquire a deep understanding of maths concepts, structures and procedures, step by step. Complex mathematical concepts are built on simpler conceptual components and when children understand every step in the learning sequence, maths becomes transparent and makes logical sense. Interactive lessons establish deep understanding in small steps, as well as effortless fluency in key facts such as tables and number bonds. The whole class works on the same content and no child is left behind.

Power Maths

- Builds every concept in small, progressive steps
- Is built with interactive, whole-class teaching in mind
- Provides the tools you need to develop growth mindsets
- Helps you check understanding and ensure that every child is keeping up
- Establishes core elements such as intelligent practice and reflection

The *Power Maths* approach

Everyone can!
Founded on the conviction that every child can achieve, *Power Maths* enables children to build number fluency, confidence and understanding, step by step.

Child-centred learning
Children master concepts one step at a time in lessons that embrace a concrete-pictorial-abstract (C-P-A) approach, avoid overload, build on prior learning and help them see patterns and connections. Same-day intervention ensures sustained progress.

Continuing professional development
Embedded teacher support and development offer every teacher the opportunity to continually improve their subject knowledge and manage whole-class teaching for mastery.

Whole-class teaching
An interactive, whole-class teaching model encourages thinking and precise mathematical language and allows children to deepen their understanding as far as they can.

What's different in the new edition?

If you have previously used the first editions of *Power Maths*, you might be interested to know how this edition is different. All of the improvements described below are based on feedback from *Power Maths* customers.

Changes to units and the progression

- The order of units has been slightly adjusted, creating closer alignment between adjacent year groups, which will be useful for mixed age teaching.
- The flow of lessons has been improved within units to optimise the pace of the progression and build in more recap where needed. For key topics, the sequence of lessons gives more opportunities to build up a solid base of understanding. Other units have fewer lessons than before, where appropriate, making it possible to fit in all the content.
- Overall, the lessons put more focus on the most essential content for that year, with less time given to non-statutory content.
- The progression of lessons matches the steps in the new White Rose Maths schemes of learning.

Lesson resources

- There is a Quick recap for each lesson in the Teacher Guide, which offers an alternative lesson starter to the Power Up for cases where you feel it would be more beneficial to surface prerequisite learning than general number fluency.
- In the **Discover** and **Share** sections there is now more of a progression from 1 a) to 1 b). Whereas before, 1 b) was mainly designed as a separate question, now 1 a) leads directly into 1 b). This means that there is an improved whole-class flow, and also an opportunity to focus on the logic and skills in more detail. As a teacher, you will be using 1 a) to lead the class into the thinking, then 1 b) to mould that thinking into the core new learning of the lesson.
- In the **Share** section, for KS1 in particular, the number of different models and representations has been reduced, to support the clarity of thinking prompted by the flow from 1 a) into 1 b).
- More fluency questions have been built into the guided and independent practice.
- Pupil pages are as easy as possible for children to access independently. The pages are less full where this supports greater focus on key ideas and instructions. Also, more freedom is offered around answer format, with fewer boxes scaffolding children's responses; squared paper backgrounds are used in the Practice Books where appropriate. Artwork has also been revisited to ensure the highest standards of accessibility.

New components

480 Individual Practice Games are available in *ActiveLearn* for practising key facts and skills in Years 1 to 6. These are designed in an arcade style, to feel like fun games that children would choose to play outside school. They can be accessed via the Pupil World for homework or additional practice in school – and children can earn rewards. There are Support, Core and Extend levels to allocate, with Activity Reporting available for the teacher. There is a Quick Guide on *ActiveLearn* and you can use the Help area for support in setting up child accounts.

There is also a new set of lesson video resources on the Professional Development tile, designed for in-school training in 10- to 20-minute bursts. For each part of the *Power Maths* lesson sequence, there is a slide deck with embedded video, which will facilitate discussions about how you can take your *Power Maths* teaching to the next level.

Your *Power Maths* resources

Pupil Textbooks

Discover, **Share** and **Think together** sections promote discussion and introduce mathematical ideas logically, so that children understand more easily.

Using a Concrete-Pictorial-Abstract approach, clear mathematical models help children to make connections and grasp concepts.

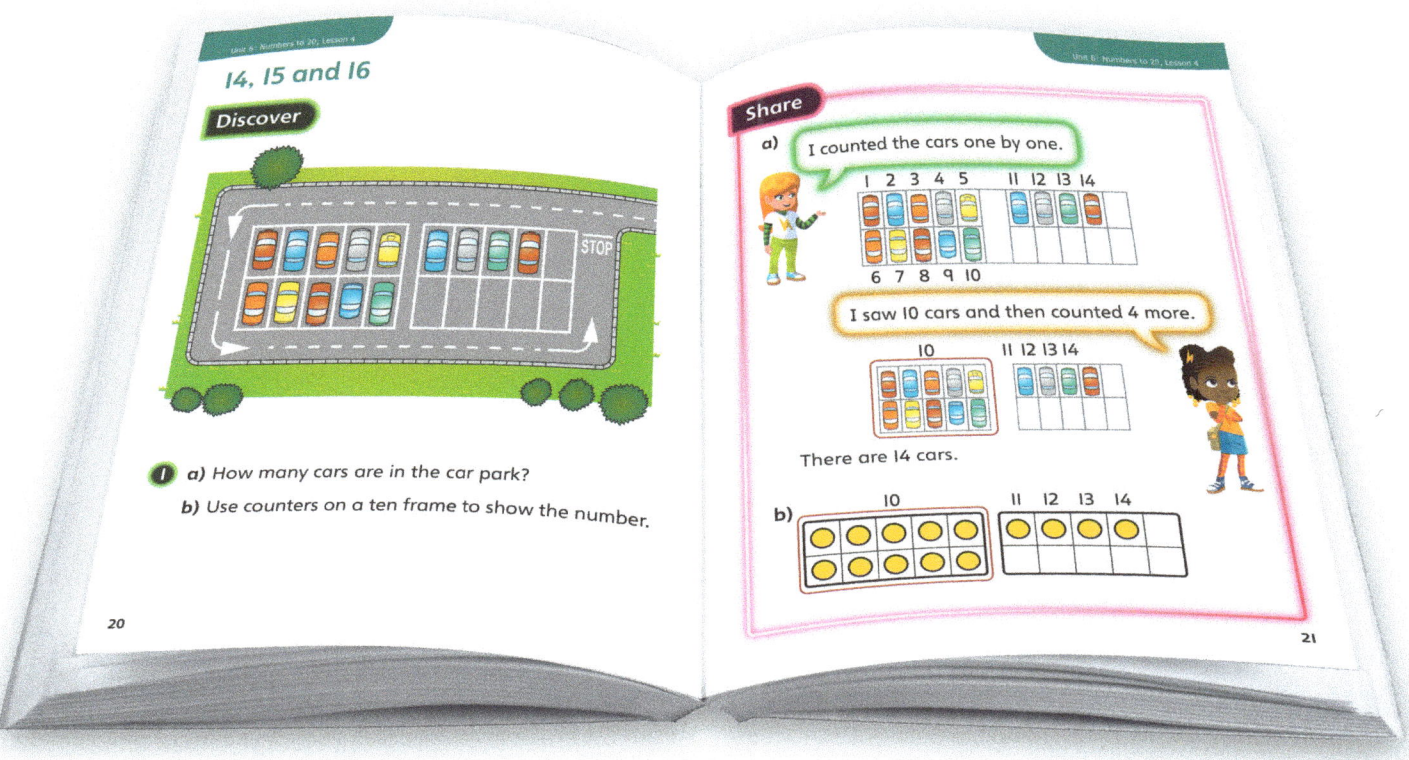

Appealing scenarios stimulate curiosity, helping children to identify the maths problem and discover patterns and relationships for themselves.

Friendly, supportive characters help children develop a growth mindset by prompting them to think, reason and reflect.

To help you teach for mastery, *Power Maths* comprises a variety of high-quality resources.

The coherent *Power Maths* lesson structure carries through into the vibrant, high-quality textbooks. Setting out the core learning objectives for each class, the lesson structure follows a carefully mapped journey through the curriculum and supports children on their journey to deeper understanding.

Pupil Practice Books

The Practice Books offer just the right amount of intelligent practice for children to complete independently in the final section of each lesson.

Practice questions are finely tuned to move children forward in their thinking and to reveal misconceptions.

The **practice questions** are for everyone – each question varies one small element to move children on in their thinking.

Calculations are connected so that children think about the underlying concept.

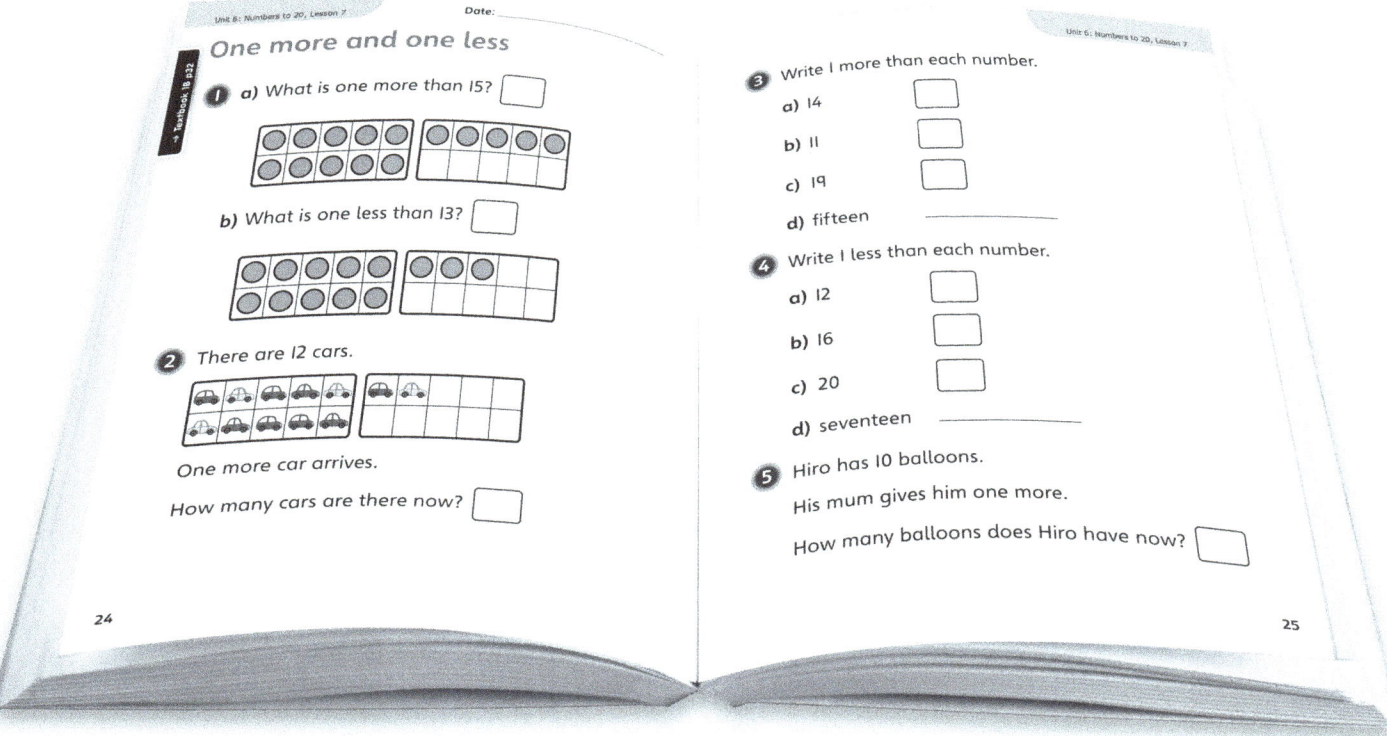

Challenge questions allow children to delve deeper into a concept.

Think differently questions encourage children to use reasoning as well as their mathematical knowledge to reach a solution.

Reflect questions reveal the depth of each child's understanding before they move on.

The *Power Maths* characters support and encourage children to think and work in different ways.

Online subscription

The online subscription will give you access to additional resources and answers from the Textbook and Practice Book.

eTextbooks

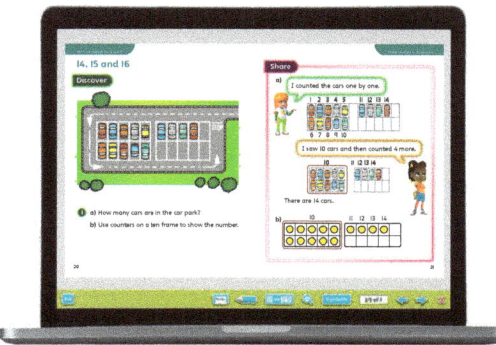

Digital versions of *Power Maths* Textbooks allow class groups to share and discuss questions, solutions and strategies. They allow you to project key structures and representations at the front of the class, to ensure all children are focusing on the same concept.

Teaching tools

Here you will find interactive versions of key *Power Maths* structures and representations.

Power Ups

Use this series of daily activities to promote and check number fluency.

Online versions of Teacher Guide pages

PDF pages give support at both unit and lesson levels. You will also find help with key strategies and templates for tracking progress.

Unit videos

Watch the professional development videos at the start of each unit to help you teach with confidence. The videos explore common misconceptions in the unit, and include intervention suggestions as well as suggestions on what to look out for when assessing mastery in your students.

End of unit Strengthen and Deepen materials

The Strengthen activity at the end of every unit addresses a key misconception and can be used to support children who need it. The Deepen activities are designed to be low ceiling/high threshold and will challenge those children who can understand more deeply. These resources will help you ensure that every child understands and will help you keep the class moving forward together. These printable activities provide an optional resource bank for use after the assessment stage.

Individual Practice Games

These enjoyable games can be used at home or at school to embed key number skills.

Professional Development videos and slides

These slides and videos of *Power Maths* lessons can be used for ongoing training in short bursts or to support new staff.

The *Power Maths* teaching model

At the heart of *Power Maths* is a clearly structured teaching and learning process that helps you make certain that every child masters each maths concept securely and deeply. For each year group, the curriculum is broken down into core concepts, taught in units. A unit divides into smaller learning steps – lessons. Step by step, strong foundations of cumulative knowledge and understanding are built.

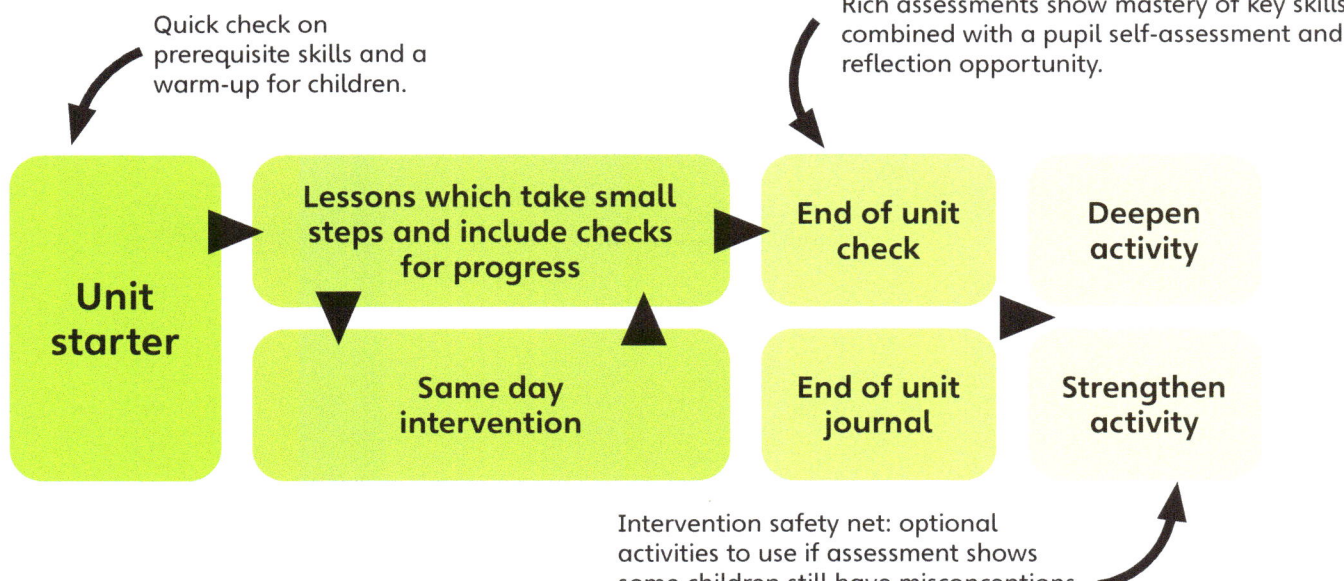

Unit starter

Each unit begins with a unit starter, which introduces the learning context along with key mathematical vocabulary and structures and representations.

- The Textbooks include a check on readiness and a warm-up task for children to complete.
- Your Teacher Guide gives support right from the start on important structures and representations, mathematical language, common misconceptions and intervention strategies.
- Unit-specific videos develop your subject knowledge and insights so you feel confident and fully equipped to teach each new unit. These are available via the online subscription.

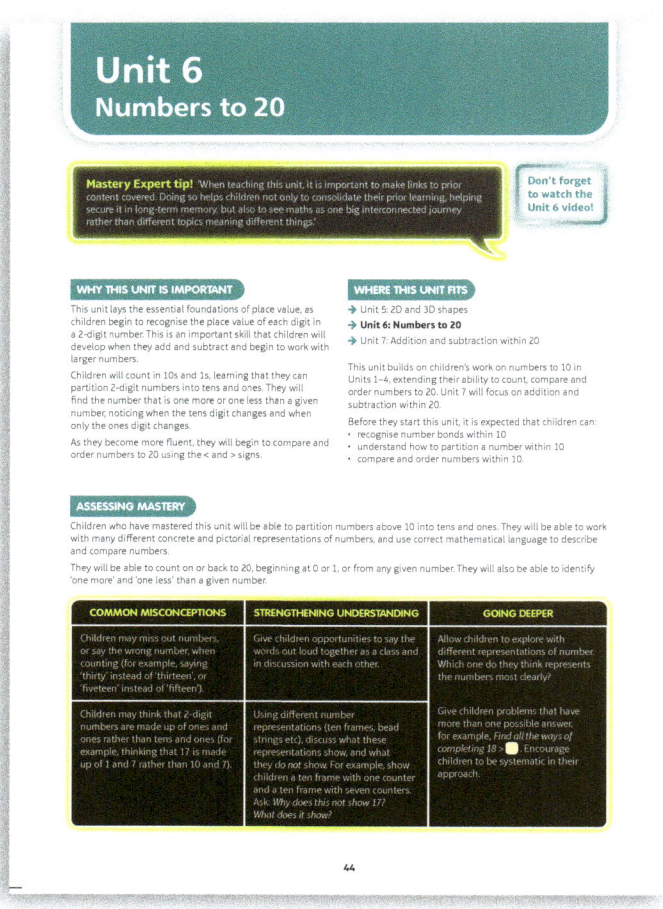

Lesson

Once a unit has been introduced, it is time to start teaching the series of lessons.

- Each lesson is scaffolded with Textbook and Practice Book activities and begins with a Power Up activity (available via online subscription) or the Quick recap activity in the Teacher Guide (see page 15).
- *Power Maths* identifies lesson by lesson what concepts are to be taught.
- Your Teacher Guide offers lots of support for you to get the most from every child in every lesson. As well as highlighting key points, tricky areas and how to handle them, you will also find question prompts to check on understanding and clarification on why particular activities and questions are used.

Same-day intervention

Same-day interventions are vital in order to keep the class progressing together. This can be during the lesson as well as afterwards (see page 27). Therefore, *Power Maths* provides plenty of support throughout the journey.

- Intervention is focused on keeping up now, not catching up later, so interventions should happen as soon as they are needed.
- Practice section questions are designed to bring misconceptions to the surface, allowing you to identify these easily as you circulate during independent practice time.
- Child-friendly assessment questions in the Teacher Guide help you identify easily which children need to strengthen their understanding.

End of unit check and journal

For each unit, the End of unit check in the Textbook lets you see which children have mastered the key concepts, which children have not and where their misconceptions lie. The Practice Books also include an End of unit journal in which children can reflect on what they have learned. Each unit also offers Strengthen and Deepen activities, available via the online subscription.

> The Teacher Guide offers different ways of managing the End of unit assessments as well as giving support with handling misconceptions.

> The End of unit check presents multiple-choice questions. Children think about their answer, decide on a solution and explain their choice.

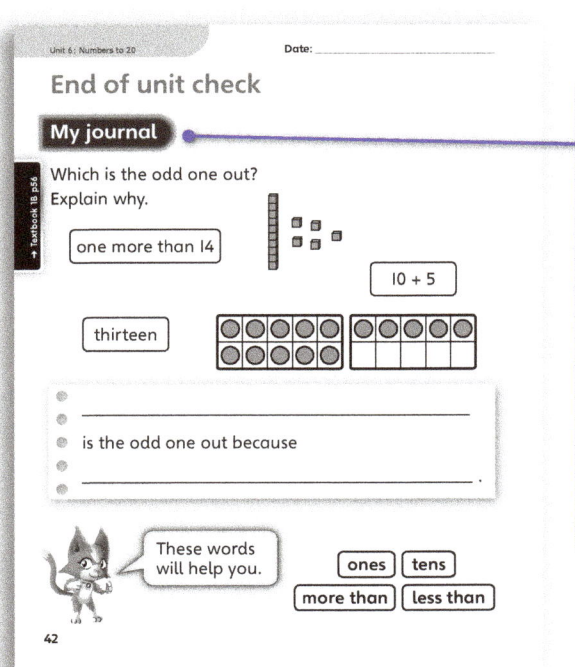

> The End of unit journal is an opportunity for children to test out their learning and reflect on how they feel about it. Tackling the 'journal' problem reveals whether a child understands the concept deeply enough to move on to the next unit.

> Your teacher will ask you these questions.

The *Power Maths* lesson sequence

At the heart of *Power Maths* is a unique lesson sequence designed to empower children to understand core concepts and grow in confidence. Embracing the National Centre for Excellence in the Teaching of Mathematics' (NCETM's) definition of mastery, the sequence guides and shapes every *Power Maths* lesson you teach.

Flexibility is built into the *Power Maths* programme so there is no one-to-one mapping of lessons and concepts and you can pace your teaching according to your class. While some children will need to spend longer on a particular concept (through interventions or additional lessons), others will reach deeper levels of understanding. However, it is important that the class moves forward together through the termly schedules.

Power Up — 5 minutes

- Each lesson begins with a Power Up activity (available via the online subscription) which supports fluency in key number facts.

- The whole-class approach depends on fluency, so the Power Up is a powerful and essential activity.

- The Quick recap is an alternative starter, for when you think some or all children would benefit more from revisiting pre-requisite work (see page 15).

TOP TIP
If the class is struggling with the task, revisit it later and check understanding.

- Power Ups reinforce the two key things that are essential for success: times-tables and number bonds.

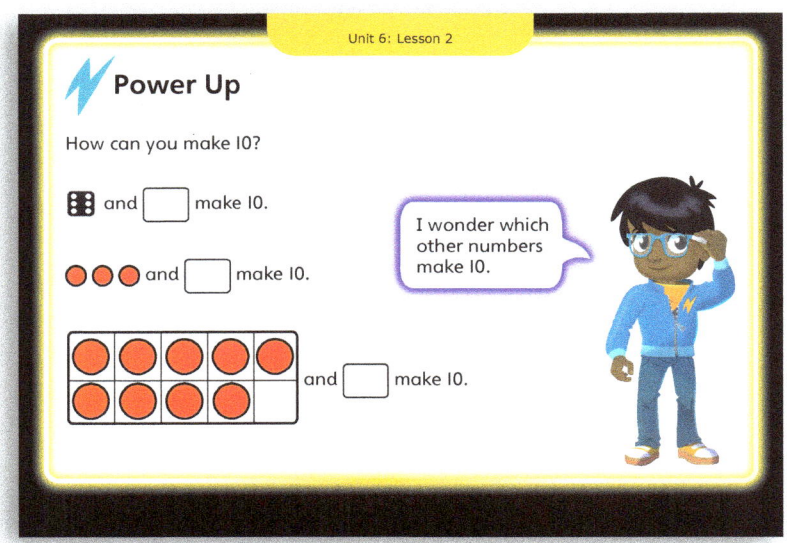

Discover — 10 minutes

- A practical, real-life problem arouses curiosity. Children find the maths through story telling.

TOP TIP
Discover works best when run at tables, in pairs with concrete objects.

- Question 1 a) tackles the key concept and question 1 b) digs a little deeper. Children have time to explore, play and discuss possible strategies.

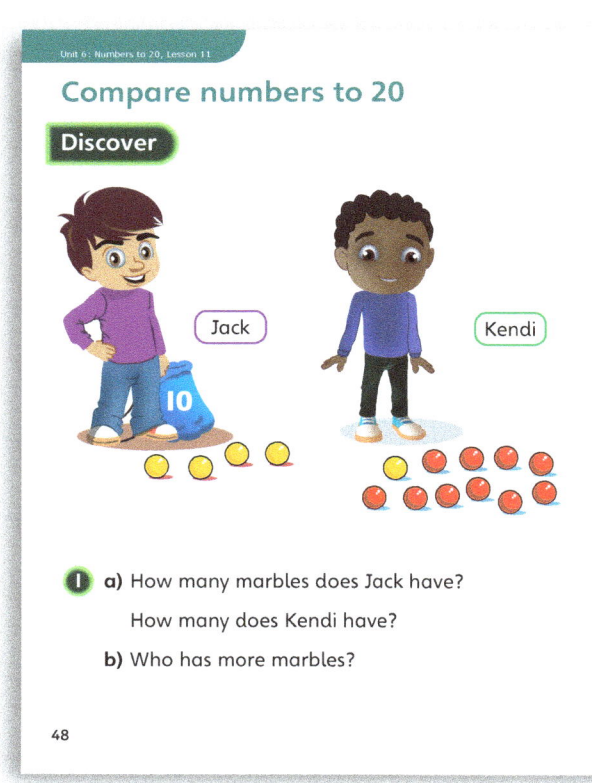

Share — 10 minutes

Teacher-led, this interactive section follows the **Discover** activity and highlights the variety of methods that can be used to solve a single problem.

TOP TIP
You can use the carpet area if you have this. Pairs sharing a textbook is a great format for **Share**!

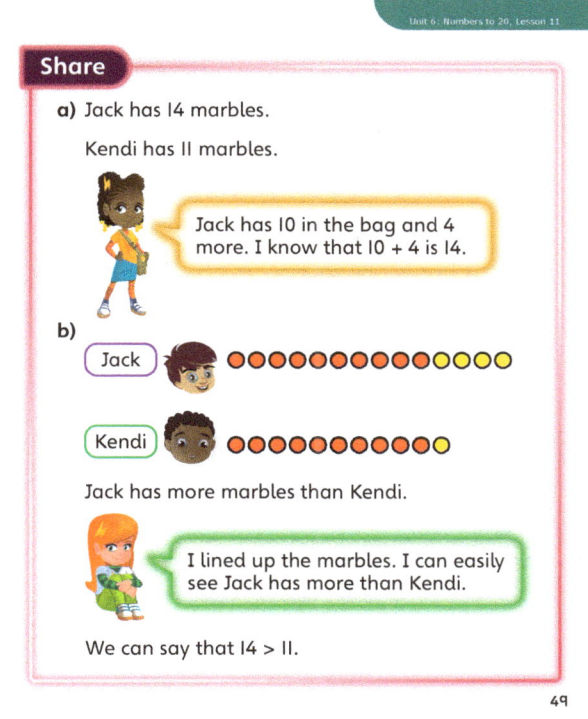

Your Teacher Guide gives target questions for children. The online toolkit provides interactive structures and representations to link concrete and pictorial to abstract concepts.

Bring children to the front to share and celebrate their solutions and strategies.

Think together

10 minutes

Children work in groups on the carpet or at tables, using their textbooks or eBooks.

TOP TIP
Make sure children have mini whiteboards or pads to write on if they are not at their tables.

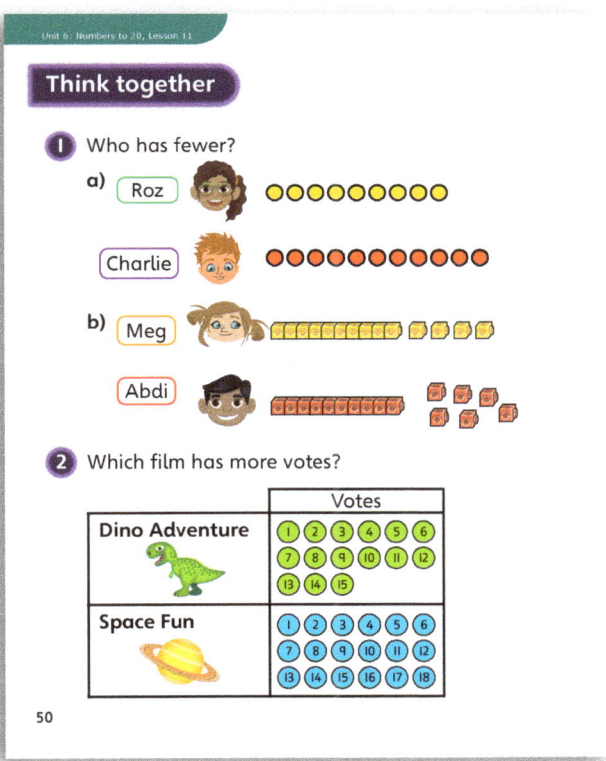

Using the Teacher Guide, model question ❶ for your class.

Question ❷ is less structured. Children will need to think together in their groups, then discuss their methods and solutions as a class.

Question ❸ – the openness of the **Challenge** question helps to check depth of understanding.

Practice ⏱ 15 minutes

Using their Practice Books, children work independently while you circulate and check on progress.

Questions follow small steps of progression to deepen learning.

TOP TIP
Some children could work separately with a teacher or assistant.

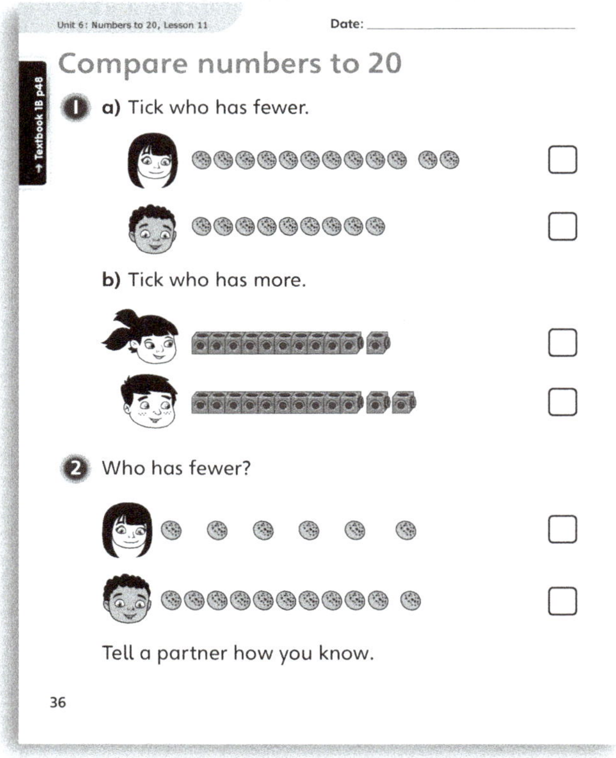

Unit 6: Numbers to 20, Lesson 11   Date: _____

Compare numbers to 20

① a) Tick who has fewer.

b) Tick who has more.

② Who has fewer?

Tell a partner how you know.

36

Are some children struggling? If so, work with them as a group, using mathematical structures and representations to support understanding as necessary.

There are no set routines: for real understanding, children need to think about the problem in different ways.

Reflect ⏱ 5 minutes

'Spot the mistake' questions are great for checking misconceptions.

The **Reflect** section is your opportunity to check how deeply children understand the target concept.

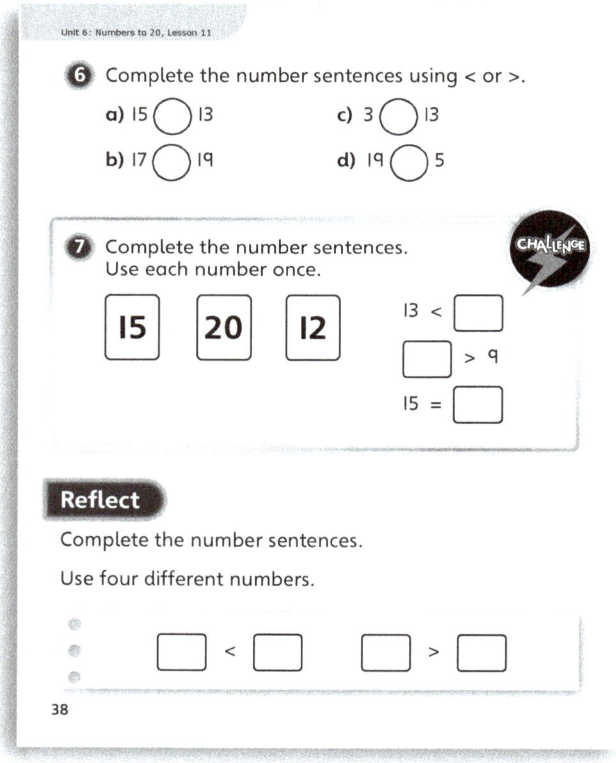

Unit 6: Numbers to 20, Lesson 11

⑥ Complete the number sentences using < or >.

a) 15 ◯ 13   c) 3 ◯ 13
b) 17 ◯ 19   d) 19 ◯ 5

⑦ Complete the number sentences. Use each number once.   **CHALLENGE**

| 15 | 20 | 12 |

13 < ☐
☐ > 9
15 = ☐

Reflect

Complete the number sentences.
Use four different numbers.

☐ < ☐   ☐ > ☐

38

The Practice Books use various approaches to check that children have fully understood each concept.

Looking like they understand is not enough! It is essential that children can show they have grasped the concept.

14

Using the *Power Maths* Teacher Guide

Think of your Teacher Guides as *Power Maths* handbooks that will guide, support and inspire your day-to-day teaching. Clear and concise, and illustrated with helpful examples, your Teacher Guides will help you make the best possible use of every individual lesson. They also provide wrap-around professional development, enhancing your own subject knowledge and helping you to grow in confidence about moving your children forward together.

- There is a Teacher Guide per year group for every term, with unit and lesson level guidance and support.
- Never feel stuck! You will find ideas for introducing every unit and lesson and questions to encourage teacher reflection before and after each lesson.
- Tips and advice on key elements such as C-P-A approaches, misconceptions, language, modelling growth mindsets and same day intervention.
- Annotations for every Textbook and Practice Book page, providing prompts for key questions to ask to expose understanding and explanations as to why key questions have been chosen.

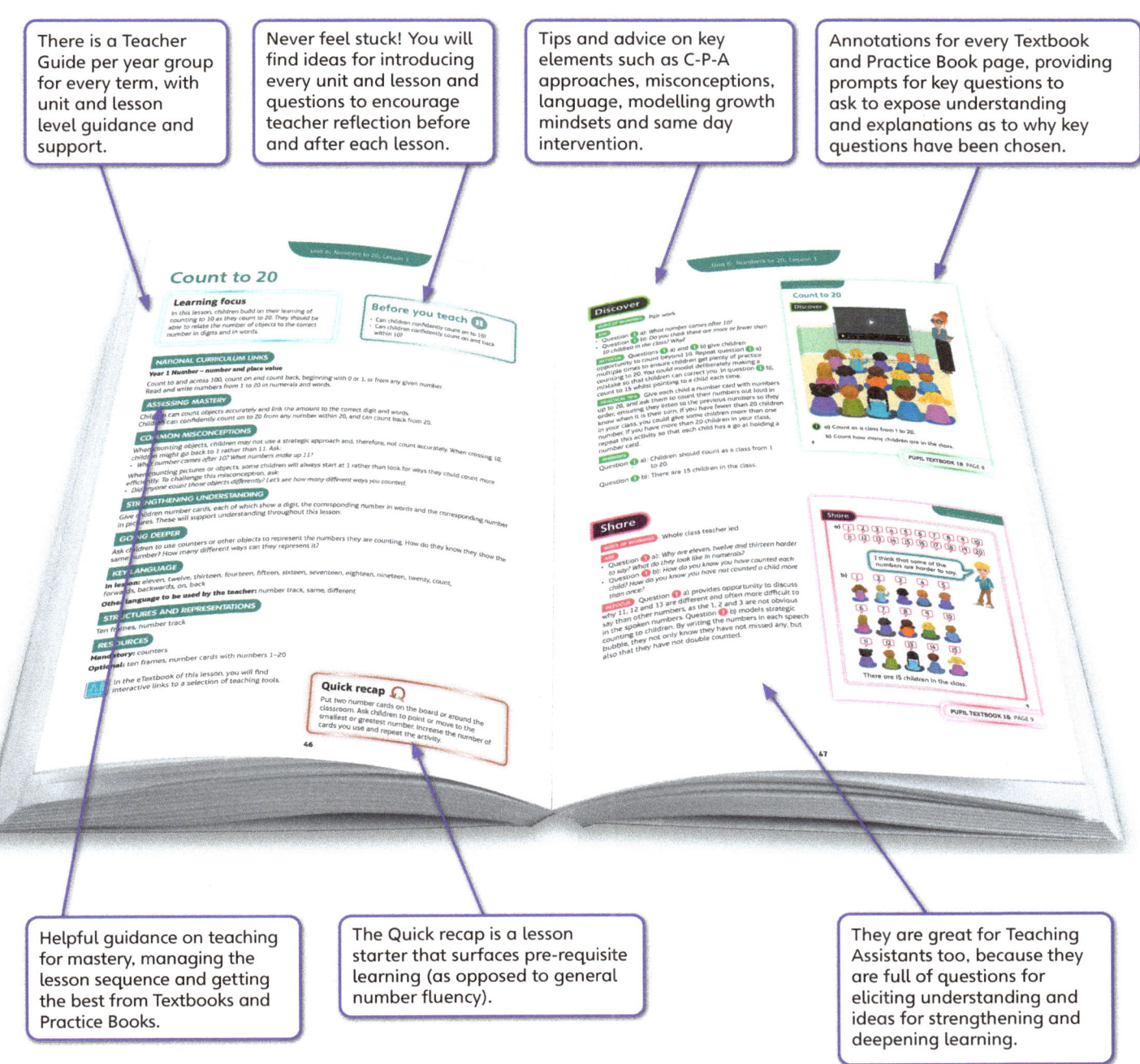

- Helpful guidance on teaching for mastery, managing the lesson sequence and getting the best from Textbooks and Practice Books.
- The Quick recap is a lesson starter that surfaces pre-requisite learning (as opposed to general number fluency).
- They are great for Teaching Assistants too, because they are full of questions for eliciting understanding and ideas for strengthening and deepening learning.

At the end of each unit, your Teacher Guide helps you identify who has fully grasped the concept, who has not and how to move every child forward. This is covered later in the Assessment strategies section.

Power Maths Year 1, yearly overview

Textbook	Strand	Unit		Number of lessons
Textbook A / Practice Book A (Term 1)	Number – number and place value	1	Numbers to 10	14
	Number – addition and subtraction	2	Part-whole within 10	7
	Number – addition and subtraction	3	Addition awithin 10	4
	Number – addition and subtraction	4	Subtraction within 10	8
	Geometry – properties of shape	5	2D and 3D shapes	5
Textbook B / Practice Book B (Term 2)	Number – number and place value	6	Numbers to 20	12
	Number – addition and subtraction	7	Addition and subtraction within 20	11
	Number – number and place value	8	Numbers to 50	7
	Measurement	9	Introducing length and height	4
	Measurement	10	Introducing weight and volume	7
Textbook C / Practice Book C (Term 3)	Number – multiplication and division	11	Multiplication and division	9
	Number – fractions	12	Halves and quarters	4
	Geometry – position and direction	13	Position and direction	5
	Number – number and place value	14	Numbers to 100	6
	Measurement	15	Money	3
	Measurement	16	Time	5

Power Maths Year 1, Textbook 1B (Term 2) overview

Strand	Unit	Unit title	Lesson number	Lesson title	NC Objective 1	NC Objective 2
Number – number and place value	6	Numbers to 20	1	Count to 20	Count to and across 100, forwards and backwards, beginning with 0 or 1, or from any given number (to 20)	Read and write numbers from 1 to 20 in numerals and words.
Number – number and place value	6	Numbers to 20	2	Understand 10	Count to and across 100, forwards and backwards, beginning with 0 or 1, or from any given number (to 20)	
Number – number and place value	6	Numbers to 20	3	11, 12 and 13	Identify and represent numbers using objects and pictorial representations including the number line, and use the language of: equal to, more than, less than (fewer), most, least	Recognise the place value of each digit in a two-digit number (tens, ones) (year 2)
Number – number and place value	6	Numbers to 20	4	14, 15 and 16	Identify and represent numbers using objects and pictorial representations including the number line, and use the language of: equal to, more than, less than (fewer), most, least	Recognise the place value of each digit in a two-digit number (tens, ones) (year 2)
Number – number and place value	6	Numbers to 20	5	17, 18 and 19	Identify and represent numbers using objects and pictorial representations including the number line, and use the language of: equal to, more than, less than (fewer), most, least	Recognise the place value of each digit in a two-digit number (tens, ones) (year 2)
Number – number and place value	6	Numbers to 20	6	Understand 20	Identify and represent numbers using objects and pictorial representations including the number line, and use the language of: equal to, more than, less than (fewer), most, least	Read and write numbers from 1 to 20 in numerals and words
Number – number and place value	6	Numbers to 20	7	One more and one less	Identify and represent numbers using objects and pictorial representations including the number line, and use the language of: equal to, more than, less than (fewer), most, least	Given a number, identify one more and one less
Number – number and place value	6	Numbers to 20	8	The number line to 20	Identify and represent numbers using objects and pictorial representations including the number line, and use the language of: equal to, more than, less than (fewer), most, least	
Number – number and place value	6	Numbers to 20	9	Label number lines	Identify and represent numbers using objects and pictorial representations including the number line, and use the language of: equal to, more than, less than (fewer), most, least	
Number – number and place value	6	Numbers to 20	10	Estimate on a number line	Identify and represent numbers using objects and pictorial representations including the number line, and use the language of: equal to, more than, less than (fewer), most, least	
Number – number and place value	6	Numbers to 20	11	Compare numbers to 20	Identify and represent numbers using objects and pictorial representations including the number line, and use the language of: equal to, more than, less than (fewer), most, least	
Number – number and place value	6	Numbers to 20	12	Order numbers to 20	Count to and across 100, forwards and backwards, beginning with 0 or 1, or from any given number (to 20)	Read and write numbers from 1 to 20 in numerals and words

Strand	Unit	Unit title	Lesson number	Lesson title	NC Objective 1	NC Objective 2
Number – addition and subtraction	7	Addition and subtraction within 20	1	Add by counting on within 20	Add and subtract one-digit and two-digit numbers to 20, including zero	
Number – addition and subtraction	7	Addition and subtraction within 20	2	Add ones using number bonds	Represent and use number bonds and related subtraction facts within 20 (within 10)	Add and subtract one-digit and two-digit numbers to 20, including zero
Number – addition and subtraction	7	Addition and subtraction within 20	3	Find and make number bonds to 20	Represent and use number bonds and related subtraction facts within 20 (within 10)	
Number – addition and subtraction	7	Addition and subtraction within 20	4	Doubles	Represent and use number bonds and related subtraction facts within 20 (within 10)	
Number – addition and subtraction	7	Addition and subtraction within 20	5	Near doubles	Represent and use number bonds and related subtraction facts within 20 (within 10)	
Number – addition and subtraction	7	Addition and subtraction within 20	6	Subtract ones using number bonds	Add and subtract one-digit and two-digit numbers to 20, including zero	Represent and use number bonds and related subtraction facts within 20 (within 10)
Number – addition and subtraction	7	Addition and subtraction within 20	7	Subtraction – count back	Solve one-step problems that involve addition and subtraction, using concrete objects and pictorial representations, and missing number problems such as 7 = – 9	Add and subtract one-digit and two-digit numbers to 20, including zero
Number – addition and subtraction	7	Addition and subtraction within 20	8	Subtraction - find the difference	Solve one-step problems that involve addition and subtraction, using concrete objects and pictorial representations, and missing number problems such as 7 = – 9	
Number – addition and subtraction	7	Addition and subtraction within 20	9	Related facts – fact families	Represent and use number bonds and related subtraction facts within 20 (within 10)	
Number – addition and subtraction	7	Addition and subtraction within 20	10	Missing number problems	Solve one-step problems that involve addition and subtraction, using concrete objects and pictorial representations, and missing number problems such as 7 = – 9	
Number – addition and subtraction	7	Addition and subtraction within 20	11	Solve word and picture problems – addition and subtraction	Solve one-step problems that involve addition and subtraction, using concrete objects and pictorial representations, and missing number problems such as 7 = – 9	
Number – number and place value	8	Numbers to 50	1	Count to 50	Count to and across 100, forwards and backwards, beginning with 0 or 1, or from any given number	Count, read and write numbers to 100 in numerals; count in multiples of twos, fives and tens
Number – number and place value	8	Numbers to 50	2	Numbers to 50	Count to and across 100, forwards and backwards, beginning with 0 or 1, or from any given number	Count, read and write numbers to 100 in numerals; count in multiples of twos, fives and tens
Number – number and place value	8	Numbers to 50	3	20, 30, 40 and 50	Identify and represent numbers using objects and pictorial representations including the number line, and use the language of: equal to, more than, less than (fewer), most, least	Recognise the place value of each digit in a two-digit number (tens, ones) (Year 2)
Number – number and place value	8	Numbers to 50	4	Count by making groups of 10s	Identify and represent numbers using objects and pictorial representations including the number line, and use the language of: equal to, more than, less than (fewer), most, least	

Strand	Unit	Unit title	Lesson number	Lesson title	NC Objective 1	NC Objective 2
Number – number and place value	8	Numbers to 50	5	Groups of 10s and 1s	Identify and represent numbers using objects and pictorial representations including the number line, and use the language of: equal to, more than, less than (fewer), most, least	
Number – number and place value	8	Numbers to 50	6	Partition into 10s and 1s	Identify and represent numbers using objects and pictorial representations including the number line, and use the language of: equal to, more than, less than (fewer), most, least	
Number – number and place value	8	Numbers to 50	7	One more, one less	Given a number, identify one more and one less	
Measurement	9	Introducing length and height	1	Compare lengths and heights	Compare, describe and solve practical problems for: lengths and heights [for example, long/short, longer/shorter, tall/short, double/half]	
Measurement	9	Introducing length and height	2	Measure length (non-standard units of measure)	Measure and begin to record the following: lengths and heights	
Measurement	9	Introducing length and height	3	Measure length (using a ruler)	Measure and begin to record the following: lengths and heights	
Measurement	9	Introducing length and height	4	Solve word problems – length	Compare, describe and solve practical problems for: lengths and heights [for example, long/short, longer/shorter, tall/short, double/half]	
Measurement	10	Introducing mass and capacity	1	Heavier and lighter	Compare, describe and solve practical problems for: mass/weight [for example, heavy/light, heavier than, lighter than]	
Measurement	10	Introducing mass and capacity	2	Measure mass	Measure and begin to record the following: mass/weight	
Measurement	10	Introducing mass and capacity	3	Compare mass	Compare, describe and solve practical problems for: mass/weight [for example, heavy/light, heavier than, lighter than]	
Measurement	10	Introducing mass and capacity	4	Full and empty	Compare, describe and solve practical problems for: capacity and volume [for example, full/empty, more than, less than, half, half full, quarter]	Measure and begin to record the following: capacity and volume
Measurement	10	Introducing mass and capacity	5	Measure capacity	Measure and begin to record the following: capacity and volume	
Measurement	10	Introducing mass and capacity	6	Compare capacity	Compare, describe and solve practical problems for: capacity and volume [for example, full/empty, more than, less than, half, half full, quarter]	
Measurement	10	Introducing mass and capacity	7	Solve word problems – mass and capacity	Compare, describe and solve practical problems for: capacity and volume [for example, full/empty, more than, less than, half, half full, quarter]	

Mindset: an introduction

Global research and best practice deliver the same message: learning is greatly affected by what learners perceive they can or cannot do. What is more, it is also shaped by what their parents, carers and teachers perceive they can do. Mindset – the thinking that determines our beliefs and behaviours – therefore has a fundamental impact on teaching and learning.

Everyone can!

Power Maths and mastery methods focus on the distinction between 'fixed' and 'growth' mindsets (Dweck, 2007).[1] Those with a fixed mindset believe that their basic qualities (for example, intelligence, talent and ability to learn) are pre-wired or fixed: 'If you have a talent for maths, you will succeed at it. If not, too bad!' By contrast, those with a growth mindset believe that hard work, effort and commitment drive success and that 'smart' is not something you are or are not, but something you become. In short, everyone can do maths!

Key mindset strategies

A growth mindset needs to be actively nurtured and developed. *Power Maths* offers some key strategies for fostering healthy growth mindsets in your classroom.

It is okay to get it wrong

Mistakes are valuable opportunities to re-think and understand more deeply. Learning is richer when children and teachers alike focus on spotting and sharing mistakes as well as solutions.

Praise hard work

Praise is a great motivator, and by focusing on praising effort and learning rather than success, children will be more willing to try harder, take risks and persist for longer.

Mind your language!

The language we use around learners has a profound effect on their mindsets. Make a habit of using growth phrases, such as, 'Everyone can!', 'Mistakes can help you learn' and 'Just try for a little longer'. The king of them all is one little word, 'yet'… I can't solve this…yet!' Encourage parents and carers to use the right language too.

Build in opportunities for success

The step-by-small-step approach enables children to enjoy the experience of success. In addition, avoid ability grouping and encourage every child to answer questions and explain or demonstrate their methods to others.

[1] Dweck, C (2007) *The New Psychology of Success*, Ballantine Books: New York

The *Power Maths* characters

The *Power Maths* characters model the traits of growth mindset learners and encourage resilience by prompting and questioning children as they work. Appearing frequently in the Textbooks and Practice Books, they are your allies in teaching and discussion, helping to model methods, alternatives and misconceptions, and to pose questions. They encourage and support your children, too: they are all hardworking, enthusiastic and unafraid of making and talking about mistakes.

Meet the team!

Creative Flo is open-minded and sometimes indecisive. She likes to think differently and come up with a variety of methods or ideas.

Determined Dexter is resolute, resilient and systematic. He concentrates hard, always tries his best and he'll never give up – even though he doesn't always choose the most efficient methods!

'Let's try again.'
'Mistakes are cool!'
'Have I found all of the solutions?'

'Let's try it this way…'
'Can we do it differently?'
'I've got another way of doing this!'

'I'm going to try this!'
'I know how to do that!'
'Want to share my ideas?'

Curious Ash is eager, interested and inquisitive, and he loves solving puzzles and problems. Ash asks lots of questions but sometimes gets distracted.

'What if we tried this…?'
'I wonder…'
'Is there a pattern here?'

Sparks the Cat

'Miaow!'

Brave Astrid is confident, willing to take risks and unafraid of failure. She's never scared to jump straight into a problem or question, and although she often makes simple mistakes, she's happy to talk them through with others.

Mathematical language

Traditionally, we in the UK have tended to try simplifying mathematical language to make it easier for young children to understand. By contrast, evidence and experience show that by diluting the correct language, we actually mask concepts and meanings for children. We then wonder why they are confused by new and different terminology later down the line! *Power Maths* is not afraid of 'hard' words and avoids placing any barriers between children and their understanding of mathematical concepts. As a result, we need to be deliberate, precise and thorough in building every child's understanding of the language of maths. Throughout the Teacher Guides you will find support and guidance on how to deliver this, as well as individual explanations throughout the pupil Textbooks.

Use the following key strategies to build children's mathematical vocabulary, understanding and confidence.

Precise and consistent

Everyone in the classroom should use the correct mathematical terms in full, every time. For example, refer to 'equal parts', not 'parts'. Used consistently, precise maths language will be a familiar and non-threatening part of children's everyday experience.

Full sentences

Teachers and children alike need to use full sentences to explain or respond. When children use complete sentences, it both reveals their understanding and embeds their knowledge.

Stem sentences

These important sentences help children express mathematical concepts accurately, and are used throughout the *Power Maths* books. Encourage children to repeat them frequently, whether working independently or with others. Examples of stem sentences are:

'4 is a part, 5 is a part, 9 is the whole.'

'There are …. groups. There are …. in each group.'

Key vocabulary

The unit starters highlight essential vocabulary for every lesson. In the pupil books, characters flag new terminology and the Teacher Guide lists important mathematical language for every unit and lesson. New terms are never introduced without a clear explanation.

Symbolic language

Symbols are used early on so that children quickly become familiar with them and their meaning. Often, the *Power Maths* characters will highlight the connection between language and particular symbols.

The role of talk and discussion

When children learn to talk purposefully together about maths, barriers of fear and anxiety are broken down and they grow in confidence, skills and understanding. Building a healthy culture of 'maths talk' empowers their learning from day one.

Explanation and discussion are integral to the *Power Maths* structure, so by simply following the books, your lessons will stimulate structured talk. The following key 'maths talk' strategies will help you strengthen that culture and ensure that every child is included.

Sentences, not words

Encourage children to use full sentences when reasoning, explaining or discussing maths. This helps both speaker and listeners to clarify their own understanding. It also reveals whether or not the speaker truly understands, enabling you to address misconceptions as they arise.

Working together

Working with others in pairs, groups or as a whole class is a great way to support maths talk and discussion. Use different group structures to add variety and challenge. For example, children could take timed turns for talking, work independently alongside a 'discussion buddy', or perhaps play different *Power Maths* character roles within their group.

Think first – then talk

Provide clear opportunities within each lesson for children to think and reflect, so that their talk is purposeful, relevant and focused.

Give every child a voice

Where the 'hands up' model allows only the more confident child to shine, *Power Maths* involves everyone. Make sure that no child dominates and that even the shyest child is encouraged to contribute – and praised when they do.

Assessment strategies

Teaching for mastery demands that you are confident about what each child knows and where their misconceptions lie; therefore, practical and effective assessment is vitally important.

Formative assessment within lessons

The **Think together** section will often reveal any confusions or insecurities; try ironing these out by doing the first **Think together** question as a class. For children who continue to struggle, you or your Teaching Assistant should provide support and enable them to move on.

▶ Performance in practice can be very revealing: check Practice Books and listen out both during and after practice to identify misconceptions.

▶ The **Reflect** section is designed to check on the all-important depth of understanding. Be sure to review how the children performed in this final stage before you teach the next lesson.

End of unit check – Textbook

Each unit concludes with a summative check to help you assess quickly and clearly each child's understanding, fluency, reasoning and problem solving skills. Your Teacher Guide will suggest ideal ways of organising a given activity and offer advice and commentary on what children's responses mean. For example, 'What misconception does this reveal?'; 'How can you reinforce this particular concept?'

For Year 1 and Year 2 children, assess in small, teacher-led groups, giving each child time to think and respond while also consolidating correct mathematical language. Assessment with young children should always be an enjoyable activity, so avoid one-to-one individual assessments, which they may find threatening or scary. If you prefer, the End of unit check can be carried out as a whole-class group using whiteboards and Practice Books.

End of unit check – Practice Book

The Practice Book contains further opportunities for assessment, and can be completed by children independently whilst you are carrying out diagnostic assessment with small groups. Your Teacher Guide will advise you on what to do if children struggle to articulate an explanation – or perhaps encourage you to write down something they have explained well. It will also offer insights into children's answers and their implications for next learning steps. It is split into three main sections, outlined below.

My journal is designed to allow children to show their depth of understanding of the unit. It can also serve as a way of checking that children have grasped key mathematical vocabulary. The question children should answer is first presented in the Textbook in the Think! section. This provides an opportunity for you to discuss the question first as a class to ensure children have understood their task. Children should have some time to think about how they want to answer the question, and you could ask them to talk to a partner about their ideas. Then children should write their answer in their Practice Book, using the word bank provided to help them with vocabulary.

The **Power check** allows pupils to self-assess their level of confidence on the topic by colouring in different smiley faces. You may want to introduce the faces as follows:

Each unit ends with either a Power play or a Power puzzle. This is an activity, puzzle or game that allows children to use their new knowledge in a fun, informal way.

Progress Tests

There are *Power Maths* Progress Tests for each half term and at the end of the year, including an Arithmetic test and Reasoning test in each case. You can enter results in the online markbook to track and analyse results and see the average for all schools' results. The tests use a 6-step scale to show results against age-related expectation.

How to ask diagnostic questions

The diagnostic questions provided in children's Practice Books are carefully structured to identify both understanding and misconceptions (if children answer in a particular way, you will know why). The simple procedure below may be helpful:

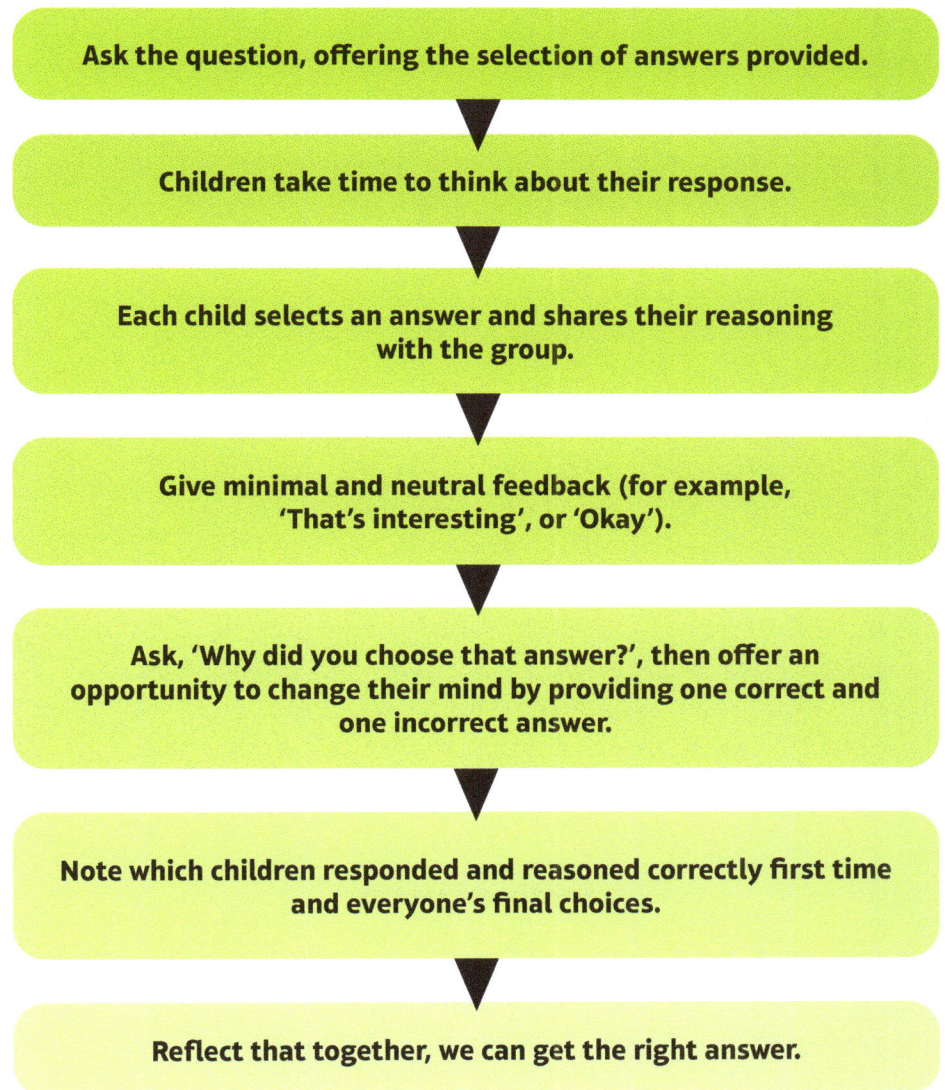

Keeping the class together

Traditionally, children who learn quickly have been accelerated through the curriculum. As a consequence, their learning may be superficial and will lack the many benefits of enabling children to learn with and from each other.

By contrast, *Power Maths'* mastery approach values real understanding and richer, deeper learning above speed. It sees all children learning the same concept in small, cumulative steps, each finding and mastering challenge at their own level. Remember that when you teach for mastery, EVERYONE can do maths! Those who grasp a concept easily have time to explore and understand that concept at a deeper level. The whole class therefore moves through the curriculum at broadly the same pace via individual learning journeys.

For some teachers, the idea that a whole class can move forward together is revolutionary and challenging. However, the evidence of global good practice clearly shows that this approach drives engagement, confidence, motivation and success for all learners, and not just the high flyers. The strategies below will help you keep your class together on their maths journey.

Mix it up

Do not stick to set groups at each table. Every child should be working on the same concept, and mixing up the groupings widens children's opportunities for exploring, discussing and sharing their understanding with others.

Recycling questions

Reuse the Textbook and Practice Book questions with concrete materials to allow children to explore concepts and relationships and deepen their understanding. This strategy is especially useful for reinforcing learning in same-day interventions.

Strengthen at every opportunity

The next lesson in a *Power Maths* sequence always revises and builds on the previous step to help embed learning. These activities provide golden opportunities for individual children to strengthen their learning with the support of Teaching Assistants.

Prepare to be surprised!

Children may grasp a concept quickly or more slowly. The 'fast graspers' won't always be the same individuals, nor does the speed at which a child understands a concept predict their success in maths. Are they struggling or just working more slowly?

Same-day intervention

Since maths competence depends on mastering concepts one by one in a logical progression, it is important that no gaps in understanding are ever left unfilled. Same-day interventions – either within or after a lesson – are a crucial safety net for any child who has not fully made the small step covered that day. In other words, intervention is always about keeping up, not catching up, so that every child has the skills and understanding they need to tackle the next lesson. That means presenting the same problems used in the lesson, with a variety of concrete materials to help children model their solutions.

We offer two intervention strategies below, but you should feel free to choose others if they work better for your class.

Within-lesson intervention

The **Think together** activity will reveal those who are struggling, so when it is time for practice, bring these children together to work with you on the first practice questions. Observe these children carefully, ask questions, encourage them to use concrete models and check that they reach and can demonstrate their understanding.

After-lesson intervention

You might like to use the **Think together** questions to recap the lesson with children who are working behind expectations during assembly time. Teaching Assistants could also work with these children at other convenient points in the school day. Some children may benefit from revisiting work from the same topic in the previous year group. Note also the suggestion for recycling questions from the Textbook and Practice Book with concrete materials on page 26.

The role of practice

Practice plays a pivotal role in the *Power Maths* approach. It takes place in class groups, smaller groups, pairs, and independently, so that children always have the opportunities for thinking as well as the models and support they need to practise meaningfully and with understanding.

Intelligent practice

In *Power Maths*, practice never equates to the simple repetition of a process. Instead we embrace the concept of intelligent practice, in which all children become fluent in maths through varied, frequent and thoughtful practice that deepens and embeds conceptual understanding in a logical, planned sequence. To see the difference, take a look at the following examples.

Traditional practice

- Repetition can be rote – no need for a child to think hard about what they are doing
- Praise may be misplaced
- Does this prove understanding?

Intelligent practice

- Varied methods – concrete, pictorial and abstract
- Equation expressed in different ways, requiring thought and understanding
- Constructive feedback

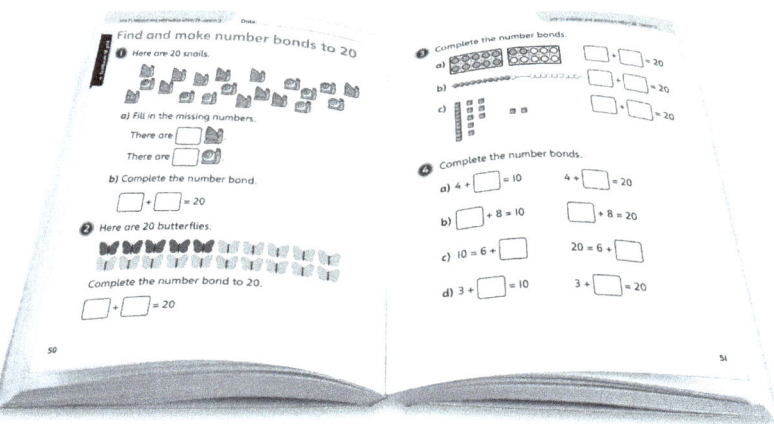

All practice questions are designed to move children on and reveal misconceptions.

Simple, logical steps build onto earlier learning.

C-P-A runs throughout – different ways of modelling and understanding the same concept.

Conceptual variation – children work on different representations of the same maths concept.

Friendly characters offer support and encourage children to try different approaches.

A carefully designed progression

The Practice Books provide just the right amount of intelligent practice for children to complete independently in the final sections of each lesson. It is really important that all children are exposed to the practice questions, and that children are not directed to complete different sections. That is because each question is different and has been designed to challenge children to think about the maths they are doing. The questions become more challenging so children grasping concepts more quickly will start to slow down as they progress. Meanwhile, you have the chance to circulate and spot any misconceptions before they become barriers to further learning.

Homework and the role of parents and carers

While *Power Maths* does not prescribe any particular homework structure, we acknowledge the potential value of practice at home. For example, practising fluency in key facts, such as number bonds and times-tables, is an ideal homework task. You can share the Individual Practice Games for homework (see pages 6 and 9), or parents and carers could work through uncompleted Practice Book questions with children at either primary stage.

However, it is important to recognise that many parents and carers may themselves lack confidence in maths, and few, if any, will be familiar with mastery methods. A Parents' and Carers' evening that helps them understand the basics of mindsets, mastery and mathematical language is a great way to ensure that children benefit from their homework. It could be a fun opportunity for children to teach their families that everyone can do maths!

Structures and representations

Unlike most other subjects, maths comprises a wide array of abstract concepts – and that is why children and adults so often find it difficult. By taking a concrete-pictorial-abstract (C-P-A) approach, *Power Maths* allows children to tackle concepts in a tangible and more comfortable way.

Non-linear stages

Concrete

Replacing the traditional approach of a teacher working through a problem in front of the class, the concrete stage introduces real objects that children can use to 'do' the maths – any familiar object that a child can manipulate and move to help bring the maths to life. It is important to appreciate, however, that children must always understand the link between models and the objects they represent. For example, children need to first understand that three cakes could be represented by three pretend cakes, and then by three counters or bricks. Frequent practice helps consolidate this essential insight. Although they can be used at any time, good concrete models are an essential first step in understanding.

Pictorial

This stage uses pictorial representations of objects to let children 'see' what particular maths problems look like. It helps them make connections between the concrete and pictorial representations and the abstract maths concept. Children can also create or view a pictorial representation together, enabling discussion and comparisons. The *Power Maths* teaching tools are fantastic for this learning stage, and bar modelling is invaluable for problem solving throughout the primary curriculum.

Abstract

Our ultimate goal is for children to understand abstract mathematical concepts, symbols and notation and of course, some children will reach this stage far more quickly than others. To work with abstract concepts, a child must be comfortable with the meaning of and relationships between concrete, pictorial and abstract models and representations. The C-P-A approach is not linear, and children may need different types of models at different times. However, when a child demonstrates with concrete models and pictorial representations that they have grasped a concept, we can be confident that they are ready to explore or model it with abstract symbols such as numbers and notation.

Use at any time and with any age to support understanding

Variation helps visualisation

Children find it much easier to visualise and grasp concepts if they see them presented in a number of ways, so be prepared to offer and encourage many different representations.

For example, the number six could be represented in various ways:

Practical aspects of *Power Maths*

One of the key underlying elements of *Power Maths* is its practical approach, allowing you to make maths real and relevant to your children, no matter their age.

Manipulatives are essential resources for both key stages and *Power Maths* encourages teachers to use these at every opportunity, and to continue the Concrete-Pictorial-Abstract approach right through to Year 6.

The Textbooks and Teacher Guides include lots of opportunities for teaching in a practical way to show children what maths means in real life.

Discover and Share

The **Discover** and **Share** sections of the Textbook give you scope to turn a real-life scenario into a practical and hands-on section of the lesson. Use these sections as inspiration to get active in the classroom. Where appropriate, use the **Discover** contexts as a springboard for your own examples that have particular resonance for your children – and allow them to get their hands dirty trying out the mathematics for themselves.

Unit videos

Every term has one unit video which incorporates real-life classroom sequences.

These videos show you how the reasoning behind mathematics can be carried out in a practical manner by showing real children using various concrete and pictorial methods to come to the solution. You can see how using these practical models, such as part-whole and bar models, helps them to find and articulate their answer.

Mastery tips

Mastery Experts give anecdotal advice on where they have used hands-on and real-life elements to inspire their children.

Mastery Expert tip! 'When teaching this unit, it is important to make links to prior content covered. Doing so helps children not only to consolidate their prior learning, helping secure it in long-term memory, but also to see maths as one big interconnected journey rather than different topics meaning different things.'

Don't forget to watch the Unit 6 video!

Concrete-Pictorial-Abstract (C-P-A) approach

Each **Share** section uses various methods to explain an answer, helping children to access abstract concepts by using concrete tools, such as counters. Remember, this isn't a linear process, so even children who appear confident using the more abstract method can deepen their knowledge by exploring the concrete representations. Encourage children to use all three methods to really solidify their understanding of a concept.

Pictorial representation – drawing the problem in a logical way that helps children visualise the maths

Concrete representation – using manipulatives to represent the problem. Encourage children to physically use resources to explore the maths.

Abstract representation – using words and calculations to represent the problem.

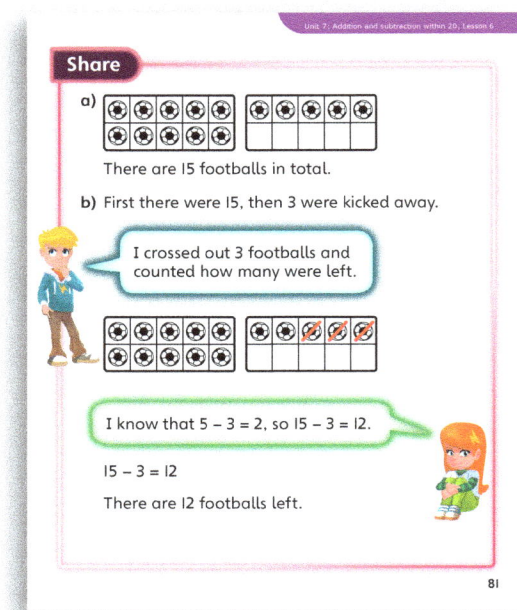

Practical tips

Every lesson suggests how to draw out the practical side of the **Discover** context.

You'll find these in the **Discover** section of the Teacher Guide for each lesson.

PRACTICAL TIPS Provide children with counters and ten frames to represent the question. Encourage them to count aloud.

Resources

Every lesson lists the practical resources you will need or might want to use. There is also a summary of all of the resources used throughout the term on page 40 to help you be prepared.

RESOURCES

Mandatory: counters

Optional: ten frames, number cards with numbers 1–20

Using *Power Maths* flexibly in Key Stage 1

Power Maths lessons have a coherent, regular structure that supports you in building up children's understanding in a series of small steps. This is something most classes will need to build up to, rather than running in from a standing start at the beginning of Year 1.

Start by using the Practice Books in small groups

In most Year 1 classes, it won't be realistic for the whole class to complete the Practice Book pages independently at the start of the year, but they will learn to do this gradually. For the Textbooks, children will need to get used to direct teaching and recording answers in their own books. And, of course, this will set them up well for the rest of Primary school.

Small teacher-led groups are likely to be the best approach for independent practice. This format allows you to talk children through the question, discuss their ideas using manipulatives (often there will be manipulatives on the page as a hint), and guide them in representing their answer. (For instance, they can tell you the answer is 5, but they may need help writing 5 or knowing that they should colour in 5 apples.)

Go through the questions one-by-one with the group. You can mark their work/give feedback there and then. As children get used to the materials, the next stage could be for the small group to work through the questions at their own pace. The style of questions in *Power Maths* is quite regular, so children will get better at knowing what they need to do.

To facilitate small group work, you are likely to need some other activities as a carousel. A good way to do this is by turning a question from the Textbook into a game (usually **Think together** question 3 will work well) and teaching this to children before you break into groups. For instance, look at the example below (pages 78–79 in Textbook 1A). You could teach children a game with a part-whole model where one child puts in the whole using counters and the other children have to put in the parts. Or they could try this with beanbags and hoops. Base the practice on the key learning from the lesson.

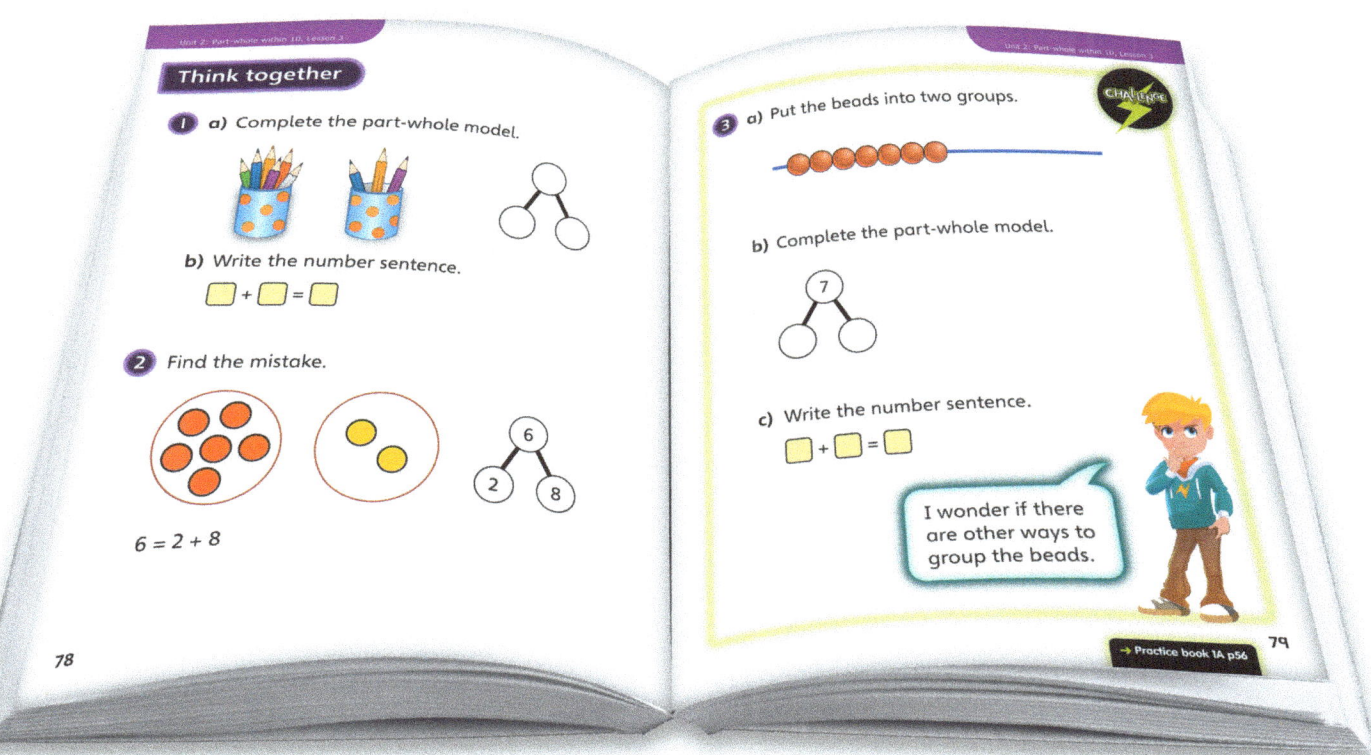

Are there any other ways to use the resources flexibly?

Don't be afraid to bring the **Discover** activity to life! Perhaps you could turn it into a game, or a role play. For instance, if the context is a teddy bear's picnic, you could share out fruit between teddies in the class. Or could you find a toy rocket to launch for the lesson below? (Textbook 1A page 32).

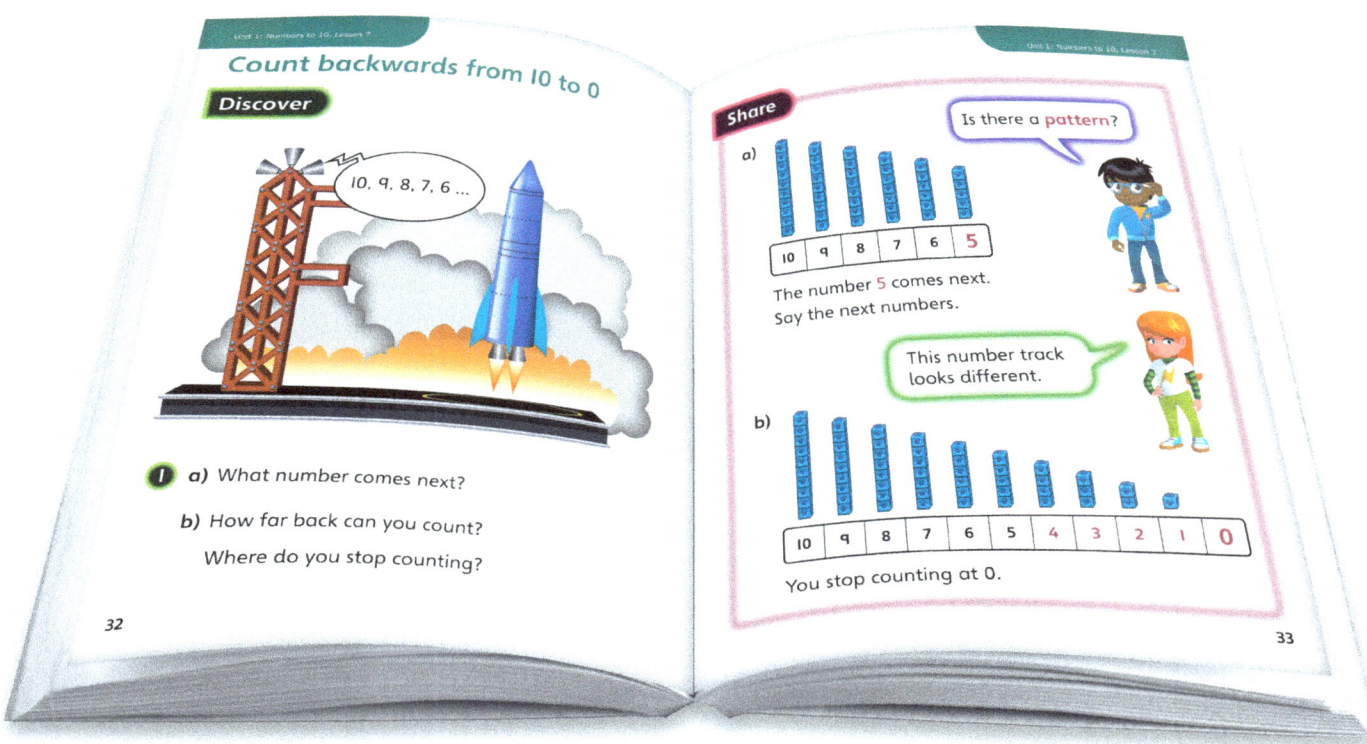

For some lessons you could consider a slightly different approach where you move backwards and forwards between the Textbook and Practice Book. If **Think together** question 1 links well with Practice Book question 1, you could do the **Think together** question together and then let children complete the Practice Book question, then the same for question 2, etc. This works better for some lessons than others, but it is one way of making practice more independent in short bursts, as a way of building up independence.

Don't forget, there isn't a *Power Maths* lesson for every lesson in the year. You can take more time where you need to, so that children's understanding is secure. In Key Stage 1, it will be all the more important to take your time, because children need to get used to the format as well as master the key learning. If using some of the ideas above means that a *Power Maths* lesson actually takes two lessons, e.g. for the first part of the year, then that's fine!

There are some further ideas for using the materials flexibly in the next section.

Working with children below age-related expectation

This section offers advice on using *Power Maths* with children who are significantly behind age-related expectation. Teacher judgement will be crucial in terms of where and why children are struggling, and in choosing the right approach. The suggestions can of course be adapted for children with special educational needs, depending on the specific details of those needs.

General approaches to support children who are struggling

Keeping the pace manageable

Remember, you have more teaching days than *Power Maths* lessons so you can cover a lesson over more than one day, and revisit key learning, to ensure all children are ready to move on. You can use the + and – buttons to adjust the time for each unit in the online planning. The NCETM's Ready-to-Progress criteria can be used to help determine what should be highest priority.

Same-day intervention

You could go over the Textbook pages or revisit the previous year's work if necessary. Remember that same-day intervention can be within the lesson, as well as afterwards (see page 29). As children start their independent practice, you can work with those who found the first part of the lesson difficult, checking understanding using manipulatives.

Fluency sessions

Fit in as much practice as you can for number bonds and times-tables, etc., at other times of the day. If you can, plan a short 'maths meeting' for this in the afternoon. You might choose to use a Power Up you haven't used already.

Pre-teaching

Find a 5- to 10-minute slot before the lesson to work with the children you feel would benefit. The afternoon before the lesson can work well, because it gives children time to think in between. Recap previous work on the topic (addressing any gaps you're aware of) and do some fluency practice, targeting number facts etc. that will help children access the learning.

Focusing on the key concepts

If children are a long way behind, it can be helpful to take a step back and think about the key concepts for children to engage with, not just the fine detail of the objective for that year group (e.g. addition with a specific number of columns). Bearing that in mind, how could children advance their understanding of the topic?

Providing extra support within the lesson

Support in the Teacher Guide
First of all, use the Strengthen support in the Teacher Guide for guided and independent work in each lesson, and share this with Teaching Assistants, where relevant. As you read through the lesson content and corresponding Teacher Guide pages before the lesson, ask yourself what key idea or nugget of understanding is at the heart of the lesson. If children are struggling, this should help you decide what's essential for all children before they move on.

Annotating pages
You can annotate questions to provide extra scaffolding or hints if you need to, but aim to build up children's ability to access questions independently wherever you can. Children tend to get used to the style of the *Power Maths* questions over time.

Quick recap as lesson starter
The Quick recap for each lesson in the Teacher Guide is an alternative starter activity to the Power Up. You might choose to use this with some or all children if you feel they will need support accessing the main lesson.

Consolidation questions
If you think some children would benefit from additional questions at the same level before moving on, write one or two similar questions on the board. (This shouldn't be at the expense of reasoning and problem-solving opportunities: take longer over the lesson if you need to.)

Hard copy Textbooks
The Textbooks help children focus in more easily on the mathematical representations, read the text more comfortably, and revisit work from a previous lesson that you are building on, as well as giving children ownership of their learning journey. In main lessons, it can work well to use the e-Textbook for **Discover** and give out the books when discussing the methods in the **Share** section.

Reading support
It's important that all children are exposed to problem solving and reasoning questions, which often involve reading. For whole-class work you can read questions together. For independent practice you could consider annotating pages to help children see what the question is asking, and stem sentences to help structure their answer. A general focus on specific mathematical language and vocabulary will help children access the questions. You could consider pairing weaker readers with stronger readers, or read questions as a group if those who need support are on the same table.

Providing extra depth and challenge with *Power Maths*

Just as prescribed in the National Curriculum, the goal of *Power Maths* is never to accelerate through a topic but rather to gain a clear, deep and broad understanding. Here are some suggestions to help ensure all children are appropriately challenged as you work with the resources.

Overall approaches

First of all, remember that the materials are designed to help you keep the class together, allowing all children to master a concept while those who grasp it quickly have time to explore it in more depth. Use the Deepen support in the Teacher Guide (see below) to challenge children who work through the questions quickly. Here are some questions and ideas to encourage breadth and depth during specific parts of the lesson, or at any time (where no part of the lesson sequence is specified):

- **Discover**: 'Can you demonstrate your solution another way?'
- **Share**: Make sure every child is encouraged to give answers and engage with the discussion, not just the most confident.
- **Think together**: 'Can you model your answers using concrete materials? Can you explain your solution to a partner?'
- Practice: Allow all children to work through the full set of questions, so that they benefit from the logical sequence.
- **Reflect**: 'Is there another way of working out the answer? And another way?'
  'Have you found all the solutions?'
  'Is that always true?'
  'What's different between this question and that question? And what's the same?'

Note that the **Challenge** questions are designed so that all children can access and attempt them, if they have worked through the steps leading up to them. There may be some children in a given lesson who don't manage to do the **Challenge**, but it is not supposed to be a distinct task for a subset of the class. When you look through the lesson materials before teaching, think about what each question is specifically asking, and compare this with the key learning point for the lesson. This will help you decide which questions you feel it's essential for all children to answer, before moving on. You can at least aim for all children to try the **Challenge**!

Deepen activities and support

The Teacher Guide provides valuable support for each stage of the lesson. This includes Deepen tips for the guided and independent practice sections, which will help you provide extra stretch and challenge within your lesson, without having to organise additional tasks. If you have a Teaching Assistant, they can also make use of this advice. There are also suggestions for the lesson as a whole in the 'Going Deeper' section on the first page of the Teacher Guide section for that lesson. Every class is different, so you can always go a bit further in the direction indicated, if appropriate, and build on the suggestions given.

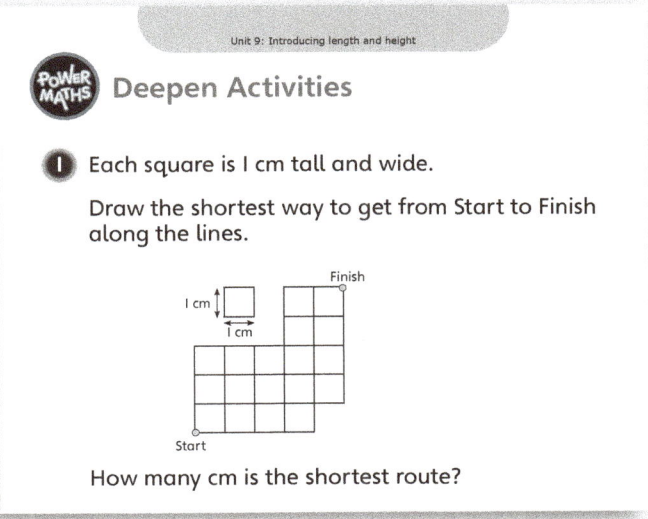

There is a Deepen activity for each unit. These are designed to follow on from the End of unit check, stretching children who have a firm understanding of the key learning from the unit. Children can work on them independently, which makes it easier for the teacher to facilitate the Strengthen activity for children who need extra support. Deepen activities could also be introduced earlier in the unit if the necessary work has been covered. The Deepen activities are on *ActiveLearn* on the Planning page for each unit, and also on the Resources page).

Using the questions flexibly to provide extra challenge

Sometimes you may want to write an extra question on the board or provide this on paper. You can usually do this by tweaking the lesson materials. The questions are designed to form a carefully structured sequence that builds understanding step by step, but, with careful thought about the purpose of each question, you can use the materials flexibly where you need to. Sometimes you might feel that children would benefit from another similar question for consolidation before moving on to the next one, or you might feel that they would benefit from a harder example in the same style. It should be quick and easy to generate 'more of the same' type questions where this is the case.

When you see a question like this one (from Unit 2, Lesson 1), it's easy to make harder examples to do afterwards if you need them. What if there were 9 cubes? Can children write the parts and wholes and find lots of different ways?

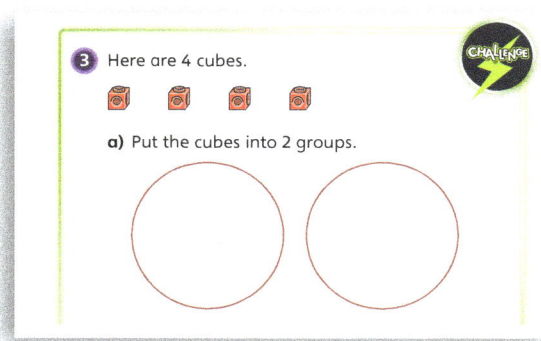

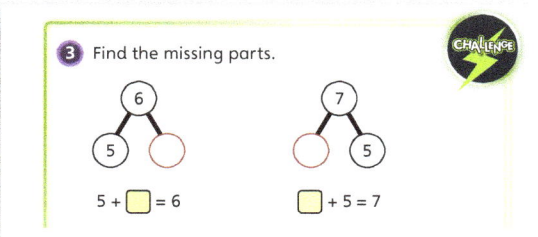

For this example (from Unit 3, Lesson 4), you could ask children to make up their own question(s) for a partner to solve. (In fact, for any of these examples you could ask early finishers to create their own question for a partner.)

Here's an example (from Unit 8, Lesson 2) where you could use the original context to provide extra challenge at the end of the lesson. For example, you could ask how far the frogs have to go to reach the next frog, or to be the winner.

Besides creating additional questions, you should be able to find a question in the lesson that you can adapt into a game or open-ended investigation, if this helps to keep everyone engaged. It could simply be that, instead of answering 5 + 6 etc. on the page, they could build a robot with 5 cubes and 6 cylinders.

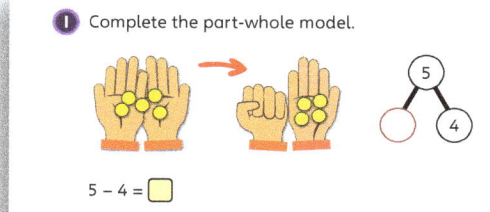

With a question like this one (from Unit 4, Lesson 4), children could play the game in pairs, taking an agreed number of counters and then showing what's in one hand. They could write a subtraction sentence each time, varying the whole and the parts.

See the bullets on the previous page for some general ideas that will help with 'opening out' questions in the books, e.g. 'Can you find all the solutions?' type questions.

Other suggestions

Another way of stretching children is through mixed ability pairs, or via other opportunities for children to explain their understanding in their own way. This is a good way of encouraging children to go deeper into the learning, rather than, for instance, tackling questions that are computationally more challenging but conceptually equivalent in level.

Using *Power Maths* with mixed age classes

Overall approaches

There are many variables between schools that would make it inadvisable to recommend a one-size-fits-all approach to mixed age teaching with *Power Maths*. These include how year groups are merged, availability of Teaching Assistants, experience and preference of teaching staff, range in pupil attainment across years, classroom space and layout, level of flexibility around timetables, and overall organisational structure (whether the school is part of a trust).

Some schools will find it best to timetable separate maths lessons for the different year groups. Others will aim to teach the class together as much as possible using the mixed age planning support on *ActiveLearn* (see the lesson exemplars for ways of organising lessons with strong/medium/weak correlation between year groups). There will also be ways of adapting these general approaches. For example, offset lessons where Year A start their lesson with the teacher, while Year B work independently on the practice from the previous lesson, and then start the next lesson with the teacher while Year A work independently; or teachers may choose to base their provision around the lesson from one year group and tweak the content up/down for the other group.

Key strategies for mixed age teaching

The mixed age teaching webinar on *ActiveLearn* provides advice on all aspects of mixed age teaching, including more detail on the ideas below.

Developing independence over time
Investing time in building up children's independence will pay off in the medium term.

Clear rationale
If someone asked, 'Why did you teach both Unit 3 and 4 in the same lesson/separate lessons?', what would your answer be?

Designing a lesson
1. Identify the core learning for each group
2. Identify any number skills necessary to access the core
3. Consider the flow of concepts and how one core leads to the other

Challenging all children
The questions are designed to build understanding step by step, but with careful thought about the purpose of each question you can tweak them to increase the challenge.

Multiple years combined
With more than two years together, teachers will inevitably need to use the resources flexibly if delivering a single lesson.

Enjoy the positives!

Comparison deepens understanding and there will be lots of opportunities for children, as well as misconceptions to explore. There is also in-built pre-teaching and the chance to build up a concept from its foundations. For teachers there is double the material to draw on! Mixed age teachers require a strong understanding of the progression of ideas across year groups, which is highly valuable for all teachers. Also, it is necessary to engage deeply with the lesson to see how to use the materials flexibly – this is recommended for all teachers and will help you bring your lesson to life!

List of practical resources

Year 1B Mandatory resources

Resource	Lesson
100 square	**Unit 8** Lesson 2
Balance scales	**Unit 10** Lesson 1
Base 10 equipment	**Unit 8** Lesson 1
Bead strings	**Unit 7** Lessons 2, 9
Boxes (of scissors or pencils)	**Unit 7** Lesson 2
Counters	**Unit 6** Lessons 1, 2, 3, 4, 5, 6, 7, 11 **Unit 7** Lessons 1, 2, 4, 5, 6, 7, 8, 9, 11 **Unit 8** Lessons 3, 4, 5, 6, 7
Counters (double-sided)	**Unit 7** Lesson 3
Counters (or cubes)	**Unit 7** Lesson 3
Cubes	**Unit 6** Lessons 11, 12 **Unit 7** Lessons 1, 5, 8
Dice	**Unit 7** Lesson 5
Multilink cubes	**Unit 8** Lesson 3 **Unit 9** Lessons 2, 3
Number lines	**Unit 7** Lessons 1, 6
Number lines (blank)	**Unit 6** Lesson 8 **Unit 7** Lessons 8, 10
Number lines (going up in 1s)	**Unit 6** Lesson 9
Number tracks	**Unit 7** Lesson 1 **Unit 8** Lesson 2
Objects (classroom, various: to compare heights and lengths)	**Unit 9** Lesson 1
Objects (to measure)	**Unit 9** Lessons 3, 4 **Unit 9** Lesson 2
Paper strips	**Unit 9** Lesson 3
Rulers	**Unit 9** Lessons 3, 4
Straws	**Unit 7** Lesson 7 **Unit 8** Lesson 3
Ten frames	**Unit 6** Lessons 2, 3, 4, 5, 6, 7 **Unit 7** Lessons 1, 2, 3, 4, 5, 6 **Unit 8** Lessons 3, 4, 5, 6, 7

Year 1B Optional resources

Resource	Lesson
100 square	**Unit 8** Lesson 1
Balance scales	**Unit 10** Lessons 2, 3
Base 10 equipment	**Unit 8** Lessons 2, 4, 6
Bead strings	**Unit 6** Lessons 4, 7 **Unit 7** Lessons 3, 6 **Unit 8** Lesson 4
Beakers	**Unit 10** Lesson 4
Cardboard tubes	**Unit 7** Lesson 11
Chalk	**Unit 6** Lessons 9, 10
Classroom stationery	**Uit 7** Lesson 8
Containers (different sizes)	**Unit 10,** Lessons 6, 7
Counters	**Unit 6** Lessons 8, 12 **Unit 7** Lesson 10 **Unit 8** Lesson 1 **Unit 10** Lesson 7
Cubes	**Unit 6** Lessons 2, 6 **Unit 7** Lesson 10 **Unit 8** Lessons 1, 5, 6, 7 **Unit 10** Lessons 2, 7
Cubes (or similar to use as non-standard units of weight)	**Unit 10** Lesson 3
Cups (to hide counters in)	**Unit 8** Lesson 5
Dice	**Unit 6** Lessons 2, 6 **Unit 7** Lesson 4
Digit cards	**Unit 8** Lesson 1
Dominoes	**Unit 6** Lesson 6
Footballs	**Unit 7** Lesson 6
Interlocking cubes	**Unit 10** Lesson 6
Jugs	**Unit 10** Lesson 4
Marbles (tub)	**Unit 8** Lesson 1
Measuring jugs and cylinders	**Unit 10** Lesson 4
Number cards	**Unit 6** Lesson 12
Number cards (0–20)	**Unit 6** Lesson 8
Number cards (1–20)	**Unit 6** Lesson 1
Number cards (11–20)	**Unit 6** Lesson 9
Number lines	**Unit 6** Lesson 9 **Unit 7** Lesson 7
Number lines (blank)	**Unit 6** Lesson 10
Number lines (blank, wipe-clean)	**Unit 6** Lesson 8
Number lines (marked in ones)	**Unit 8** Lesson 1
Number tracks	**Unit 6** Lessons 6, 7, 8, 11 **Unit 7** Lessons 6, 7
Number tracks (blank)	**Unit 8** Lesson 1
Number track (large: on playground)	**Unit 7** Lesson 1
Objects (classroom: general)	**Unit 6** Lesson 6
Objects (to weigh, small)	**Unit 10** Lesson 3
Objects (to weigh: a selection)	**Unit 10** Lesson 2
Objects (to weigh: a selection, e.g. teddy bear, soft toy, toy car or lorries)	**Unit 10** Lesson 1

Year 1B Optional resources – *continued*

Resource	Lesson
Objects (with same measurements as the objects in the questions)	**Unit 9** Lesson 4
Pegs	**Unit 6** Lesson 8
Pencil cases (to measure)	**Unit 9** Lesson 3
Plastic glasses	**Unit 10** Lesson 4
Practical apparatus and materials (for model measuring including sand, rice, water, beakers, buckets, scoops, spoons)	**Unit 10** Lesson 5
Rekenreks	**Unit 6** Lessons 2, 3, 4, 5, 6, 7 **Unit 7** Lessons 2, 4, 5 **Unit 8** Lessons 5, 6, 7
Sentence scaffolds (blank) and number sentences	**Unit 7** Lesson 8
Skipping ropes (3: of different lengths)	**Unit 9** Lesson 1
Straws	**Unit 8** Lessons 4, 5 **Unit 9** Lesson 4
String (to make a number line)	**Unit 6** Lesson 8
Sweets	**Unit 6** Lesson 12
Ten frames (blank)	**Unit 6** Lessons 1, 8, 11, 12 **Unit 7** Lesson 10 **Unit 10** Lesson 7
Tin	**Unit 7** Lesson 9
Toys (or classroom objects – for counting)	**Unit 6** Lesson 12
Toys (cuddly toys)	**Unit 7** Lesson 8
Toys (teddy bears or toy cars)	**Unit 6** Lesson 4
Water	**Unit 10** Lessons 6, 7
Water (or squash; to fill the containers)	**Unit 10** Lesson 4
Whiteboard pens	**Unit 6** Lesson 8
Whiteboards (mini)	**Unit 6** Lesson 8

Getting started with *Power Maths*

As you prepare to put *Power Maths* into action, you might find the tips and advice below helpful.

STEP 1: Train up!

A practical, up-front full day professional development course will give you and your team a brilliant head-start as you begin your *Power Maths* journey. You will learn more about the ethos, how it works and why.

STEP 2: Check out the progression

Take a look at the yearly and termly overviews. Next take a look at the unit overview for the unit you are about to teach in your Teacher Guide, remembering that you can match your lessons and pacing to your class.

STEP 3: Explore the context

Take a little time to look at the context for this unit: what are the implications for the unit ahead? (Think about key language, common misunderstandings and intervention strategies, for example.) If you have the online subscription, don't forget to watch the corresponding unit video.

STEP 4: Prepare for your first lesson

Familiarise yourself with the objectives, essential questions to ask and the resources you will need. The Teacher Guide offers tips, ideas and guidance on individual lessons to help you anticipate children's misconceptions and challenge those who are ready to think more deeply.

STEP 5: Teach and reflect

Deliver your lesson – and enjoy!

Afterwards, reflect on how it went … Did you cover all five stages? Does the lesson need more time? How could you improve it?

Unit 6
Numbers to 20

> **Mastery Expert tip!** 'When teaching this unit, it is important to make links to prior content covered. Doing so helps children not only to consolidate their prior learning, helping secure it in long-term memory, but also to see maths as one big interconnected journey rather than different topics meaning different things.'

Don't forget to watch the Unit 6 video!

WHY THIS UNIT IS IMPORTANT

This unit lays the essential foundations of place value, as children begin to recognise the place value of each digit in a 2-digit number. This is an important skill that children will develop when they add and subtract and begin to work with larger numbers.

Children will count in 10s and 1s, learning that they can partition 2-digit numbers into tens and ones. They will find the number that is one more or one less than a given number, noticing when the tens digit changes and when only the ones digit changes.

As they become more fluent, they will begin to compare and order numbers to 20 using the < and > signs.

WHERE THIS UNIT FITS

→ Unit 5: 2D and 3D shapes
→ **Unit 6: Numbers to 20**
→ Unit 7: Addition and subtraction within 20

This unit builds on children's work on numbers to 10 in Units 1–4, extending their ability to count, compare and order numbers to 20. Unit 7 will focus on addition and subtraction within 20.

Before they start this unit, it is expected that children can:
- recognise number bonds within 10
- understand how to partition a number within 10
- compare and order numbers within 10.

ASSESSING MASTERY

Children who have mastered this unit will be able to partition numbers above 10 into tens and ones. They will be able to work with many different concrete and pictorial representations of numbers, and use correct mathematical language to describe and compare numbers.

They will be able to count on or back to 20, beginning at 0 or 1, or from any given number. They will also be able to identify 'one more' and 'one less' than a given number.

COMMON MISCONCEPTIONS	STRENGTHENING UNDERSTANDING	GOING DEEPER
Children may miss out numbers, or say the wrong number, when counting (for example, saying 'thirty' instead of 'thirteen', or 'fiveteen' instead of 'fifteen').	Give children opportunities to say the words out loud together as a class and in discussion with each other.	Allow children to explore with different representations of number. Which one do they think represents the numbers most clearly?
Children may think that 2-digit numbers are made up of ones and ones rather than tens and ones (for example, thinking that 17 is made up of 1 and 7 rather than 10 and 7).	Using different number representations (ten frames, bead strings etc), discuss what these representations show, and what they *do not* show. For example, show children a ten frame with one counter and a ten frame with seven counters. Ask: *Why does this not show 17? What does it show?*	Give children problems that have more than one possible answer, for example, *Find all the ways of completing 18 > ☐*. Encourage children to be systematic in their approach.

Unit 6: Numbers to 20

UNIT STARTER PAGES

Introduce the unit using teacher-led discussion. Give children time to discuss each question in small groups or pairs and then discuss their ideas as a class.

STRUCTURES AND REPRESENTATIONS

Ten frame: Ten frames play a key role in helping children to recognise the structure of 2-digit numbers.

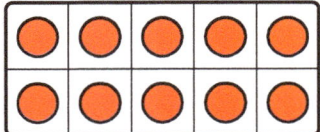

Number line: Number lines can be used to support children in counting on or back and in comparing numbers.

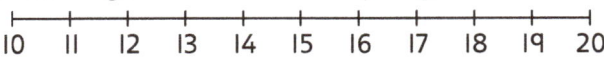

Bead string: Bead strings can be used to help children partition numbers into tens and ones, and to support them in finding one more or one less than a number.

Straws: Straws help children see the importance of 10: it is much easier to count in 10s when counting large numbers and using straws emphasises this.

KEY LANGUAGE

There is some key language that children will need to know as part of the learning in this unit.

- numbers 11–20
- count, count on, count back
- tens, ones
- one more, one less
- greatest, larger, smallest, smaller, fewer, fewest, most, least
- order, compare
- equal to, more than, less than, fewer than, greater than

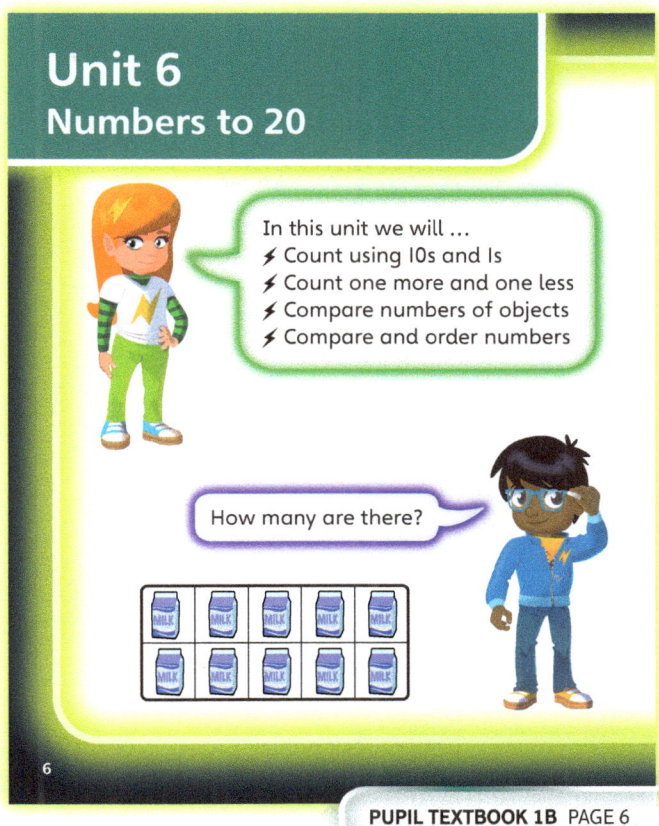

PUPIL TEXTBOOK 1B PAGE 6

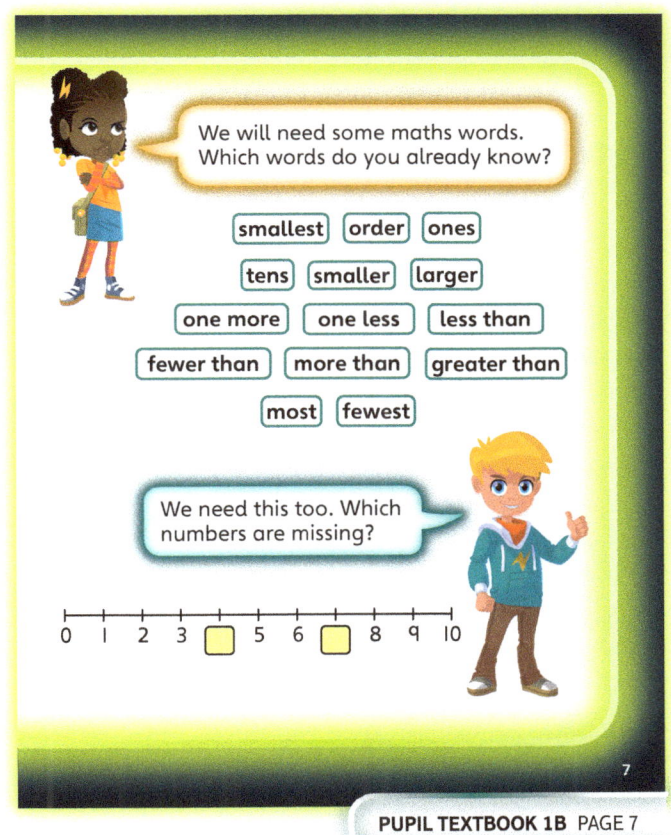

PUPIL TEXTBOOK 1B PAGE 7

Unit 6: Numbers to 20, Lesson 1

Count to 20

Learning focus
In this lesson, children build on their learning of counting to 10 as they count to 20. They should be able to relate the number of objects to the correct number in digits and in words.

Before you teach
- Can children confidently count on to 10?
- Can children confidently count on and back within 10?

NATIONAL CURRICULUM LINKS

Year 1 Number – number and place value

Count to and across 100, count on and count back, beginning with 0 or 1, or from any given number.
Read and write numbers from 1 to 20 in numerals and words.

ASSESSING MASTERY

Children can count objects accurately and link the amount to the correct digit and words.
Children can confidently count on to 20 from any number within 20, and can count back from 20.

COMMON MISCONCEPTIONS

When counting objects, children may not use a strategic approach and, therefore, not count accurately. When crossing 10, children might go back to 1 rather than 11. Ask:
- *What number comes after 10? What numbers make up 11?*

When counting pictures or objects, some children will always start at 1 rather than look for ways they could count more efficiently. To challenge this misconception, ask:
- *Did anyone count those objects differently? Let's see how many different ways you counted.*

STRENGTHENING UNDERSTANDING

Give children number cards, each of which show a digit, the corresponding number in words and the corresponding number in pictures. These will support understanding throughout this lesson.

GOING DEEPER

Ask children to use counters or other objects to represent the numbers they are counting. How do they know they show the same number? How many different ways can they represent it?

KEY LANGUAGE

In lesson: eleven, twelve, thirteen, fourteen, fifteen, sixteen, seventeen, eighteen, nineteen, twenty, count, forwards, backwards, on, back

Other language to be used by the teacher: number track, same, different

STRUCTURES AND REPRESENTATIONS

Ten frames, number track

RESOURCES

Mandatory: counters

Optional: ten frames, number cards with numbers 1–20

 In the eTextbook of this lesson, you will find interactive links to a selection of teaching tools.

Quick recap
Put two number cards on the board or around the classroom. Ask children to point or move to the smallest or greatest number. Increase the number of cards you use and repeat the activity.

Unit 6: Numbers to 20, Lesson 1

Discover

WAYS OF WORKING Pair work

ASK

- Question 1 a): *What number comes after 10?*
- Question 1 b): *Do you think there are more or fewer than 10 children in the class? Why?*

IN FOCUS Questions 1 a) and 1 b) give children opportunity to count beyond 10. Repeat question 1 a) multiple times to ensure children get plenty of practice counting to 20. You could model deliberately making a mistake so that children can correct you. In question 1 b), count to 15 whilst pointing to a child each time.

PRACTICAL TIPS Give each child a number card with numbers up to 20, and ask them to count their numbers out loud in order, ensuring they listen to the previous numbers so they know when it is their turn. If you have fewer than 20 children in your class, you could give some children more than one number. If you have more than 20 children in your class, repeat this activity so that each child has a go at holding a number card.

ANSWERS

Question 1 a): Children should count as a class from 1 to 20.

Question 1 b): There are 15 children in the class.

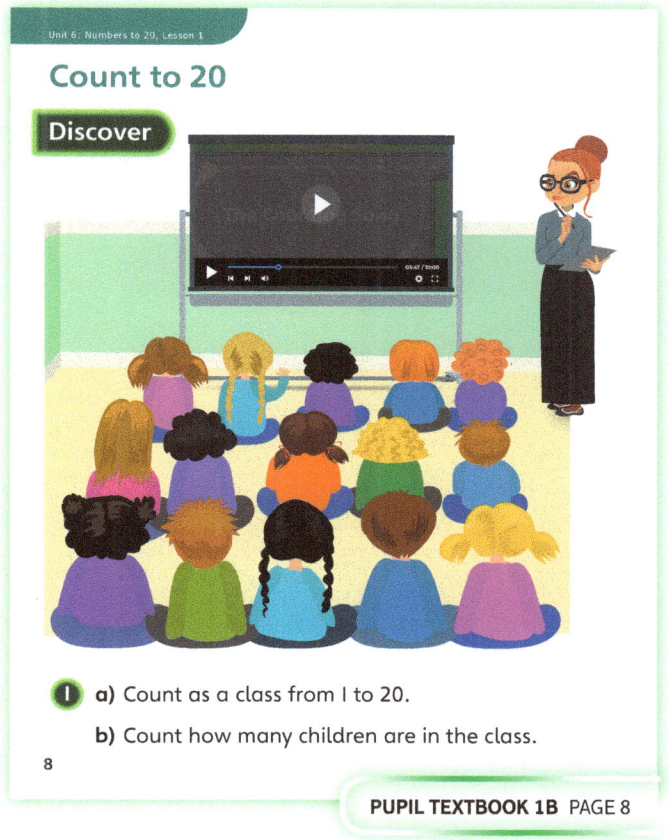

Share

WAYS OF WORKING Whole class teacher led

ASK

- Question 1 a): *Why are eleven, twelve and thirteen harder to say? What do they look like in numerals?*
- Question 1 b): *How do you know you have counted each child? How do you know you have not counted a child more than once?*

IN FOCUS Question 1 a) provides opportunity to discuss why 11, 12 and 13 are different and often more difficult to say than other numbers, as the 1, 2 and 3 are not obvious in the spoken numbers. Question 1 b) models strategic counting to children. By writing the numbers in each speech bubble, they not only know they have not missed any, but also that they have not double counted.

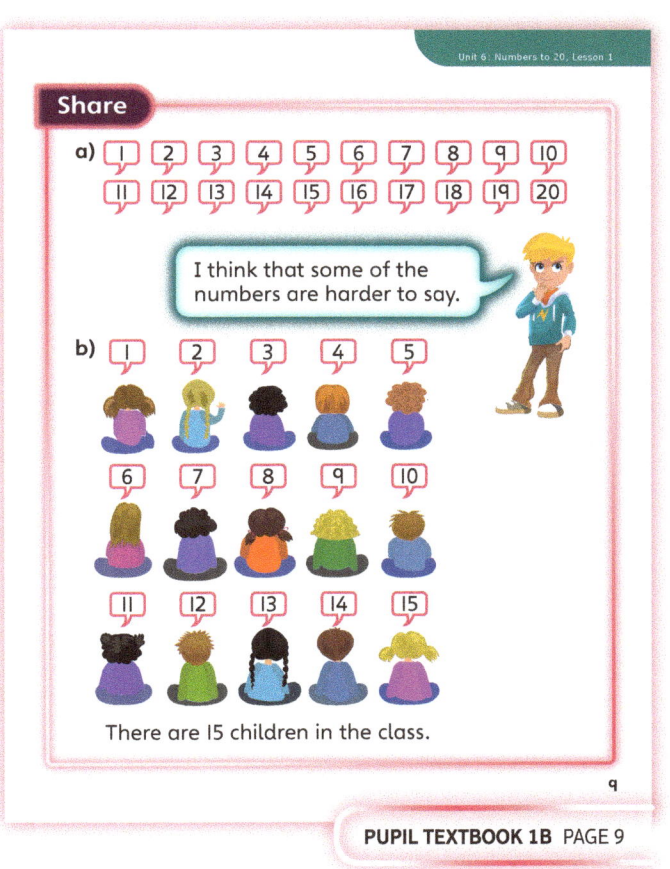

47

Think together

WAYS OF WORKING Whole class teacher led (I do, We do, You do)

ASK
- Question ❶: *What number comes next? How do you know? Do you need to start counting from 1?*
- Question ❷: *How do you know you have counted them all?*

IN FOCUS These questions provide opportunity to count beyond 10. In question ❶, children could count on from 1 to help them. In question ❷, discuss with children what they can do to make sure they do not miss out any ladybirds.

STRENGTHEN Provide children with counters or other physical resources to support their counting. They could place a counter over each ladybird in question ❷ to make sure they have not missed any or counted any twice.

DEEPEN Question ❸ requires children to count without starting at 1 and also to count back from 20. When counting back, ask: *How do you know what number comes next? Can you use counters to show me?*

ASSESSMENT CHECKPOINT Use questions ❶ and ❷ to assess whether children can confidently count on within 20 from 1. Use question ❸ to assess whether children can count within 20 from any number and whether they can count back from 20.

ANSWERS

Question ❶: 9, 10, 11, 12, 13, 14, 15, 16, 17, 18, 19, 20

Question ❷: There are 18 ladybirds.

Question ❸ a): 8, 9, 10, 11, 12, 13, 14, 15

Question ❸ b): 12, 13, 14, 15, 16, 17, 18, 19

Question ❸ c): 20, 19, 18, 17, 16, 15, 14, 13

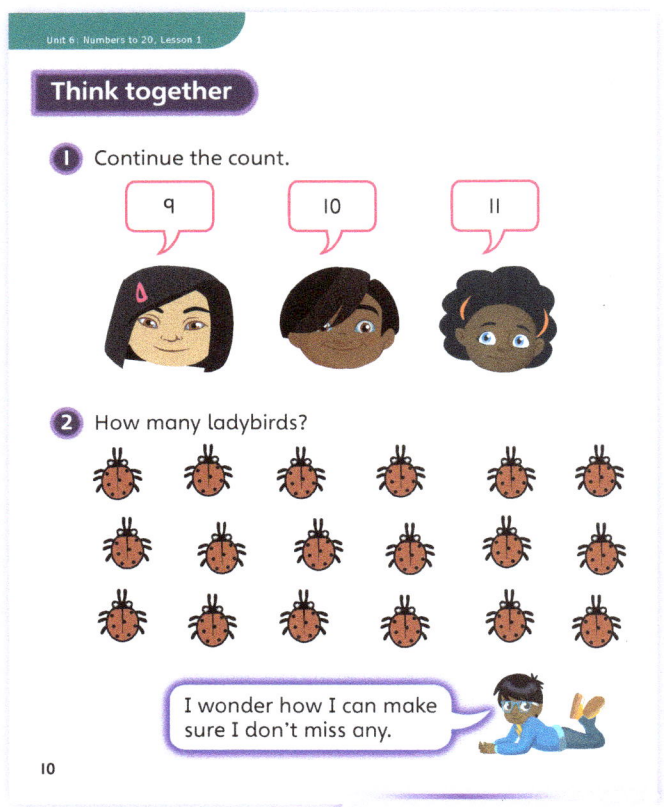

PUPIL TEXTBOOK 1B PAGE 10

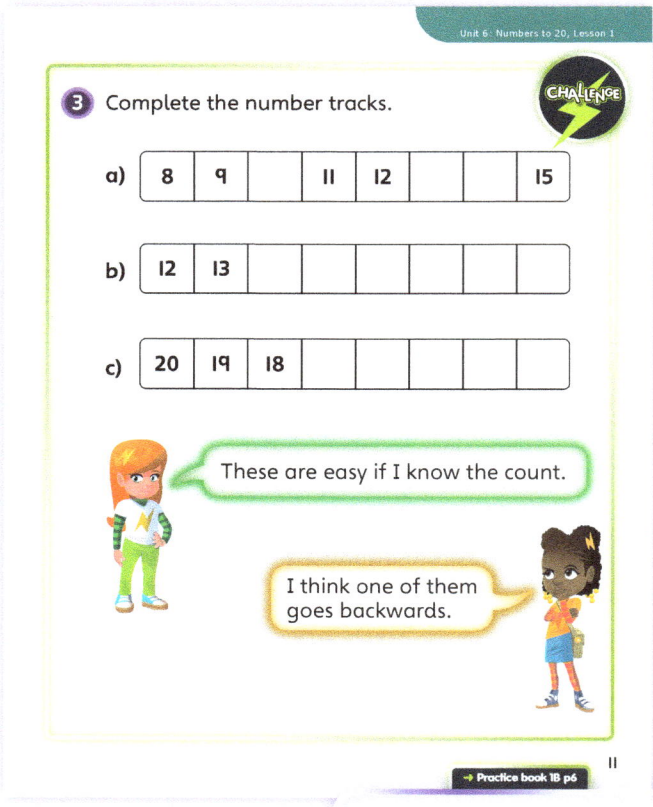

PUPIL TEXTBOOK 1B PAGE 11

Unit 6: Numbers to 20, Lesson 1

Practice

WAYS OF WORKING Independent thinking

IN FOCUS Questions ❶ and ❷ provide opportunity for children to consolidate counting within 20, starting from any number within 20. Questions ❸ and ❹ require children to use a strategic method of counting to ensure that they do not miss any objects out. The number track in question ❺ b) requires children to cross 10. This question will allow you to ensure they do not go back to 1.

STRENGTHEN Provide children with counters or other physical resources to support their counting. They could start by always counting on from 1, before building up to starting from any number. Encourage them to count aloud to support their thinking.

DEEPEN In question ❼ b), ask: *What is the same about the numbers in the top row of the number track and the numbers in the bottom row? What is different?*

THINK DIFFERENTLY Question ❻ shows a dot-to-dot activity, which provides children with opportunity to count from 1 to 20 within a context they should be familiar with. Ask: *Where do you need to start? How do you know?*

ASSESSMENT CHECKPOINT Use questions ❶, ❷ and ❺ to assess whether children can count on from any number within 20. Use questions ❸ and ❹ to assess whether children can count amounts within 20.

ANSWERS Answers for the **Practice** part of the lesson can be found in the *Power Maths* online subscription.

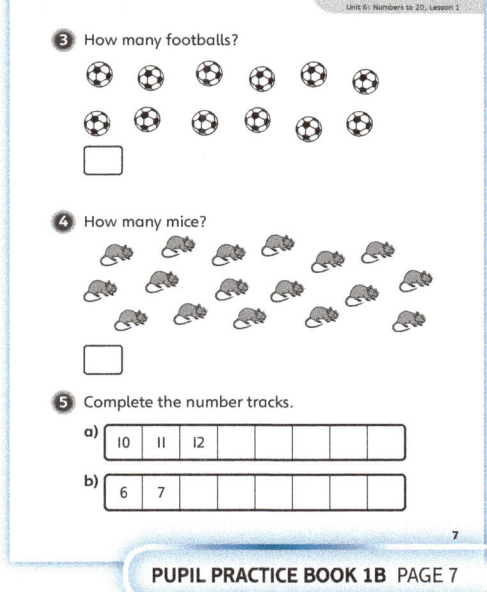

Reflect

WAYS OF WORKING Whole class

IN FOCUS This activity provides opportunity to recap the learning from this lesson, as children count both on and back to and from 20.

ASSESSMENT CHECKPOINT Children should be able to confidently count both on to and back from 20.

ANSWERS Answers for the **Reflect** part of the lesson can be found in the *Power Maths* online subscription.

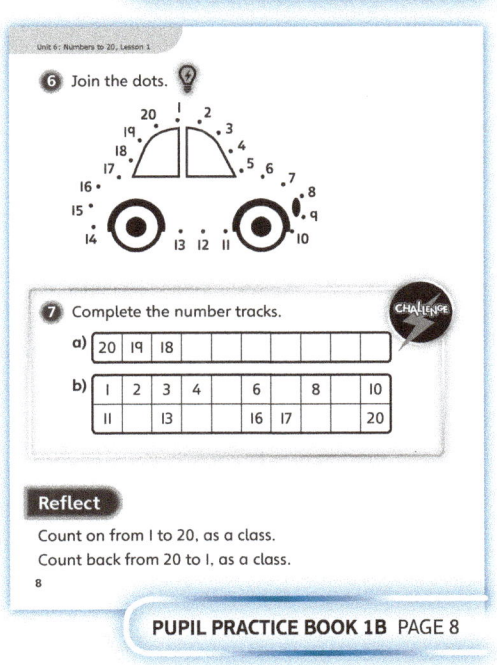

After the lesson
- Can children confidently count on to 20?
- Can children confidently count back from 20?
- Can children correctly identify a number of objects where the number is within 20?

49

Unit 6: Numbers to 20, Lesson 2

Understand 10

Learning focus
In this lesson, children recap learning from earlier in the year as they develop their understanding of 10. It is important that children understand the structure of 10 and can subitise when it is presented in a standard representation.

Before you teach
- Can children confidently count to 10?
- Can children count out ten items?

NATIONAL CURRICULUM LINKS

Year 1 Number – number and place value

Count to and across 100, count on and count back, beginning with 0 or 1, or from any given number (to 20).

ASSESSING MASTERY

Children can instantly recognise 10 without needing to count. They can recognise various different representations of 10, and they can represent 10 in multiple different ways.

COMMON MISCONCEPTIONS

Children may always rely on counting on from 1 to decide whether the representation shows 10 or not, rather than using their understanding to instantly recognise, or subitise, 10. Ask:
- *Do you have to count each object to know how many there are?*

STRENGTHENING UNDERSTANDING

When presented in different ways, provide children with counters and ten frames to represent the objects to support them in building 10. Encourage them to show how they counted.

GOING DEEPER

Explore the idea that 10 ones are equivalent to 1 ten. Using a ten frame can support this, as one full ten frame holds 10 counters.

KEY LANGUAGE

In lesson: ten, full, count

Other language to be used by the teacher: subitise, recognise

STRUCTURES AND REPRESENTATIONS

Ten frames

RESOURCES

Mandatory: ten frames, counters

Optional: rekenreks, cubes, dice

 In the eTextbook of this lesson, you will find interactive links to a selection of teaching tools.

Quick recap

Provide children with different pieces of equipment and ask them to build 10. Ask them to count aloud as they are building it, and discuss each other's representations as a class.

Unit 6: Numbers to 20, Lesson 2

Discover

WAYS OF WORKING Pair work

ASK

- Question 1 a): *How many counters are in the ten frame? How do you know?*

IN FOCUS Questions 1 a) and 1 b) give children opportunity to develop their understanding of 10. In question 1 a), they should recognise that, as the ten frame is full, they know there are 10 without needing to count. Question 1 b) provides opportunity for children to make 10 in different ways using different equipment in the classroom.

PRACTICAL TIPS Other equipment could be linked to ten frames to support understanding. For example, if children are making 10 using cubes, they could place the cubes in a ten frame. Do they need to count to know when they have made 10?

ANSWERS

Question 1 a): Children should know there are 10 counters without counting because the ten frame is full.

Question 1 b): There are various ways children could show 10, such as 10 fingers, two 5s on dice, or a rod of 10 cubes. The rekenreks show 10 in different ways: there are 10 red beads and 10 white beads; there are also 10 beads on each row.

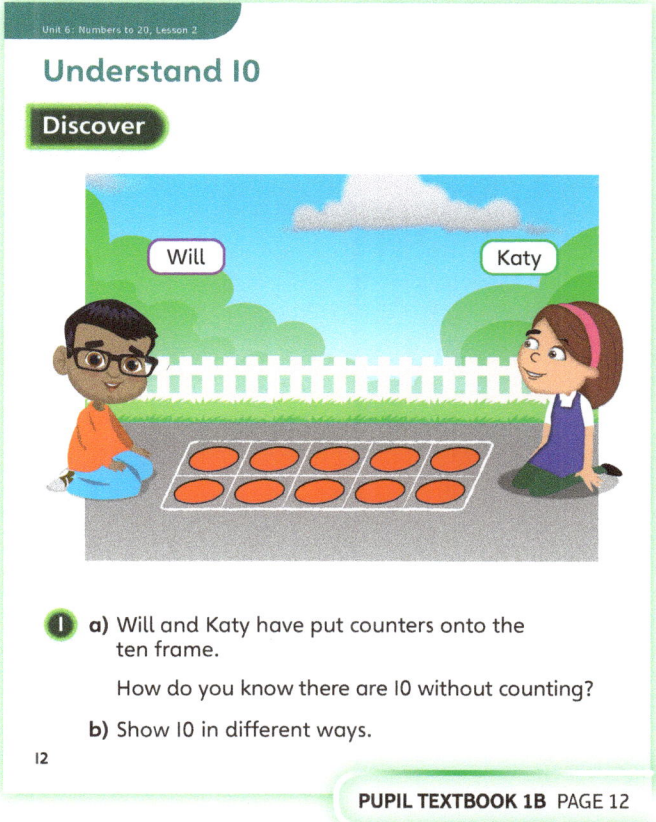

PUPIL TEXTBOOK 1B PAGE 12

Share

WAYS OF WORKING Whole class teacher led

ASK

- Question 1 a): *Did you need to count each counter in the ten frame to know how many there were?*
- Question 1 b): *How do each of these representations show 10? Did you find any other ways to make 10?*

IN FOCUS In question 1 a), children should know that, because the ten frame is full, they do not need to count. If the ten frame was not full, then they should know that there are not 10 counters. In question 1 b), take time to explore each of the representations and how they show 10. Ask children how they counted each time. Did they always need to count each item or could they subitise, for example, 5 on a dice and know that the two 5s make 10?

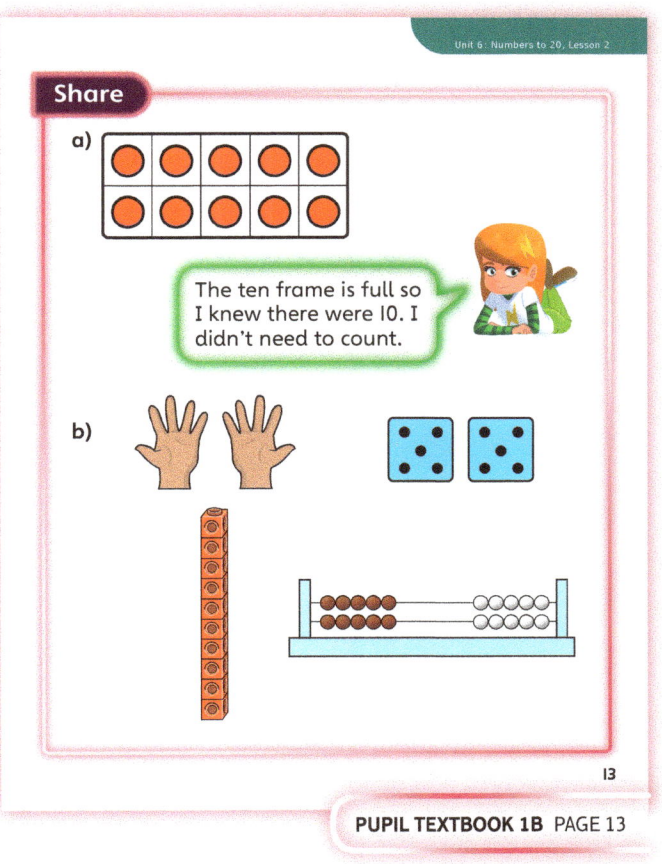

PUPIL TEXTBOOK 1B PAGE 13

Think together

WAYS OF WORKING Whole class teacher led (I do, We do, You do)

ASK

- Question ①: *How do you know that filling the ten frame will show 10?*
- Question ②: *After children have suggested one way to make 10, ask: Can you see 10 made up in any other ways?*

IN FOCUS Questions ① and ② provide opportunity for children to develop their understanding of 10. In question ②, encourage children to see 10 in different ways, for example, on each individual row or also as the two rows of red beads.

STRENGTHEN Provide children with a ten frame that they can use to build 10. Give them a rekenrek and ask them first to spot the 5s and then use their understanding of two 5s making 10 to recognise 10 in different ways.

DEEPEN In question ③, encourage children to explain their reasoning of how they decided which representation showed 10. Did they need to count every object each time? If so, how did they count? Did they always start at 1 or could they subitise some of the objects and then count on?

ASSESSMENT CHECKPOINT Use questions ① and ② to assess children's understanding of 10.

ANSWERS

Question ①: Children should use cubes or counters to fill all 10 cells in the ten frame.

Question ②: There are 10 beads on each row.
There are 10 white and 10 red beads.

Question ③ a):

Question ③ b):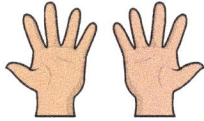

Question ③ c):

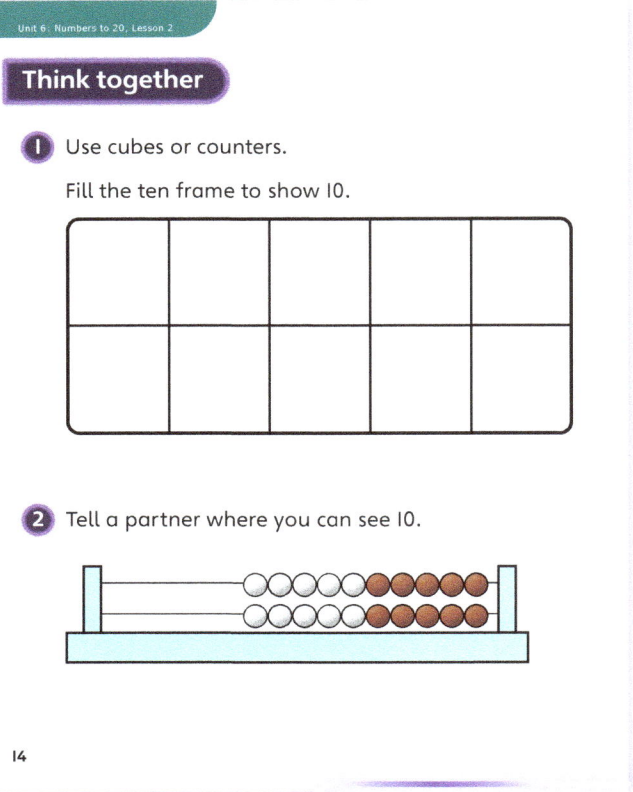

PUPIL TEXTBOOK 1B PAGE 14

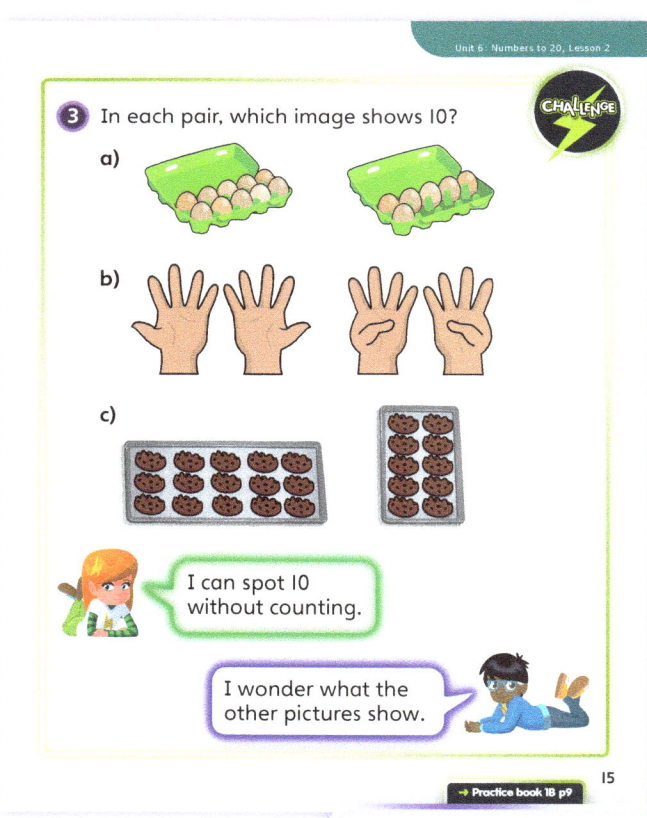

PUPIL TEXTBOOK 1B PAGE 15

Unit 6: Numbers to 20, Lesson 2

Practice

WAYS OF WORKING Independent thinking

IN FOCUS Questions ❶ and ❷ further build on children's understanding of a full ten frame showing 10. In question ❸, encourage children to explain their reasoning; did they need to count each object each time? Encourage children to subitise 10 rather than count on from 1 each time.
In question ❹, children should be able to recognise rather than count 5 on each dice, and know that they need to shade all of the dots to show 10.

STRENGTHEN Provide children with physical representations that they can use to count on and build their understanding. These should be gradually removed as children move towards being able to instantly recognise 10.

DEEPEN In question ❺, encourage children to find 10 in as many different ways as possible. Encourage children to discuss, with a partner, what is the same and what is different about each 10.

ASSESSMENT CHECKPOINT Use questions ❶ and ❷ to assess whether children understand that a full ten frame represents 10. Use questions ❸ and ❹ to assess whether children can recognise 10 rather than always having to rely on counting.

ANSWERS Answers for the **Practice** part of the lesson can be found in the *Power Maths* online subscription.

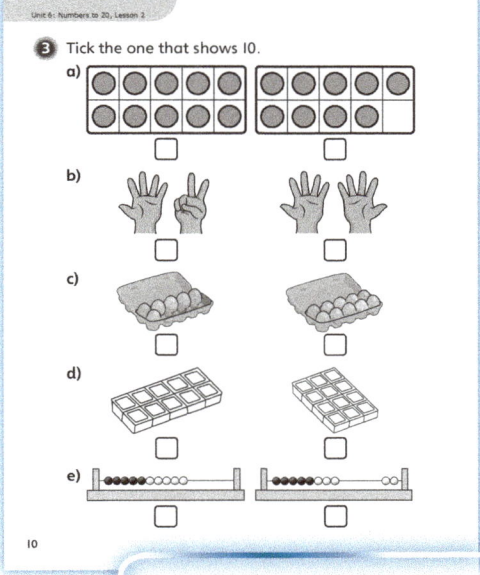

PUPIL PRACTICE BOOK 1B PAGE 9

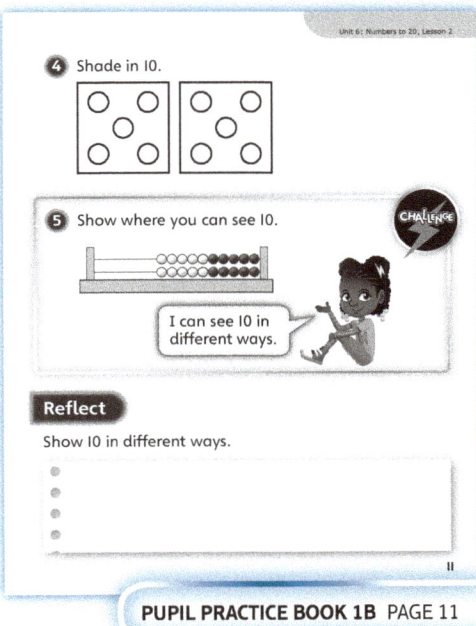

PUPIL PRACTICE BOOK 1B PAGE 10

Reflect

WAYS OF WORKING Independent thinking

IN FOCUS This activity provides children with opportunity to represent 10 in different ways. This could be drawn in their book or made physically in the classroom.

ASSESSMENT CHECKPOINT Children should be able to represent 10 in different ways and explain how they know their representation shows 10 without counting from 1 each time.

ANSWERS Answers for the **Reflect** part of the lesson can be found in the *Power Maths* online subscription.

After the lesson

- Do children know that a full ten frame shows 10?
- Can children instantly recognise 10 when presented in a standard way?
- Can children represent 10 in different ways?

Unit 6: Numbers to 20, Lesson 3

11, 12 and 13

Learning focus

In this lesson, children develop their understanding of the numbers 11, 12 and 13 as 1 ten and some ones using the '10 and a bit' structure. They recognise that each number is made up as 1 whole ten and some extra ones.

Before you teach

- Do children have a deep understanding of the structure of 10?
- Can children confidently count beyond 10 to 20?

NATIONAL CURRICULUM LINKS

Year 1 Number – number and place value

Identify and represent numbers using objects and pictorial representations including the number line, and use the language of: equal to, more than, less than (fewer), most, least.

Year 2 Number – number and place value

Recognise the place value of each digit in a two-digit number (tens, ones).

ASSESSING MASTERY

Children can recognise 11, 12 and 13 when these numbers are presented in different ways. Children understand that 11, 12 and 13 are made up of 1 whole ten and some ones. They can represent 11, 12 and 13 in different ways.

COMMON MISCONCEPTIONS

Children may write, for example, 103 instead of 13, because they can see 10 and 3 in their representations. Ask:
- *Can you put counters into ten frames to make the number 13? How many counters are there in each ten frame? How many tens are there? How many ones are there?*

STRENGTHENING UNDERSTANDING

11, 12 and 13 can often be difficult numbers for children to understand as they cannot hear the 1, 2 or 3 within the word. Spend time counting beyond 10 and making the numbers to ensure children relate the spoken word with the visual representation and the written numeral.

GOING DEEPER

Ask children to partition 11, 12 and 13 and record these partitions as a number sentence. They should be able to correctly identify the ten and the ones in any representation and show how this links to the number sentence.

KEY LANGUAGE

In lesson: 11, 12, 13, 10, ones, **more than**

Other language to be used by the teacher: partition

STRUCTURES AND REPRESENTATIONS

Ten frames

RESOURCES

Mandatory: counters, ten frames

Optional: rekenrek

 In the eTextbook of this lesson, you will find interactive links to a selection of teaching tools.

Quick recap

Count on to 20 and back from 20 as a class. Pay particular attention to the numbers 11, 12 and 13. Ask: *What number comes after 10, 11 or 12?* This allows children to understand the position of these numbers when counting.

Unit 6: Numbers to 20, Lesson 3

Discover

WAYS OF WORKING Pair work

ASK
- Question 1 a): *How many eggs fit in the box? How do you know?*
- Question 1 b): *Will there be any eggs left over? How do you know?*
- Question 1 b): *How many eggs are there in total?*

IN FOCUS This activity provides opportunity for children to understand the '10 and a bit' structure of 11. They fill their ten frame with 10 counters to represent the eggs and the egg box. They use their knowledge of counting beyond 10 to identify that, since there is 1 more egg that will not fit in the egg box, there are 11 eggs in total.

PRACTICAL TIPS Provide children with counters and ten frames to represent the question. Encourage them to count aloud.

ANSWERS
Question 1 a): 10 eggs will fill the box.
Question 1 b): There are 11 eggs on the tray.

PUPIL TEXTBOOK 1B PAGE 16

Share

WAYS OF WORKING Whole class teacher led

ASK
- Question 1 a): *How many counters are in a full ten frame? How do you know? How many eggs are in a full box?*
- Question 1 b): *Are there more than 10 eggs or fewer than 10 eggs? How do you know?*
- Question 1 b): *What number comes after 10? How many eggs are there in total?*

IN FOCUS Questions 1 a) and 1 b) provide opportunity for children to develop their understanding of 11. They use their learning from the previous lesson in question 1 a) as they fill a ten frame, or an egg box, to show 10. They then make connections between previous learning from counting as they see that the 1 egg left over means there is one more than 10, or 11, eggs in total.

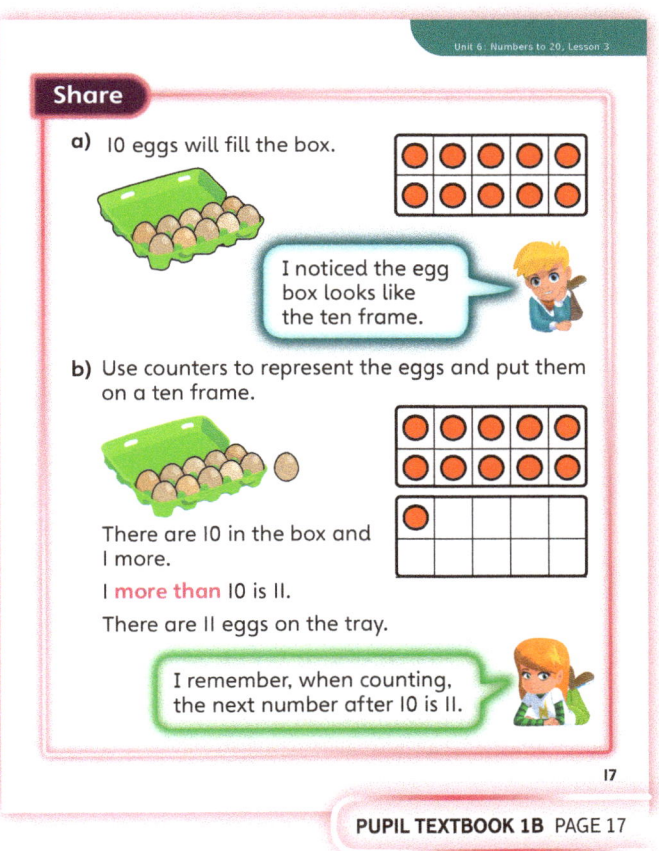

PUPIL TEXTBOOK 1B PAGE 17

Unit 6: Numbers to 20, Lesson 3

Think together

WAYS OF WORKING Whole class teacher led (I do, We do, You do)

ASK
- Question ①: *How many eggs are in a full egg box? How many more eggs are there? How many eggs are there altogether?*
- Question ②: *How many counters are in a full ten frame? How many more counters are there? How many counters are there altogether?*
- Question ②: *Did you need to count on from 1? Why or why not?*

IN FOCUS Questions ① and ② provide opportunity for children to develop their understanding of the '10 and a bit' structure of 12 and 13. They should be able to clearly see the full 10 within each representation and then count on to find the total amount.

STRENGTHEN Provide children with a ten frame and counters to represent the eggs in question ①. Once the ten frame is full, ask: *How many eggs are in the box? How do you know?* Encourage children to recognise rather than count the 10.

DEEPEN Question ③ formalises the '10 and a bit' structure and gives children opportunity to partition 11, 12 and 13 using number sentences. Encourage them to explain the representations and which bit of the number sentences they can see.

ASSESSMENT CHECKPOINT Use questions ① and ② to assess whether children understand how the numbers 12 and 13 follow on from 11. Use question ③ to assess whether children can partition 11, 12 and 13 into a ten and some ones.

ANSWERS

Question ①: There are 12 eggs.

Question ②: The number 13 is shown.

Question ③: Explanations will vary but should mention that a complete ten frame shows 10, with extra counters added on to 10.

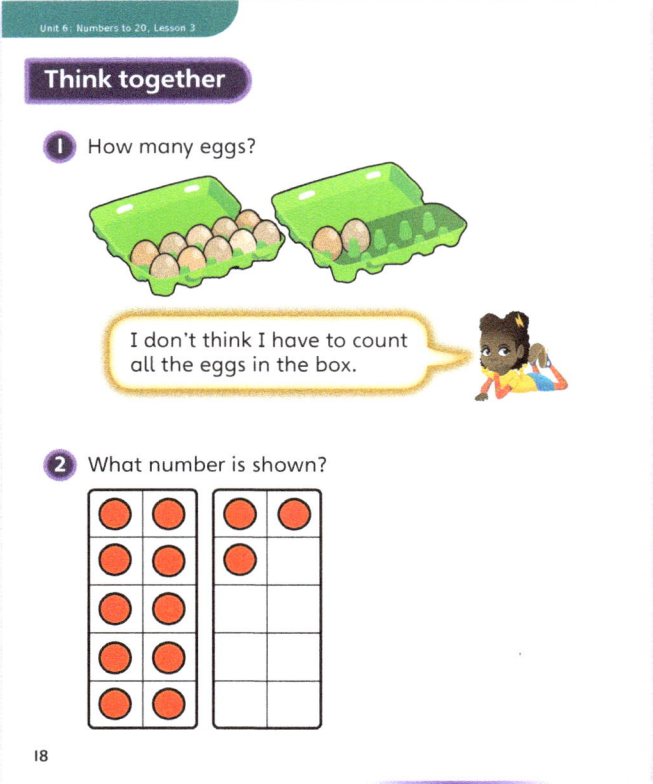

PUPIL TEXTBOOK 1B PAGE 18

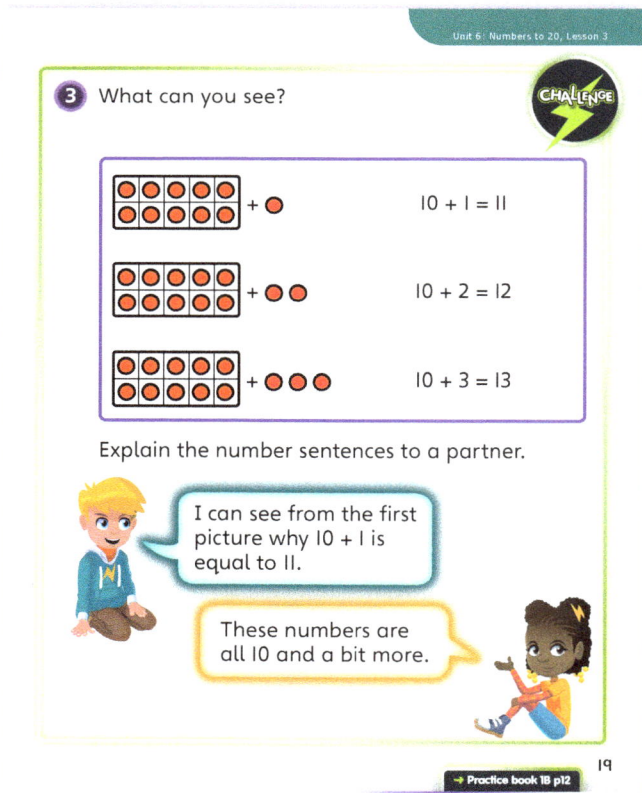

PUPIL TEXTBOOK 1B PAGE 19

Unit 6: Numbers to 20, Lesson 3

Practice

WAYS OF WORKING Independent thinking

IN FOCUS Question ❶ provides opportunity for children to represent the numbers 11, 12 and 13 with the whole 10 given. Ask: *How many counters do you need to draw in the empty ten frame? How do you know?* This will develop children's understanding of these numbers. Questions ❷, ❸ and ❹ provide opportunity for children to recognise the numbers 11, 12 and 13 from different pictorial representations. They should be encouraged to explain their reasoning.

STRENGTHEN Remind children that a full ten frame shows 10 and model counting on from 10 to support understanding, ensuring that children do not revert to counting on from 1 each time.

DEEPEN In question ❺, children identify that there are 10 black beads on each rekenrek because they are arranged in a similar way to 10 counters on a ten frame. From this, they should see that adding 2 more white beads gives 12.

In question ❻, children match the numbers to their partitioned form. They should be able to explain why each number sentence matches each number, and could be encouraged to represent the number sentences using counters or other resources.

ASSESSMENT CHECKPOINT Use questions ❶, ❸ and ❹ to assess whether children understand the structure of the numbers 11, 12 and 13. Use question ❷ to assess whether children can identify the correct representation of a number.

ANSWERS Answers for the **Practice** part of the lesson can be found in the *Power Maths* online subscription.

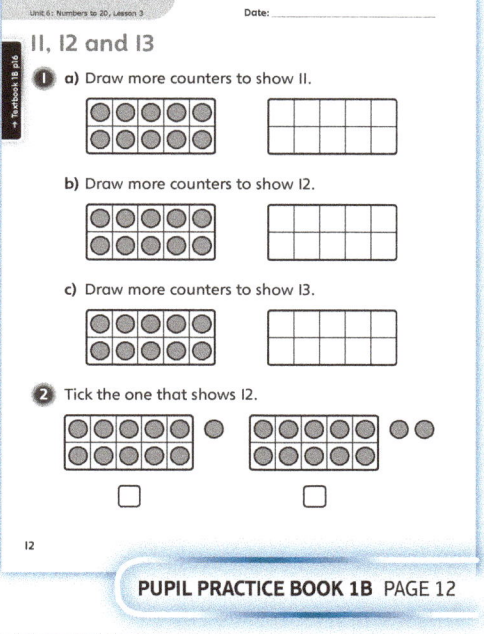

PUPIL PRACTICE BOOK 1B PAGE 12

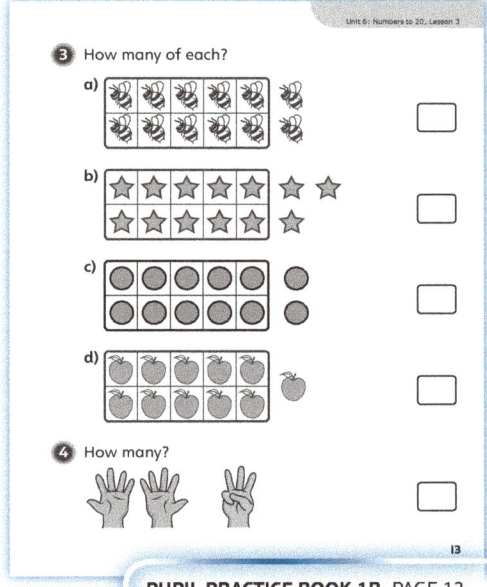

PUPIL PRACTICE BOOK 1B PAGE 13

Reflect

WAYS OF WORKING Independent thinking

IN FOCUS This activity provides opportunity for children to build the number 12 and asks them to reflect on their approach to the task.

ASSESSMENT CHECKPOINT Children should start by making 10 and realise that they only need to count on 2 more.

ANSWERS Answers for the **Reflect** part of the lesson can be found in the *Power Maths* online subscription.

After the lesson

- Do children understand that 11, 12 and 13 are made up of 1 whole ten and some ones?
- Can children recognise representations of 11, 12 and 13?
- Can children represent 11, 12 and 13 in different ways?

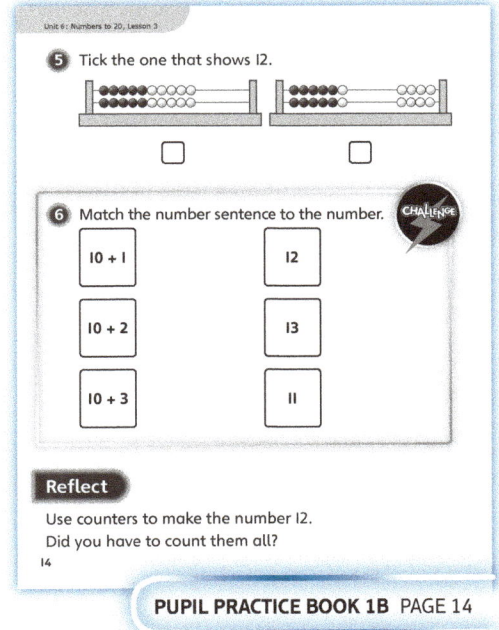

PUPIL PRACTICE BOOK 1B PAGE 14

57

Unit 6: Numbers to 20, Lesson 4

14, 15 and 16

Learning focus

In this lesson, children develop their understanding of the numbers 14, 15 and 16 as 1 ten and some ones, using the '10 and a bit' structure they saw in the previous lesson. They recognise that each number is made up of 1 whole ten and some extra ones.

Before you teach

- Do children have a deep understanding of the structure of 10?
- Can children confidently count beyond 10 to 20?
- Do children understand the structure of the numbers 11, 12 and 13?

NATIONAL CURRICULUM LINKS

Year 1 Number – number and place value

Identify and represent numbers using objects and pictorial representations including the number line, and use the language of: equal to, more than, less than (fewer), most, least.

Year 2 Number – number and place value

Recognise the place value of each digit in a two-digit number (tens, ones).

ASSESSING MASTERY

Children can recognise the numbers 14, 15 and 16 when presented in different ways. Children understand that 14, 15 and 16 are made up of 1 whole ten and some ones. They can represent 14, 15 and 16 in different ways.

COMMON MISCONCEPTIONS

Children may write 14 as 41, because they hear the 4 first in the spoken word. Similarly, children may write 15 as 105, because they can see 10 and 5 in the representations. Ask:
- Can you show 14/15 on ten frames? How many tens are there? How many ones are there? Can you write this as a number?

STRENGTHENING UNDERSTANDING

Spend time counting beyond 10 and making the numbers to ensure children relate the oral word with the visual representation and the written numeral. Model to children counting on from 10 rather than always reverting back to counting from 1.

GOING DEEPER

Ask children to partition the numbers 14, 15 and 16 and record these partitions as number sentences. They should be able to correctly identify the ten and the ones in any representation and show how this links to the number sentence.

KEY LANGUAGE

In lesson: 14, 15, 16, 10, ones

Other language to be used by the teacher: partition

STRUCTURES AND REPRESENTATIONS

Ten frames

RESOURCES

Mandatory: counters, ten frames

Optional: rekenrek, teddy bears or toy cars, bead strings

 In the eTextbook of this lesson, you will find interactive links to a selection of teaching tools.

Quick recap

Provide children with two ten frames and some counters. Ask them to show 10, encouraging them to instantly fill a ten frame rather than counting from 1. Then ask them to show 11, 12 and 13, building on their understanding from the previous lesson. Ask: *What comes next?*

Unit 6: Numbers to 20, Lesson 4

Discover

WAYS OF WORKING Pair work

ASK

- Question 1 a): *How many cars are in the car park? How do you know?*
- Question 1 b): *Where can you see 10 cars? How can you show this on a ten frame? How many more cars are there?*

IN FOCUS This activity provides opportunity for children to understand the '10 and a bit' structure of 14.

In question 1 b), they should recognise the block of 10 cars in the car park and know this will fill a ten frame. They can then count on from 10 as they add counters to the second ten frame, to recognise that there are 14 cars in total.

PRACTICAL TIPS Model counting aloud from 10 to children, adding 1 counter to the ten frame each time you count.

ANSWERS

Question 1 a): There are 14 cars in the car park.

Question 1 b): Children should show one complete 10 frame and another with 4 cells filled.

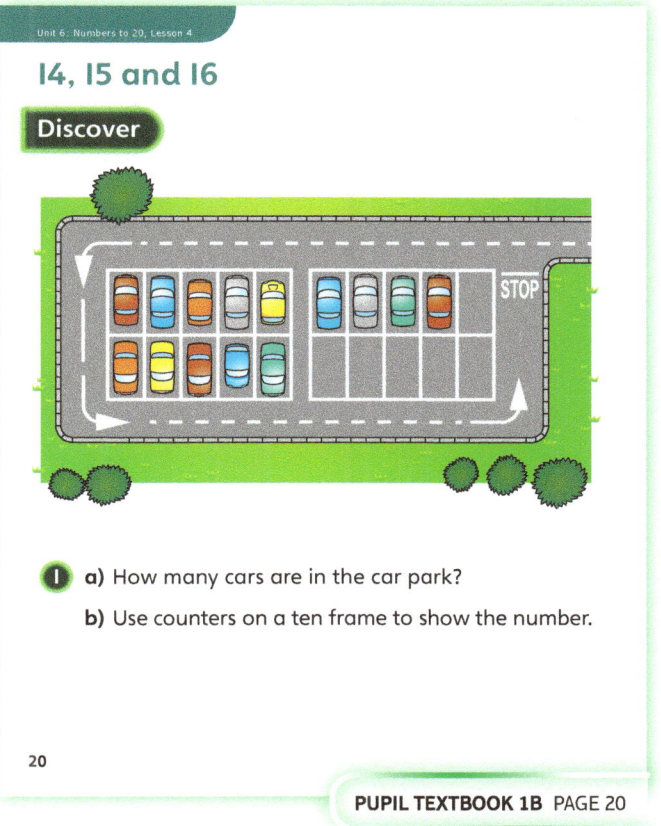

PUPIL TEXTBOOK 1B PAGE 20

Share

WAYS OF WORKING Whole class teacher led

ASK

- Question 1 a): *How did you count the cars? Whose method do you think is more efficient?*
- Question 1 b): *Why is there 1 full ten frame? How many more counters are needed? How do you know there is 1 counter for each car?*

IN FOCUS Questions 1 a) and 1 b) provide opportunity for children to develop their understanding of the number 14. They use their learning from the previous lesson as they use the '10 and a bit' structure to represent 14.

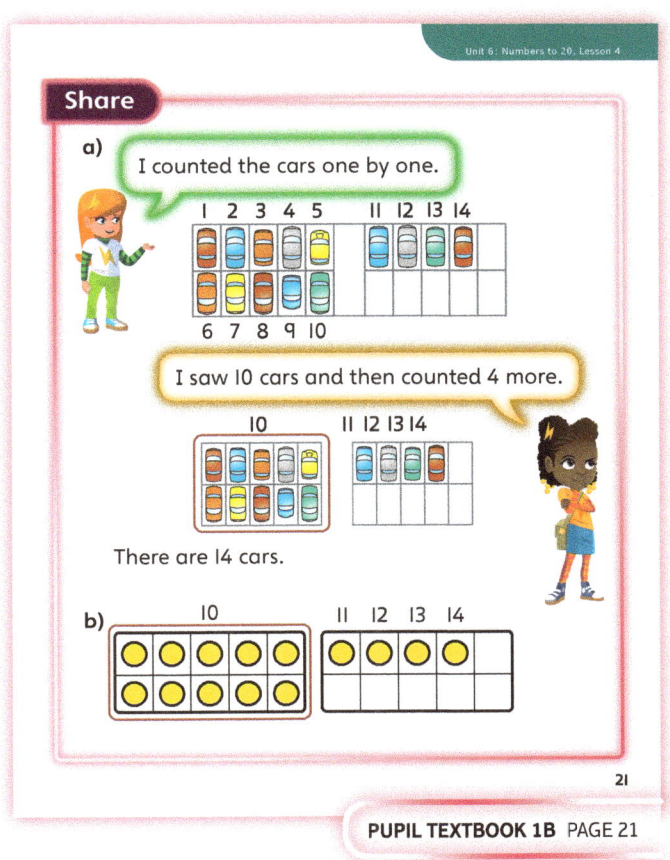

PUPIL TEXTBOOK 1B PAGE 21

Unit 6: Numbers to 20, Lesson 4

Think together

WAYS OF WORKING Whole class teacher led (I do, We do, You do)

ASK

- Question ❶: *Do you need to count on from 1? How do you know?*
- Question ❶: *How many more than 10 cars are there? How many cars are there altogether?*
- Question ❷: *Can you see a full 10? Can you see 6 more? Do they show 16?*

IN FOCUS Questions ❶ and ❷ provide opportunity for children to develop their understanding of the '10 and a bit' structure of 15 and 16. They should be able to clearly see the full 10 within each representation, and then count on to identify the total amount in each question.

STRENGTHEN Children could line up teddy bears or toy cars to practise counting one more or one less. To further reinforce their learning, children could then represent these numbers using a bead string.

DEEPEN Question ❸ formalises the '10 and a bit' structure and gives children opportunity to partition the numbers 14, 15 and 16, looking for patterns in the number sentences. Encourage them to explain any patterns they notice and make links between the written number sentences and the pictorial representations.

ASSESSMENT CHECKPOINT Use questions ❶ and ❷ to assess whether children understand the numbers 15 and 16. Use question ❸ to assess whether children can partition 14, 15 and 16 into 1 whole ten and some ones.

ANSWERS

Question ❶: There are 15 cars.

Question ❷ a): Yes, the rekenrek shows 16: 10 + 6 = 16.

Question ❷ b): Yes, the rekenrek shows 16: 8 + 8 = 16.

Question ❸ a): 10 + 4 = 14

Question ❸ b): 10 + 5 = 15

Question ❸ c): 10 + 6 = 16

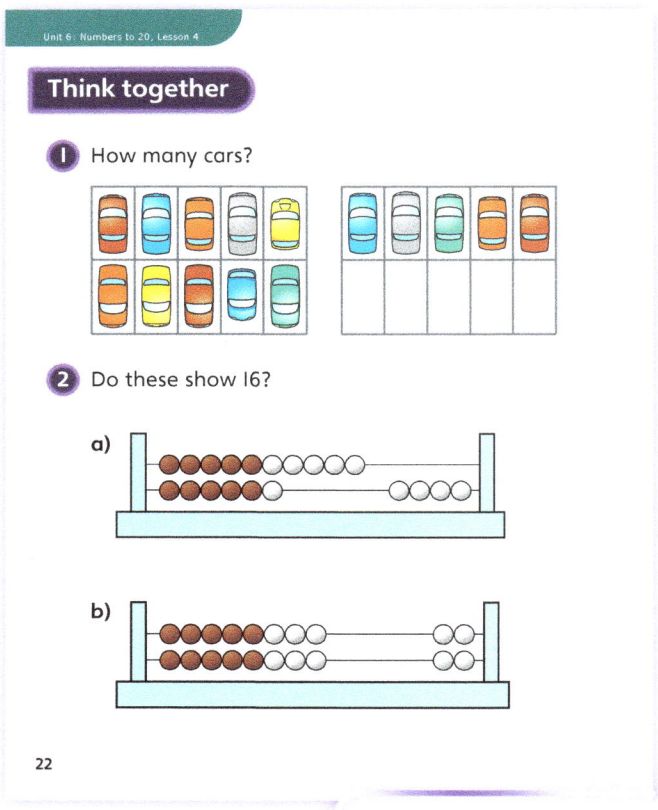

PUPIL TEXTBOOK 1B PAGE 22

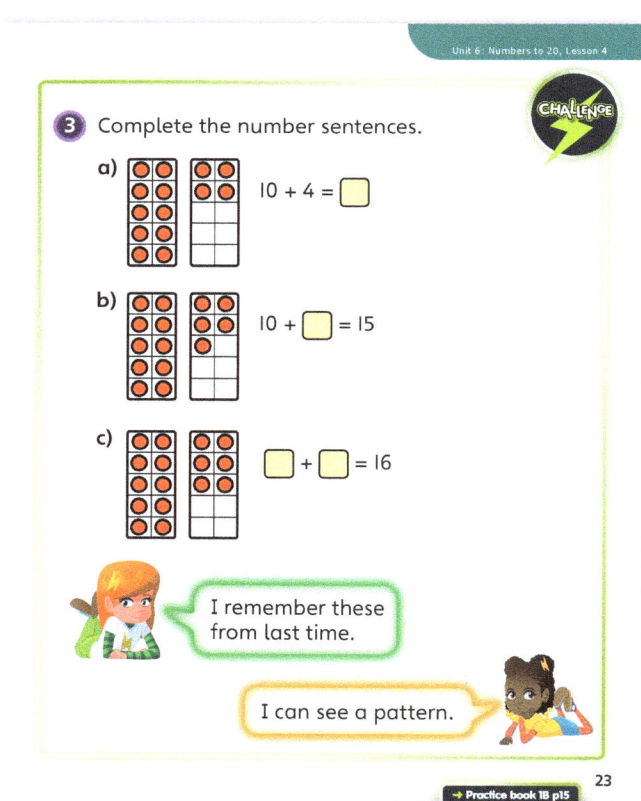

PUPIL TEXTBOOK 1B PAGE 23

60

Unit 6: Numbers to 20, Lesson 4

Practice

WAYS OF WORKING Independent thinking

IN FOCUS Question ❶ provides opportunity for children to represent the numbers 14, 15 and 16 with the whole ten given. Ask: *How many more counters do you need to draw? How do you know?* This will help to develop children's understanding of these numbers. Questions ❷, ❸ and ❹ provide opportunity for children to recognise these numbers in pictorial representations. They should be able to recognise rather than count the 10 in each representation.

STRENGTHEN Remind children that a full ten frame shows 10 and model counting on from 10 to support understanding, ensuring that children do not revert back to counting from 1 each time.

DEEPEN In question ❺, children demonstrate their understanding of the partitioning of numbers that are made up of 1 whole ten and some ones. They should be encouraged to represent each number on ten frames and demonstrate how each part of the number sentence is shown in their representation.

ASSESSMENT CHECKPOINT Use questions ❶, ❷ and ❸ to assess children's understanding of the numbers 14, 15 and 16.

ANSWERS Answers for the **Practice** part of the lesson can be found in the *Power Maths* online subscription.

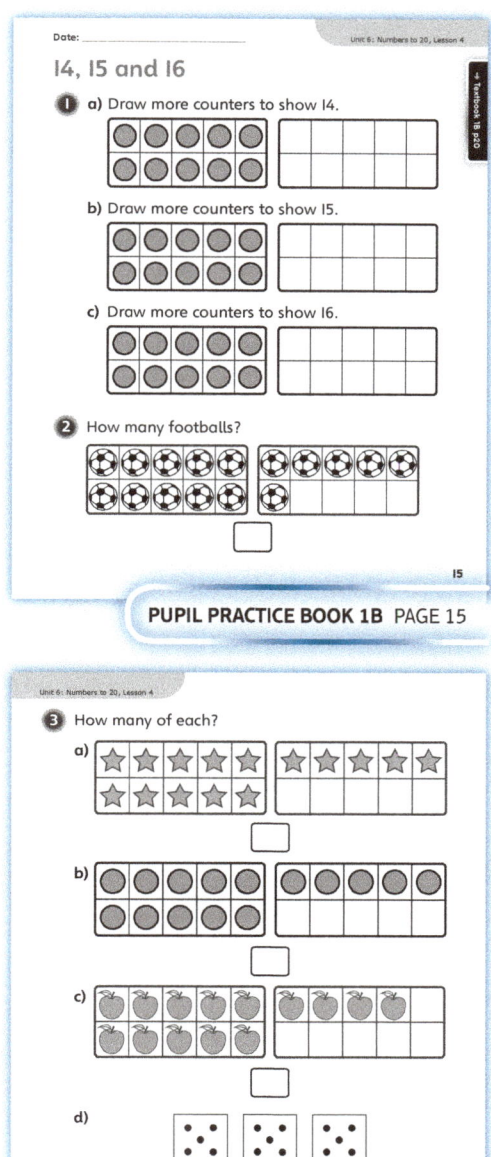

Reflect

WAYS OF WORKING Independent thinking

IN FOCUS This activity provides opportunity for children to demonstrate their understanding of the number 15.

ASSESSMENT CHECKPOINT Children should be able to build 10 by filling a ten frame and then count on 5 more. They could even use their understanding of 5 being half a full ten frame to not have to count any of the counters.

ANSWERS Answers for the **Reflect** part of the lesson can be found in the *Power Maths* online subscription.

After the lesson ⏸

- Do children understand that 14, 15 and 16 are made up of 1 whole ten and some ones?
- Can children recognise representations of 14, 15 and 16?
- Can children represent 14, 15 and 16 in different ways?

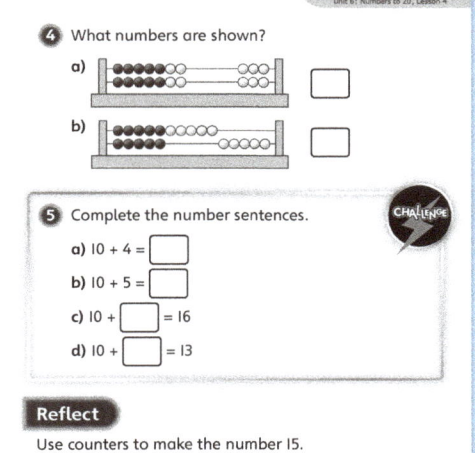

Unit 6: Numbers to 20, Lesson 5

17, 18 and 19

Learning focus
In this lesson, children develop their understanding of 17, 18 and 19 as 1 ten and some ones using the '10 and a bit' structure they saw in the previous two lessons. They recognise that each number is made up of 1 whole ten and some extra 1s.

Before you teach
- Do children have a deep understanding of the structure of 10?
- Can children confidently count beyond 10 to 20?
- Do children understand the structure of the numbers from 11 to 16?

NATIONAL CURRICULUM LINKS

Year 1 Number – number and place value

Identify and represent numbers using objects and pictorial representations including the number line, and use the language of: equal to, more than, less than (fewer), most, least.

Year 2 Number – number and place value

Recognise the place value of each digit in a two-digit number (tens, ones).

ASSESSING MASTERY

Children can recognise 17, 18 and 19 when presented in different ways. Children understand that 17, 18 and 19 are made up of 1 whole ten and some 1s. They can represent 17, 18 and 19 in different ways.

COMMON MISCONCEPTIONS

Children may write 17 as 71, because they hear the 7 first in the spoken word. Children may write 18 as 108, as they can see 10 and 8 in the representations. Ask:
- Can you show 17/18 on ten frames? How many tens are there? How many ones are there? Can you write this as a number?

STRENGTHENING UNDERSTANDING

Spend time counting beyond 10 and making the numbers to ensure children relate the oral word with the pictorial representation and the numeral. Model to children counting on from 10 rather than always reverting to counting from 1.

GOING DEEPER

Ask children to partition the numbers 17, 18 and 19 and record these partitions as number sentences. They should be able to correctly identify the ten and the ones in any representation and show how this links to the number sentence.

KEY LANGUAGE

In lesson: 17, 18, 19, 10, ones

Other language to be used by the teacher: partition

STRUCTURES AND REPRESENTATIONS

Ten frames

RESOURCES

Mandatory: counters, ten frames

Optional: rekenrek

 In the eTextbook of this lesson, you will find interactive links to a selection of teaching tools.

Quick recap

Spend time looking at the numbers from 11 to 16. Show children representations of these numbers where the 10 is clearly visible. Ask: *How many altogether?* Ask children to explain how they know, ensuring they do not count on from 1 each time.

Unit 6: Numbers to 20, Lesson 5

Discover

WAYS OF WORKING Pair work

ASK

- Question ❶ a): *How many red counters are on the ten frame? How many yellow counters are on the ten frame? How do you know?*
- Question ❶ b): *How many more counters do you need to add to make 18 and 19?*

IN FOCUS This activity provides opportunity for children to develop their understanding of the '10 and a bit' structure of 17, 18 and 19. In question ❶ a), they should recognise the 10 shown in the full ten frame, and count on to identify what number the children have made. In question ❶ b), they then use their understanding from counting earlier in the unit to see that 1 more counter makes 18, and another 1 makes 19.

PRACTICAL TIPS Model counting aloud from 10 to children, adding 1 counter to the ten frame each time you count.

ANSWERS

Question ❶ a): The children have made the number 17: 10 + 7 = 17.

Question ❶ b): Children should show one complete ten frame plus 8 in the second ten frame: 10 + 8 = 18.
Children should show one complete ten frame plus 9 in the second ten frame: 10 + 9 = 19.

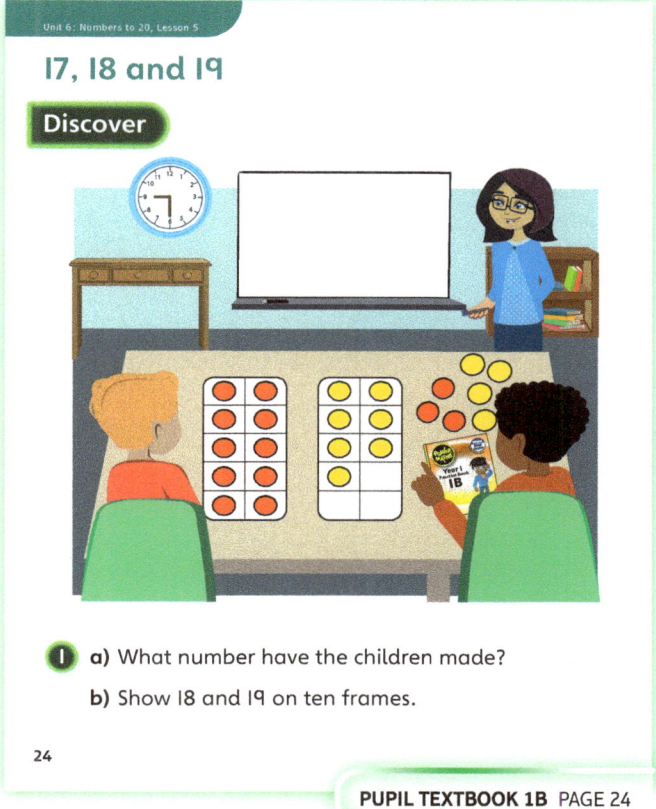

PUPIL TEXTBOOK 1B PAGE 24

Share

WAYS OF WORKING Whole class teacher led

ASK

- Question ❶ a): *Where is the full 10? Where are the 7 more? How do you write this as a number?*
- Question ❶ b): *How many more counters do you need to add to make 18? How do you know? How many more counters do you need to add to make 19? How do you know?*

IN FOCUS Questions ❶ a) and ❶ b) provide opportunity for children to develop their understanding of 17, 18 and 19. They use their learning from the previous lessons as they use the '10 and a bit' structure to understand these numbers.

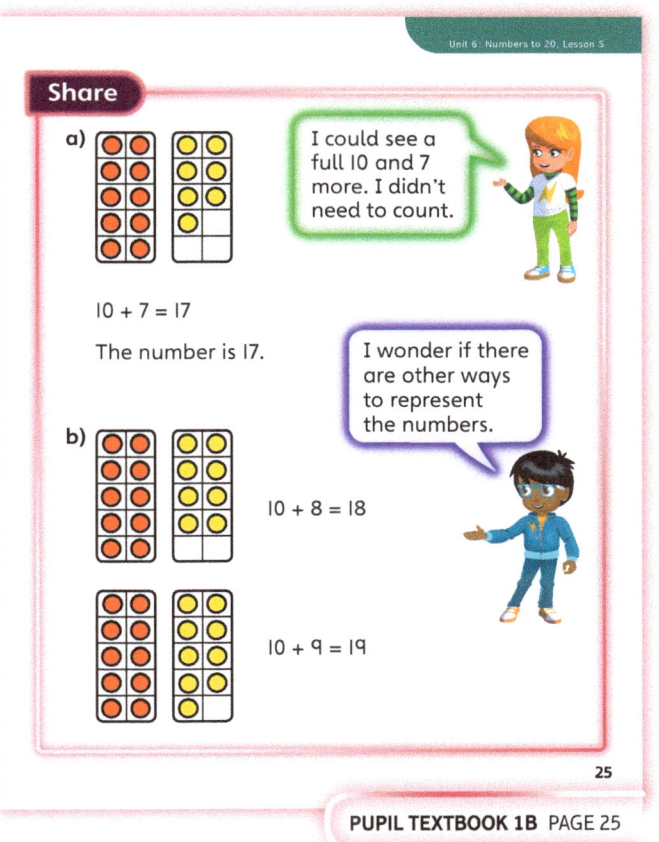

PUPIL TEXTBOOK 1B PAGE 25

Unit 6: Numbers to 20, Lesson 5

Think together

WAYS OF WORKING Whole class teacher led (I do, We do, You do)

ASK
- Question ❶: *Do you need to count from 1? How do you know? Where can you see a full 10? How many more are there? How many altogether?*
- Question ❷: *Where is the full 10? How many more are there? What number is it? When you count back from 20, what number comes next? How does that help you know how many counters there are?*

IN FOCUS Questions ❶ and ❷ provide opportunity for children to develop their understanding of the '10 and a bit' structure of 17, 18 and 19, as well as their understanding of 20 being made up of 2 whole tens. They should be able to clearly see any full 10s within a representation, and use their understanding of the '10 and a bit' structure to find how many altogether.

STRENGTHEN Provide children with a ten frame and counters to represent the questions and support their understanding. Model filling the ten frame without counting, then counting on from 10 rather than counting on from 1.

DEEPEN Question ❸ provides opportunity for children to understand the partitions of all numbers from 11 to 19, and write these as number sentences. Encourage them to use ten frames and counters to model each partition and show how the numbers are related.

ASSESSMENT CHECKPOINT Use questions ❶ and ❷ to assess whether children understand and can recognise the numbers 17, 18 and 19, as well as the number 20.

ANSWERS

Question ❶: a) 18 = 9 + 9

Question ❶: b) 17 = 10 + 7

Question ❷: Both children are correct: one fewer than 20 is 19 and 10 + 9 = 19.

Question ❸: 10 + 1 = 11
10 + 2 = 12
10 + 3 = 13
10 + 4 = 14
10 + 5 = 15
10 + 6 = 16
10 + 7 = 17
10 + 8 = 18
10 + 9 = 19
All answers have 1 ten.
The number of ones is different each time.

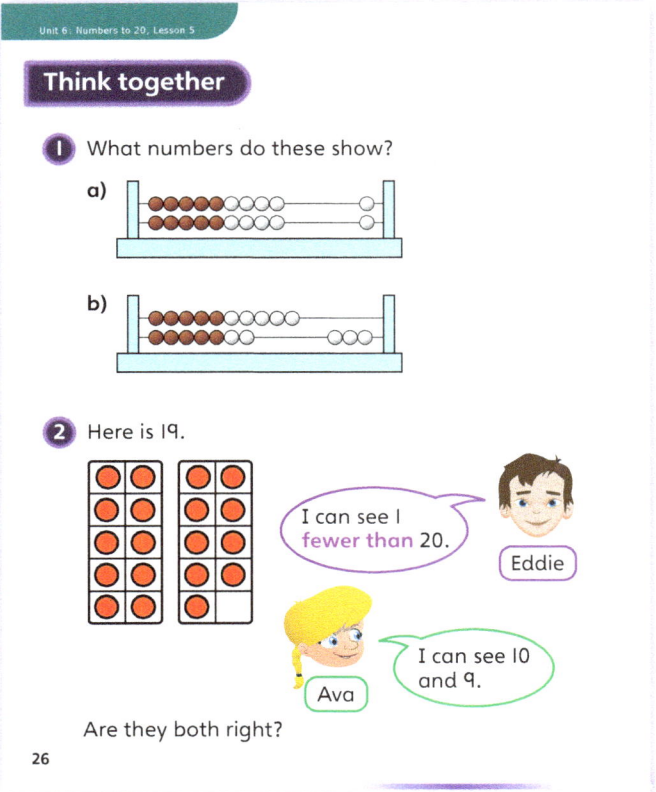

PUPIL TEXTBOOK 1B PAGE 26

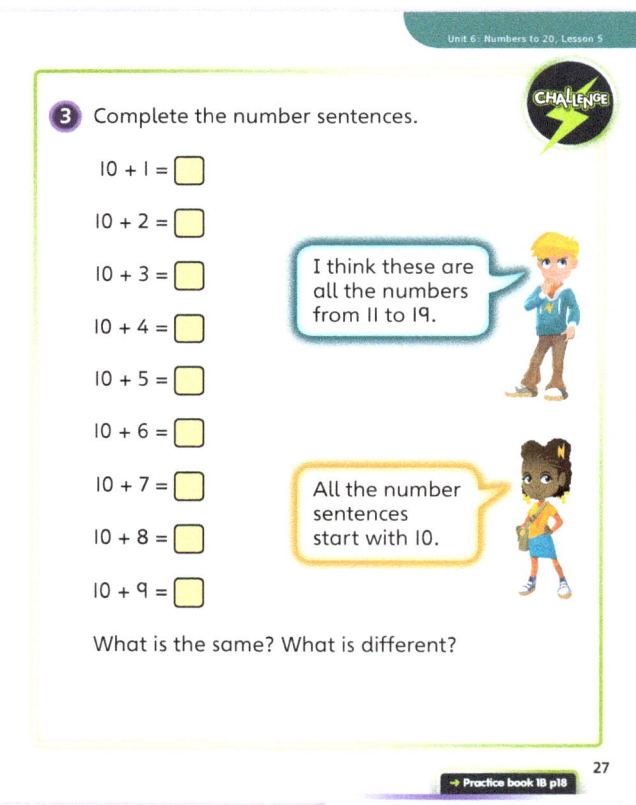

PUPIL TEXTBOOK 1B PAGE 27

Unit 6: Numbers to 20, Lesson 5

Practice

WAYS OF WORKING Independent thinking

IN FOCUS Question ① provides opportunity for children to match counters on ten frames to the correct numeral. Ask them to explain their thinking each time and whether they needed to count the 10. Question ② develops children's understanding of the '10 and a bit' structure of 18. Questions ③ and ④ provide opportunity for children to recognise 17, 18 and 19. They should be encouraged to explain how they counted each time.

STRENGTHEN Remind children that a full ten frame shows 10. Model counting on from 10 to support understanding, ensuring that children do not revert to counting from 1 each time.

DEEPEN Question ⑥ provides opportunity for children to demonstrate their understanding of the partitioning of numbers that are made up of 1 whole ten and some ones, including finding missing numbers in number sentences. Children should be encouraged to represent each number sentence and explain how each part of the number sentence can be seen in their representation.

THINK DIFFERENTLY In question ⑤, children may automatically start by identifying the numbers that are represented. Encourage them to consider other ways of deciding, such as the three spaces on the ten frames mirroring the three beads on the right in the rekenrek.

ASSESSMENT CHECKPOINT Use questions ① to ④ to assess whether children can represent and recognise the numbers 17, 18 and 19.

ANSWERS Answers for the **Practice** part of the lesson can be found in the *Power Maths* online subscription.

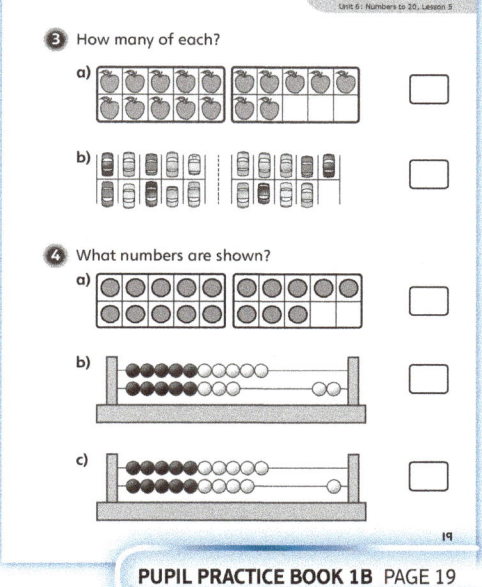

PUPIL PRACTICE BOOK 1B PAGE 18

PUPIL PRACTICE BOOK 1B PAGE 19

PUPIL PRACTICE BOOK 1B PAGE 20

Reflect

WAYS OF WORKING Pair work

IN FOCUS In this activity, children bring together their learning from this lesson and the previous two lessons as they make a number from 11 to 19. They work with a partner, asking their partner to identify the number they have made.

ASSESSMENT CHECKPOINT Children should be able to recognise their partner's number and explain the structure of it as being 1 whole ten and some ones.

ANSWERS Answers for the **Reflect** part of the lesson can be found in the *Power Maths* online subscription.

After the lesson

- Do children understand that 17, 18 and 19 are made up of 1 whole ten and some ones?
- Can children represent and recognise the numbers 17, 18 and 19?

Unit 6: Numbers to 20, Lesson 6

Understand 20

Learning focus
In this lesson, children recognise that 20 is made up of 2 tens using representations they are familiar with. Children use their fingers, ten frames and a rekenrek to instantly recognise 20, and that it is clearly made up of 2 tens.

Before you teach
- Are children confident counting to 10?
- Can children say each number to 20 fluently?
- Are children confident using ten frames and can they instantly recognise 10 when using familiar representations?

NATIONAL CURRICULUM LINKS

Year 1 Number – number and place value

Identify and represent numbers using objects and pictorial representations including the number line, and use the language of: equal to, more than, less than (fewer), most, least.

Read and write numbers from 1 to 20 in numerals and words.

ASSESSING MASTERY

Children can instantly recognise 20 when using familiar representations such as their fingers, ten frames and a rekenrek or bead string. Children understand that 20 is made up of 2 tens.

COMMON MISCONCEPTIONS

Children may want to count each time to check there are 10 objects. The aim is for children to be able to recognise 20 as 2 tens, so this should be reinforced throughout the lesson. Ask:
- *If a ten frame is full, what number does it show? If two ten frames are full, what number do they show?*

STRENGTHENING UNDERSTANDING

Use songs, rhymes and books with children to explore 20 in a fun and engaging way. Each time you use a song, rhyme or book, ask children to act it out using concrete manipulatives, such as counters and a ten frame.

GOING DEEPER

Children could begin to recognise 20 as 2 tens in numerals rather than seeing the 2 tens as representations where the 1s are visible and countable.

KEY LANGUAGE

In lesson: make, show, **tens**

STRUCTURES AND REPRESENTATIONS

Ten frames

RESOURCES

Mandatory: ten frames and counters

Optional: number tracks, rekenrek, dice, cubes, dominoes and other general objects from around the classroom

 In the eTextbook of this lesson, you will find interactive links to a selection of teaching tools.

Quick recap
Make 10 in different ways, such as on a ten frame, with a rekenrek or with a tower of cubes.

Unit 6: Numbers to 20, Lesson 6

Discover

WAYS OF WORKING Pair work

ASK

- Question 1 a): *How many fingers are on one of the children's hands? How many fingers are on two of the children's hands? How many fingers are on three of the children's hands? How many fingers are on four of the children's hands?*
- Question 1 b): *How many counters fill 1 ten frame? How many counters fill 2 ten frames?*

IN FOCUS In questions 1 a) and 1 b), children think about the composition of 20 by considering fingers and ten frames. Children are already familiar with both of these representations, so should see the link between 2 tens being the same as 20.

PRACTICAL TIPS Build 20 on ten frames together alongside a rhyme or song to 20.

ANSWERS

Question 1 a): Both children are showing 10 fingers. 10 + 10 = 20. Yes, the children are showing 20.

Question 1 b): Children should show two full ten frames.

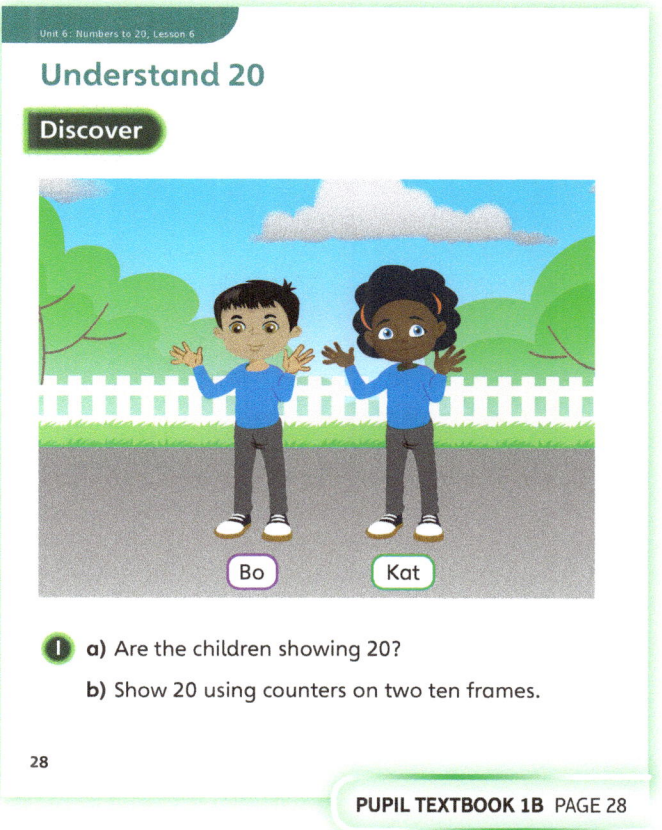

PUPIL TEXTBOOK 1B PAGE 28

Share

WAYS OF WORKING Whole class teacher led

ASK

- Question 1 a): *How many fingers are the children showing? How do you know?*
- Question 1 b): *When a ten frame is full, how many counters are there? How many ten frames do you need to show 20 counters?*

IN FOCUS Children use representations they are familiar with to see that 20 is made up of 2 tens. In question 1 b), children do not need to count to see that there are 20 counters in total.

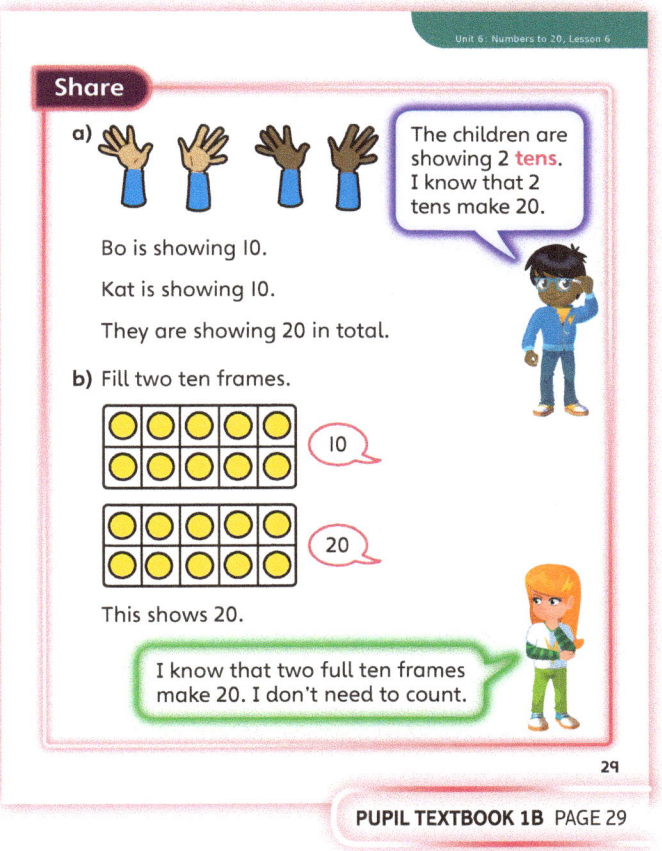

PUPIL TEXTBOOK 1B PAGE 29

Think together

WAYS OF WORKING Whole class teacher led (I do, We do, You do)

ASK

- Question ❶: *How many red counters are there? How many yellow counters are there? How many tens are there? How many tens make 20?*
- Question ❷: *What is the difference between the rekenreks? Which one shows 2 tens?*
- Question ❸: *Can you think of another way to make 20? Could you use dominoes?*

IN FOCUS Children see 20 as 2 tens using different representations. They could also start to think about 20 in different ways, such as with four dice each showing 5. Questions ❶ and ❷ allow children to become confident with representations where the ones are visible, first with ten frames and counters, rekenreks and bead strings, before moving on to question ❸ where the ones are not always visible.

STRENGTHEN Count out 20 counters onto two ten frames to reinforce the fact that each full ten frame represents 10 and two full ten frames represent 20. You could do the same with a rekenrek, counting each bead individually. This helps children see for themselves that there are 20 counters or beads rather than just being told.

DEEPEN Ask children to experiment with making 20 in different ways, such as with combinations of dice or dominoes. They could get creative and realise that two lots of a 6 and a 4 on a dice also make 20, alongside four 5s.

ASSESSMENT CHECKPOINT Check that children can instantly recognise 10 and 20 when using standard, familiar representations such as full ten frames or full rows on a rekenrek.

ANSWERS

Question ❶: Yes, two full ten frames show 20.

Question ❷: The top rekenrek shows 20 = 10 + 10. The bottom rekenrek shows 19 = 10 + 9.

Question ❸: There are various ways children could show 20 in addition to those shown on the page, such as 20 fingers or two full ten frames.

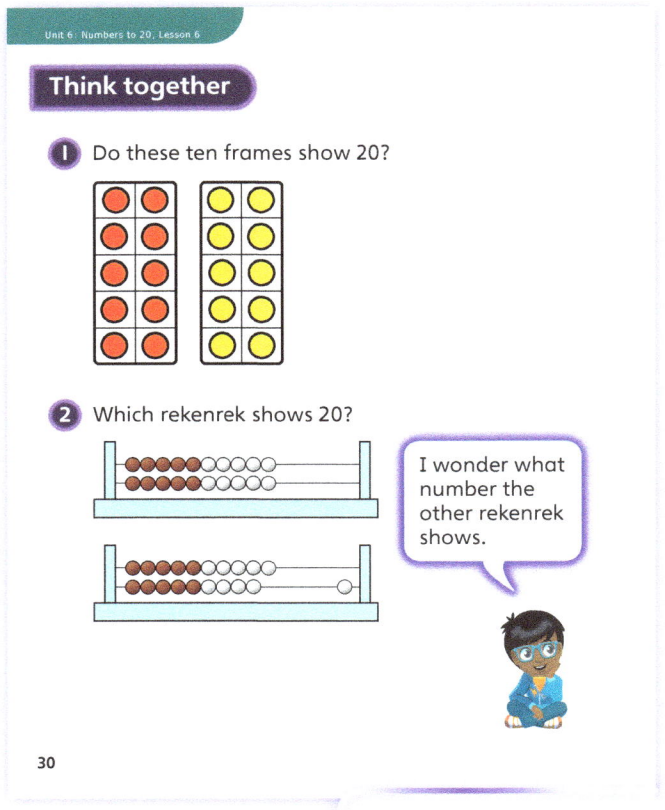

PUPIL TEXTBOOK 1B PAGE 30

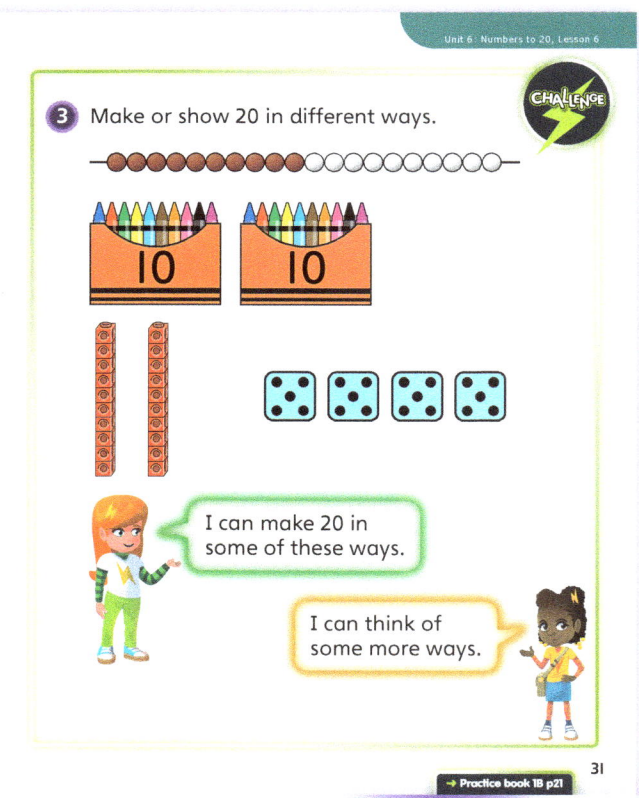

PUPIL TEXTBOOK 1B PAGE 31

Unit 6: Numbers to 20, Lesson 6

Practice

WAYS OF WORKING Independent thinking

IN FOCUS Children recognise 20 in a variety of ways and also recognise what is not 20. Questions ❶ and ❷ reinforce that 20 is made from 2 tens. Question ❸ shows what 20 is and what 20 is not. This helps children recognise 20 as 2 tens.

STRENGTHEN Use the concrete representations of the pictures used in this section and encourage children to count to check their answers.

DEEPEN Build on question ❷ b) by experimenting with dice. Can children make 20 in a different way to using 4 fives? Question ❹ starts to expose children to more abstract style questions where a pictorial representation is not included for support.

THINK DIFFERENTLY Question ❺ gives children the opportunity to recognise 20 as 2 tens, with no ones visible.

ASSESSMENT CHECKPOINT Children should be able to recognise 20 as 2 tens without counting. They should know that one full ten frame or one full row on a rekenrek is equal to 10 and, therefore, two of these is equal to 20.

ANSWERS Answers for the **Practice** part of the lesson can be found in the *Power Maths* online subscription.

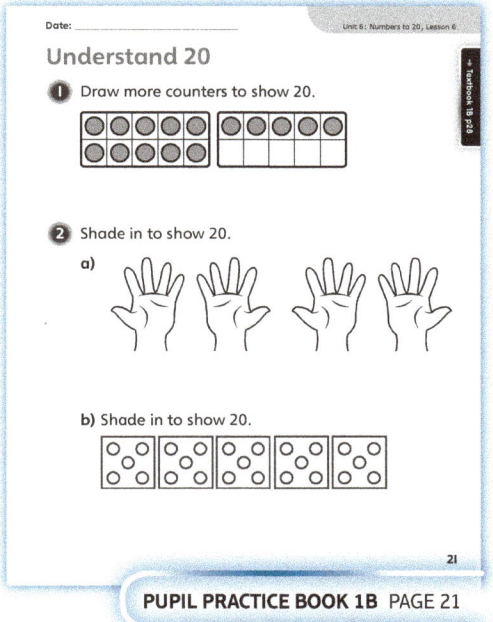
PUPIL PRACTICE BOOK 1B PAGE 21

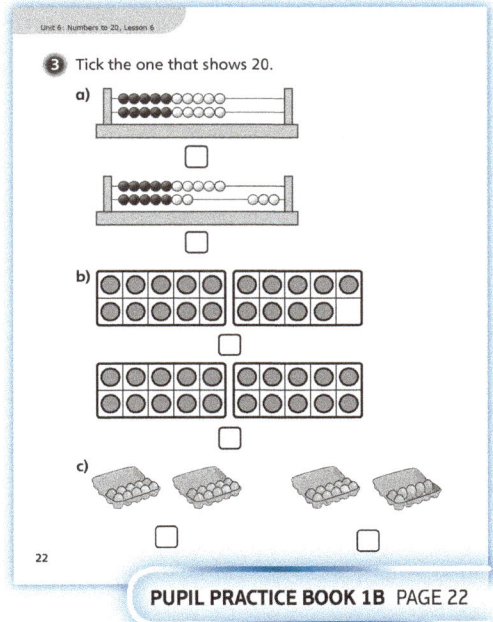
PUPIL PRACTICE BOOK 1B PAGE 22

Reflect

WAYS OF WORKING Independent thinking

IN FOCUS Children show 20 in different ways using objects or representations of their choice. Pay attention to how creative the children are and if they are able to use representations that have not been used in the lesson.

ASSESSMENT CHECKPOINT Assess what children use to show 20. Some may choose ten frames or rekenreks as they have already been using them in the lesson. Other, more confident children may choose something completely different. This would show that they have a deeper understanding of the concept.

ANSWERS Answers for the **Reflect** part of the lesson can be found in the *Power Maths* online subscription.

After the lesson

- Are children able to recognise 10 without counting?
- Do children know that 2 tens make 20?
- How many different ways can children show 20?

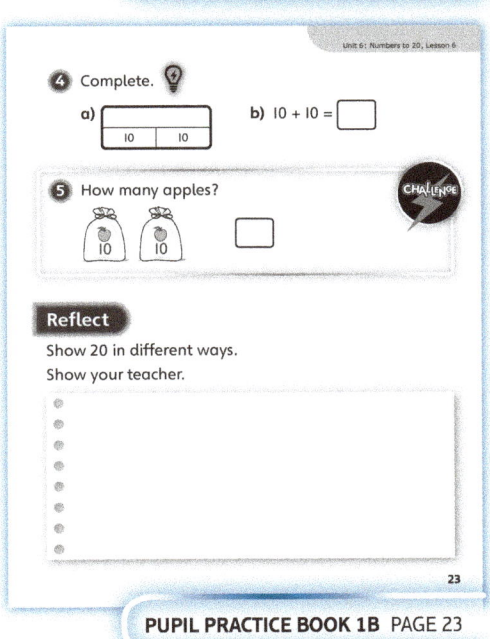
PUPIL PRACTICE BOOK 1B PAGE 23

Unit 6: Numbers to 20, Lesson 7

One more and one less

Learning focus

In this lesson, children will apply their counting skills to find one more and one less than any number within 20. Children have already been exposed to the language of more and less but this may need reinforcing with real-life examples. Representations such as ten frames are useful for showing one more and one less. Towers of cubes are particularly useful to show the one more pattern of consecutive numbers really clearly.

Before you teach

- Do children understand the language of 'more' and 'less'?
- Are children secure counting to 20? Do they miss numbers out or get them in the wrong order?
- Do children have a secure understanding of 10? Do they know that when a ten frame is full, it represents 10?

NATIONAL CURRICULUM LINKS

Year 1 Number – number and place value

Identify and represent numbers using objects and pictorial representations including the number line, and use the language of: equal to, more than, less than (fewer), most, least.

Given a number, identify one more and one less.

ASSESSING MASTERY

Children can find one more or one less than any number to 20. This includes finding one more than 0. They can show their answers using concrete manipulatives and also find one more and one less of a number abstractly.

COMMON MISCONCEPTIONS

Children may mix up the vocabulary of 'more' and 'less'. Give them plenty of practice using real-life examples in context. Some children may add one more ten rather than one more one. Concrete resources will help overcome this. Ask:
- How can you use cubes to show one more? What do you need to do to show more? Add cubes or remove some?

STRENGTHENING UNDERSTANDING

Showing a number track alongside concrete resources is a great way to help children secure their understanding of the concept. Note that children have not yet learned about the number line to 20.

GOING DEEPER

Children can be challenged with problems where one more or one less is given and they have to find the starting number.

KEY LANGUAGE

In lesson: one more, one less, add, take away, **less than**

Other language to be used by the teacher: fewer

STRUCTURES AND REPRESENTATIONS

Ten frames, bead strings, rekenreks

RESOURCES

Mandatory: ten frames, counters

Optional: bead strings, rekenreks, number tracks

 In the eTextbook of this lesson, you will find interactive links to a selection of teaching tools.

Quick recap

Ask children to show you 'more' than something to see if they understand the word. Checking 'less' is more difficult to do as children would show you 'fewer' counters, for example, not 'less' counters. Be careful with this so that children are using the correct vocabulary between 'less' and 'fewer'.

Unit 6: Numbers to 20, Lesson 7

Discover

WAYS OF WORKING Pair work

ASK
- Question 1 a): *How many ten frames will we need? How do you know?*
- Question 1 b): *How can we show one more?*

IN FOCUS Question 1 a) requires children to link concrete and pictorial representations of the number 12. In question 1 b), children see the concept of one more in a real-life context.

PRACTICAL TIPS Ask some children to line up. Say out loud the number of children in the line. Ask one child to join the line and say the number that is one more. Ask one child to leave and say the number that is one less. Half the class could be in the line while the other half models with ten frames, then they could swap.

ANSWERS

Question 1 a): There are 12 children in the line. Children should show one complete ten frame plus 2 counters in a second ten frame.

Question 1 b): If one more child joins, there are now 13 children in the line. Children should show one complete ten frame plus 3 counters in a second ten frame.

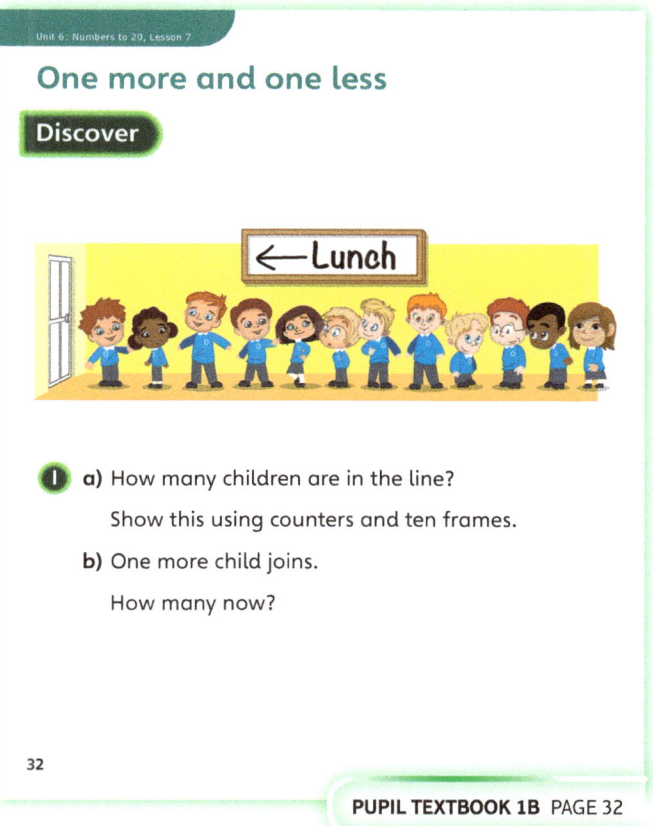

PUPIL TEXTBOOK 1B PAGE 32

Share

WAYS OF WORKING Whole class teacher led

ASK
- Question 1 a): *How many tens are there? How many more counters do you need? What is 10 add 2 equal to?*
- Question 1 b): *How many counters do we have now?*

IN FOCUS Using ten frames alongside the practical activity reinforces children's prior learning. In question 1 a), they should be able to see that they have 12 counters because they have one full ten frame and 2 ones. In question 1 b), adding one more in a different colour helps children visually see the concept of one more.

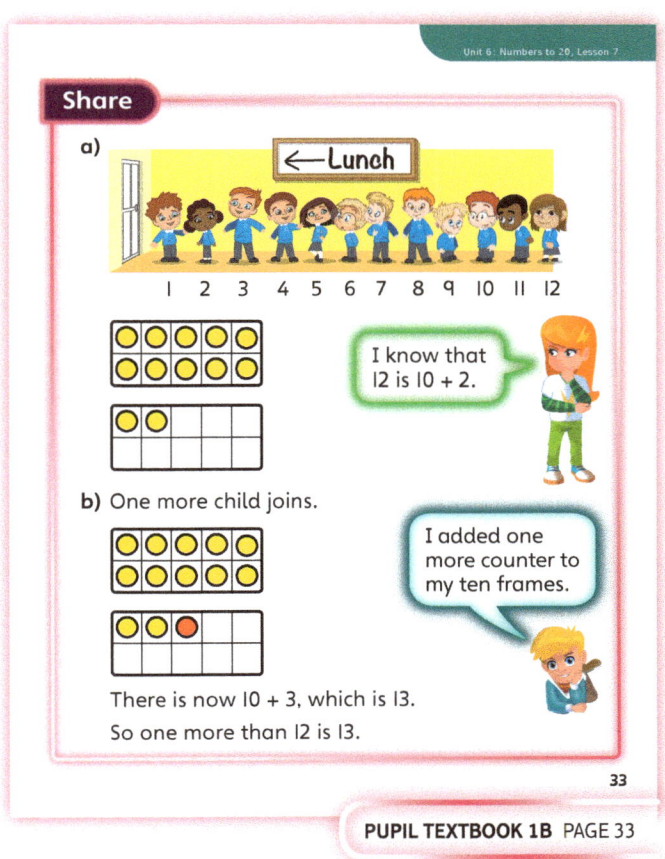

PUPIL TEXTBOOK 1B PAGE 33

Think together

WAYS OF WORKING Whole class teacher led (I do, We do, You do)

ASK
- Question 1: *How many tens and ones do you have? Add one more. How many do you have now? Is the number bigger or smaller when you add one more?*
- Question 2: *When you find one less, will the number get bigger or smaller?*
- Question 3: *Can you imagine a ten frame? How many counters would you have? Is finding one more and one less than a number just like counting?*

IN FOCUS Two pictorial representations are used in questions 1 and 2 to help children see the concept of one more and one less. Children then move on to abstract representations of one more and one less in question 3, where they should try to visualise a pictorial representation.

STRENGTHEN Use ten frames and bead strings or rekenreks alongside the questions. Use songs and rhymes to reinforce the concept.

DEEPEN Children can be challenged with a problem where one more or one less is given and they have to find the starting number.

ASSESSMENT CHECKPOINT Can children find one more and one less than a given number using practical equipment to support them? Check that they understand the vocabulary and can link it to an amount getting bigger or smaller.

ANSWERS

Question 1 a): One more than 14 is 15.

Question 1 b): One more than 17 is 18.

Question 2 a): One less than 17 is 16.

Question 2 b): One less than 11 is 10.

Question 3 a): 1 more than 15 is 16.

Question 3 b): 1 more than 19 is 20.

Question 3 c): 1 less than 13 is 12.

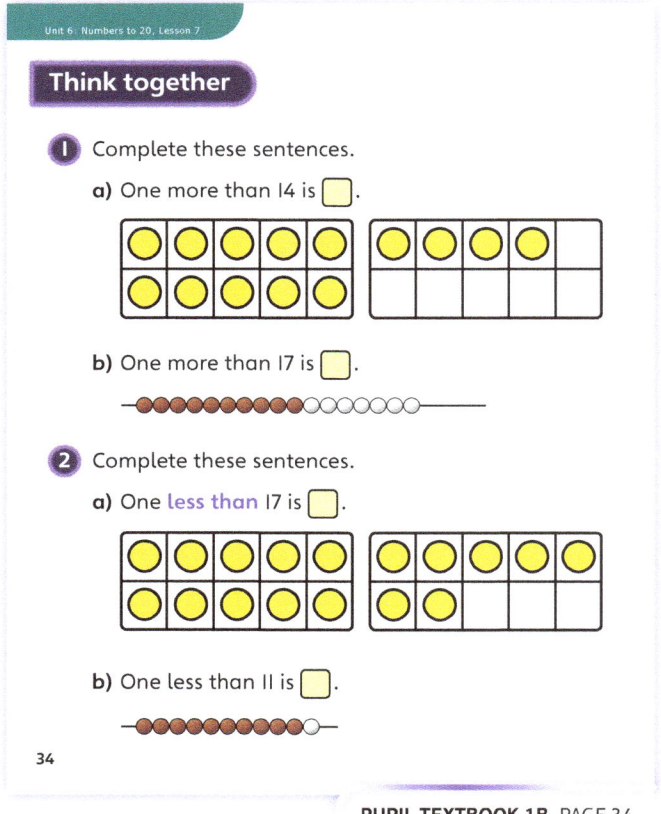

PUPIL TEXTBOOK 1B PAGE 34

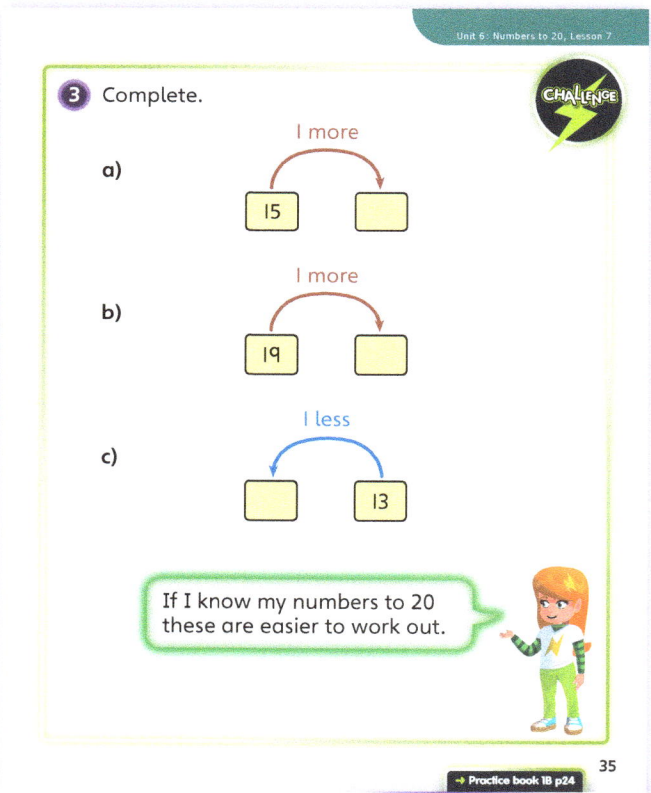

PUPIL TEXTBOOK 1B PAGE 35

Unit 6: Numbers to 20, Lesson 7

Practice

WAYS OF WORKING Independent thinking

IN FOCUS Question ❶ uses the concept of one more and one less in a familiar context, using ten frames and counters. Check children can complete these fairly easily. If not, bring them back to some practical activities. Question ❷ brings in a slightly different context and links one more with the word 'arrives'. Question ❺ is important as children explore one more than 10.

STRENGTHEN Use ten frames and counters or bead strings throughout or allow children to act out questions using the line activity in the **Discover** section.

DEEPEN Ask children to find the starting number when you give them what one more or one less is. For example, ask: *One less than my number is 12. What is my number?*

ASSESSMENT CHECKPOINT Questions ❸ and ❹ can be used as an assessment checkpoint. The last part of each question uses words rather than the numeral so children may find this a little tricky. Check they can verbally say the answer then support them with writing the numeral if needed.

ANSWERS Answers for the **Practice** part of the lesson can be found in the *Power Maths* online subscription.

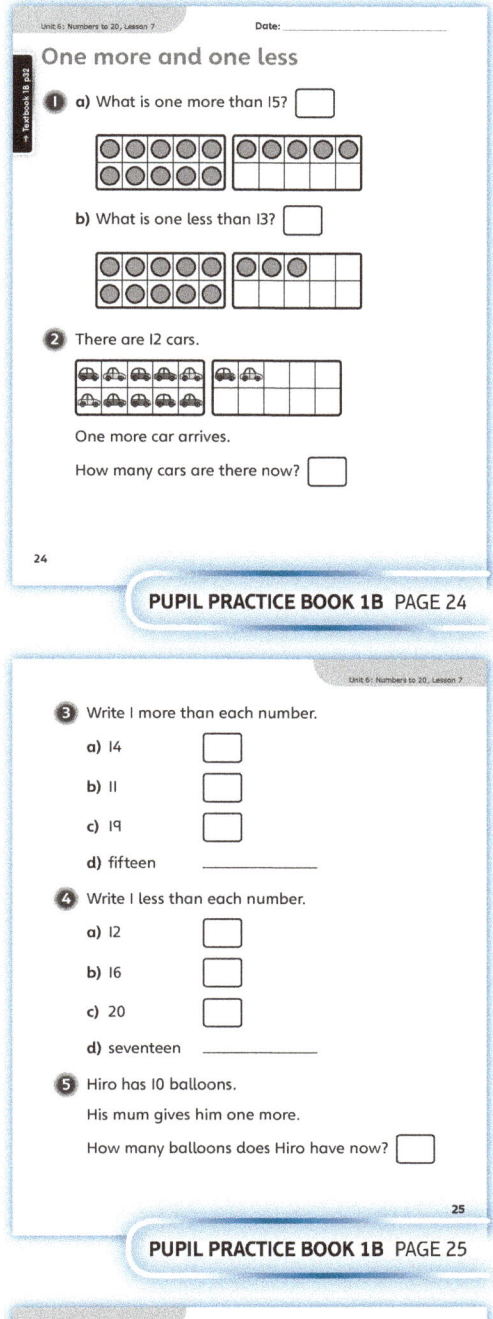

PUPIL PRACTICE BOOK 1B PAGE 24

PUPIL PRACTICE BOOK 1B PAGE 25

Reflect

WAYS OF WORKING Pair work

IN FOCUS Children choose their own number and find one more or one less. Share answers as a class. What is the most popular choice?

ASSESSMENT CHECKPOINT Check that children have got their answers the correct way round and are not mixing up one more and one less.

ANSWERS Answers for the **Reflect** part of the lesson can be found in the *Power Maths* online subscription.

After the lesson ⏸

- Can children confidently find one more or one less than any number to 20?
- Do children know that one less than 1 is 0?
- Do children understand when to use the word 'less' and when to use the word 'fewer'?

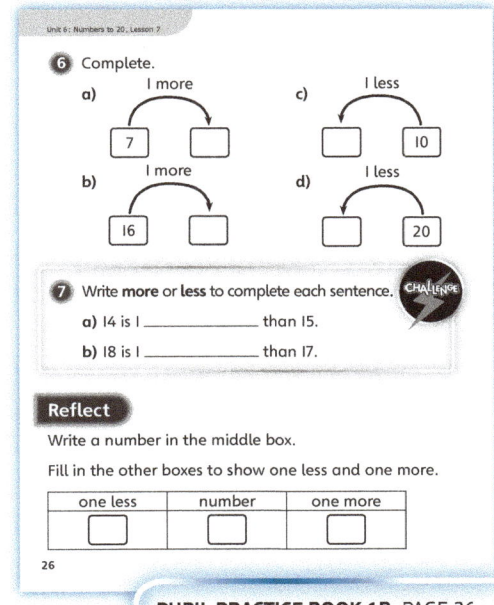

PUPIL PRACTICE BOOK 1B PAGE 26

Unit 6: Numbers to 20, Lesson 8

The number line to 20

Learning focus
In this lesson, children will learn about the number line to 20. Children learned about the number line to 10 in the autumn term. All number lines will count in 1s in this lesson. Children will recap counting on from 0 to 20 when labelling a number line and can practise counting back, if reading from right to left. They will be able to consolidate their learning in the previous lesson as they clearly see that one more is the next number along the number line, whilst one less is the previous number.

Before you teach
- How did your class do in the autumn term when they explored the number line to 10?
- Are children confident counting to 20?
- What skills could you consolidate during this lesson?

NATIONAL CURRICULUM LINKS

Year 1 Number – number and place value

Identify and represent numbers using objects and pictorial representations including the number line, and use the language of: equal to, more than, less than (fewer), most, least.

ASSESSING MASTERY

Children can confidently complete a number line to 20 from any starting number and place numbers to 20 in the correct place on a number line.

COMMON MISCONCEPTIONS

Children may write the numbers in between markers rather than on markers when labelling a number line. Children may assume that all number lines start at 0. To overcome this, expose children to number lines that do not start at 0. Ask:
- *What is the starting number on this number line? Does the start number always have to be 0?*

Children may label the number line the wrong way around. For example, starting it at 20 with the numbers decreasing. Ensure children know that number lines always start with the smaller number on the left. Ask:
- *Which number should the number line start with? The smaller number or the larger number?*

STRENGTHENING UNDERSTANDING

A number line is a great opportunity to count from 0, as children do not do this when counting physical things. Ensure you use a variety of number lines all counting in 1s, for example, a number line from 0 to 20 and also a number line from 10 to 20.

GOING DEEPER

Children can spot and correct mistakes on number lines and they can use number lines to show previous learning such as finding one more and one less.

KEY LANGUAGE

In lesson: number line, order

STRUCTURES AND REPRESENTATIONS

Number lines

RESOURCES

Mandatory: blank horizontal and vertical number lines

Optional: number tracks, pegs, string to make a number line, ten frames, counters, 0–20 number cards, wipe-clean blank number lines, mini whiteboards, whiteboard pens

 In the eTextbook of this lesson, you will find interactive links to a selection of teaching tools.

Quick recap

Explore the number line to 10 again. Draw a blank number line on the board from 0 to 10 with unlabelled marked intervals. Tell children they need to help you finish drawing a number line to 10. Ask: *What number do we start at? What are we counting up in?*

Unit 6: Numbers to 20, Lesson 8

Discover

WAYS OF WORKING Pair work

ASK
- Question 1 a): *How do you know which numbers come next?*
- Question 1 b): *What is ordering numbers the same as?*

IN FOCUS In question 1 a), children link the numbers on the back of the shirts to a number line and counting in order. Notice that the number line does not start at 0 to help children see that a number line can start with any number.

PRACTICAL TIPS Make your own line with numbers. Peg each number on the line as if it was a football shirt so that children can act out the activity.

ANSWERS

Question 1 a): The next two shirts will be numbers 14 and 15.

Question 1 b): 11, 12, 13, 14, 15, 16, 17, 18, 19, 20

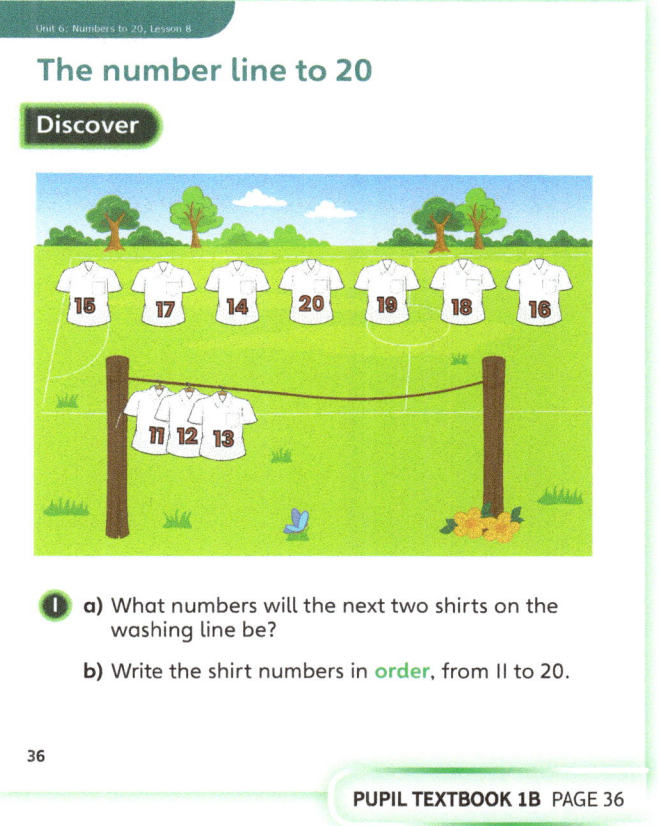

PUPIL TEXTBOOK 1B PAGE 36

Share

WAYS OF WORKING Whole class teacher led

ASK
- Question 1 a): *What comes after the number 12? What comes after that?*
- Question 1 b): *What does the line of shirts look like? Are the numbers in order?*

IN FOCUS In question 1 b), children link the numbers on the back of the shirts to a number line and counting in order. Ten frames and counters, or rekenreks, could be used alongside this activity if children need some extra support.

PUPIL TEXTBOOK 1B PAGE 37

Think together

WAYS OF WORKING Whole class teacher led (I do, We do, You do)

ASK
- Question ❶: *Where could we start? What is the smallest number?*
- Question ❷: *Can you count aloud?*
- Question ❸: *Does it matter which way we draw the number line?*

IN FOCUS In question ❷, children focus on ordering numbers and seeing them spaced out on a number line, before completing their own number lines in question ❸. Seeing number lines in a horizontal and vertical orientation in question ❸ is important for children for their later learning.

STRENGTHEN Give small groups of children number cards from 1 to 20 and ask them to recreate question ❶. Ensure you have some horizontal and vertical number lines available for children to use.

DEEPEN Ask children to order a set of number cards backwards, starting at 20. As this lesson is about number lines, ensure they start with 20 on the right and work back to the left, so that the ordering still looks like a number line. Number lines should always start with the smallest number on the left.

ASSESSMENT CHECKPOINT Can children find one more and one less using practical equipment to support them? Check that they understand the vocabulary and can link it to a number getting bigger or smaller.

ANSWERS

Question ❶: Children should point to each number card as they count up from 1 to 20.

Question ❷: Children should point to each number on the number line as they count up from 11 to 20.

Question ❸: Children should continue the horizontal number line, counting up:
0, 1, 2, 3, 4, 5, 6, 7, 8, 9, 10, 11, 12, 13, 14, 15, 16, 17, 18, 19, 20.
Children should continue the vertical number line, counting back:
20, 19, 18, 17, 16, 15, 14, 13, 12, 11, 10, 9, 8, 7, 6, 5, 4, 3, 2, 1, 0.

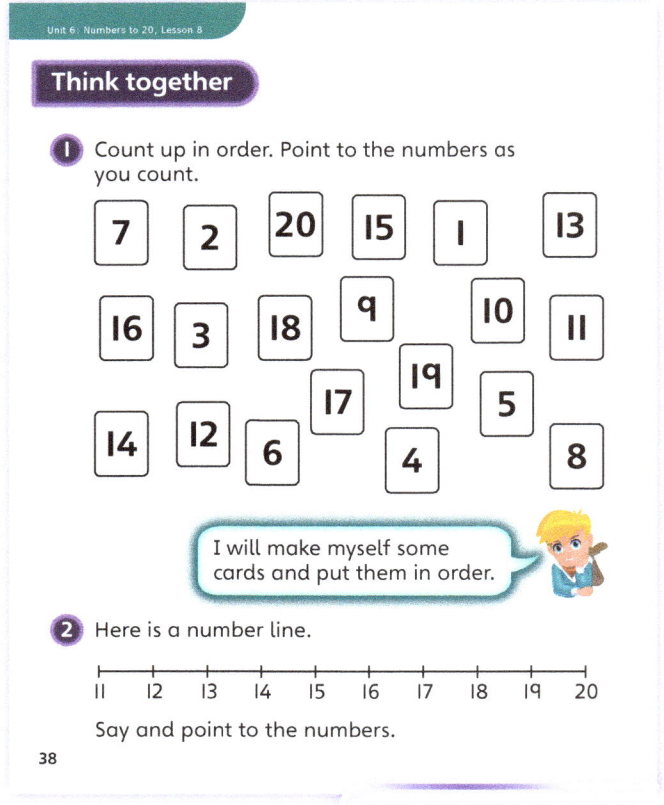

PUPIL TEXTBOOK 1B PAGE 38

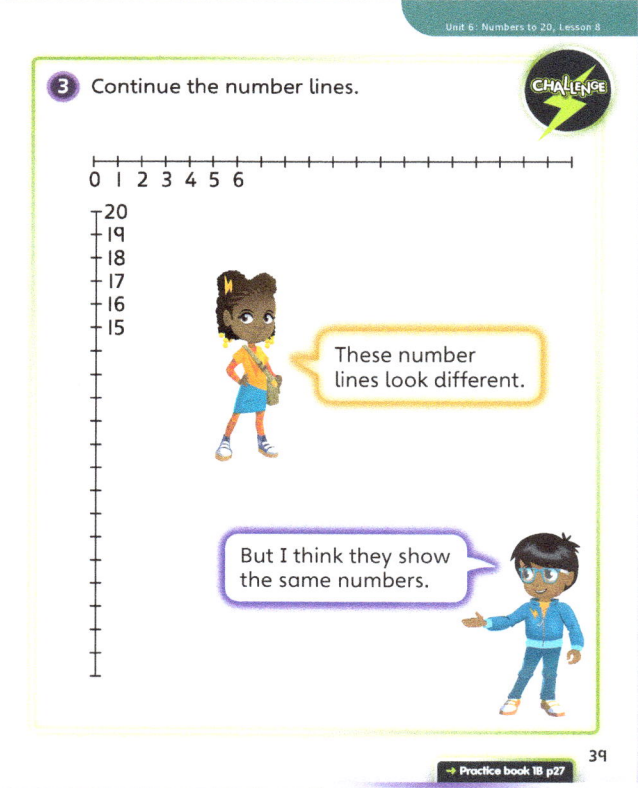

PUPIL TEXTBOOK 1B PAGE 39

Unit 6: Numbers to 20, Lesson 8

Practice

WAYS OF WORKING Independent thinking

IN FOCUS Questions 1, 2 and 3 ask children to order numbers in ways in which they are familiar, before leading into ordering numbers on number lines in question 4. In question 4, children complete number lines from a variety of different starting numbers. Question 5 challenges children to label a completely blank number line with the numbers 11 to 20. Some children may be tempted to start at 0 or they might start at 12, with this being the next number after 11. Another error children might make is to start labelling with 11 at the second marker and then run out of markers at the end.

STRENGTHEN Provide children with wipe clean number lines, so they can practise filling them in from a variety of different starting points.

DEEPEN Combine children's learning on number lines with previous learning, such as finding one more and one less. How can children show this concept on a number line?

ASSESSMENT CHECKPOINT Use questions 4 and 5 to check children's understanding of number lines. Use blank, wipeable number lines for children to use for further practice.

ANSWERS Answers for the **Practice** part of the lesson can be found in the *Power Maths* online subscription.

Reflect

WAYS OF WORKING Pair work

IN FOCUS Children draw their own number line without scaffolding. This is a difficult skill for children and they could first sketch their number line on a mini whiteboard, rather than try to draw a neat, accurate one straight in their books. Children should try to get their intervals as equally spaced out as possible.

ASSESSMENT CHECKPOINT Check if children are able to think about the key features of a number line. Have they drawn a straight line split into equal intervals?

ANSWERS Answers for the **Reflect** part of the lesson can be found in the *Power Maths* online subscription.

After the lesson

- Continue to use the number line to 20 with other learning.
- Ensure you have a 0–20 number line displayed in your classroom and refer to it regularly.
- Continue to show number lines in different orientations and number lines that do not start at 0.

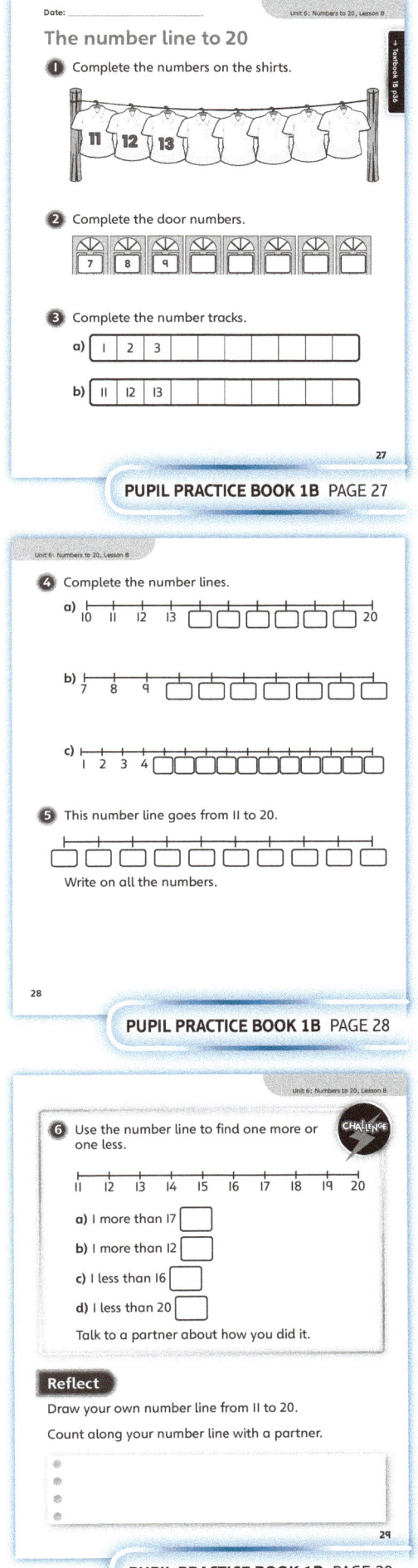

Unit 6: Numbers to 20, Lesson 9

Label number lines

Learning focus
In this lesson, children label number lines to 20. Children will recap counting from 0 to 20 when labelling a number line and can practise counting back if reading from right to left. They will be able to clearly see that one more is the next number along the number line, whilst one less is the previous number.

Before you teach
- Are children confident counting to 20?
- What skills could you consolidate during this lesson?
- Are children able to label a number line to 10?

NATIONAL CURRICULUM LINKS
Year 1 Number – number and place value

Identify and represent numbers using objects and pictorial representations including the number line, and use the language of: equal to, more than, less than (fewer), most, least.

ASSESSING MASTERY
Children can confidently complete a number line to 20 from any starting number and place numbers to 20 in their correct position on a number line.

COMMON MISCONCEPTIONS
Children may write the numbers in between markers rather than on markers when labelling a number line. Ask:
- *Are the start and end numbers on a marker or in between markers? Where should other numbers on the number line go?*

Children may assume that all number lines start at 0. To overcome this, expose children to number lines that do not start at 0. Ask:
- *What is the start number on this number line? Do number lines always have to start at 0?*

Children may label the number line the wrong way around. For example, starting it at 20 with the numbers decreasing. Ensure children know that number lines always start with the smaller number on the left. Ask:
- *What number should the number line start with? The smaller number or the larger number?*

STRENGTHENING UNDERSTANDING
A number line is a great opportunity to count from 0, as children do not do this when counting physical things. Ensure you use a variety of number lines all counting in 1s, for example, use a number line from 0 to 20 but also a number line from 10 to 20 and number lines starting at a less conventional starting point such as 7 or 11.

GOING DEEPER
Although at this point children are only expected to use number lines going up in intervals of 1, some children could explore number lines going up in intervals of 2. Can they estimate where a number would lie on a number line like this?

KEY LANGUAGE
In lesson: label, number line

Other language to be used by the teacher: interval

STRUCTURES AND REPRESENTATIONS
Number lines

RESOURCES
Mandatory: number lines starting at different points, going up in 1s

Optional: vertical number lines, chalk, number cards with the numbers 11–20

 In the eTextbook of this lesson, you will find interactive links to a selection of teaching tools.

Quick recap
Putting numbers in order from 0 to 20. As a class, ask children to chant the numbers from 0 to 20 in order. As children say each number, write the number on the whiteboard.

Unit 6: Numbers to 20, Lesson 9

Discover

WAYS OF WORKING Pair work

ASK

- Question 1 a): *Can you count from 11 to 20 in order? Does this help you see which numbers are the wrong way round on the number line?*
- Question 1 b): *When counting from 11 to 20, did you say any numbers that aren't on the number line?*

IN FOCUS Children begin to explore number lines that are partially complete. In question 1 b), they need to focus on counting in the correct order to be able to successfully complete the number line and correct the mistakes.

PRACTICAL TIPS Create the 11–20 number line outside with chalk and use number cards to partially label the line. Include some mistakes for children to correct before they complete the number line. This could be repeated with different gaps and different mistakes.

ANSWERS

Question 1 a): 15 and 16 are the wrong way around.

Question 1 b): 14, 17 and 19 are the missing numbers.

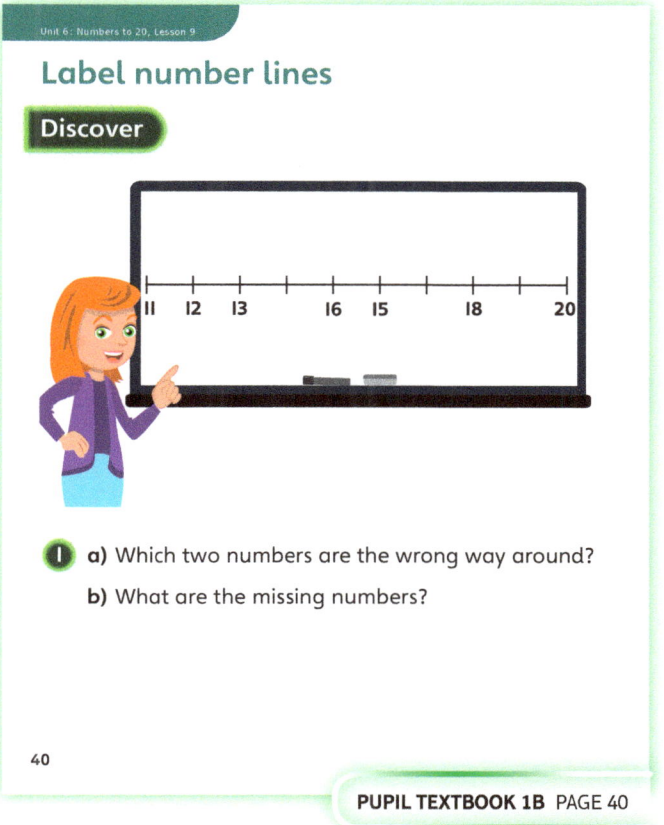

PUPIL TEXTBOOK 1B PAGE 40

Share

WAYS OF WORKING Whole class teacher led

ASK

- Question 1 a): *How did you know that 15 and 16 were the wrong way around? Did you count to check?*
- Question 1 b): *Did counting from 11 to 20 help you fill in the gaps?*

IN FOCUS In question 1 b), children should use their counting skills to 20 to complete the number line and correct the mistake. Children may find it easier to complete the number line first then find the mistake with 15 and 16 being the wrong way around.

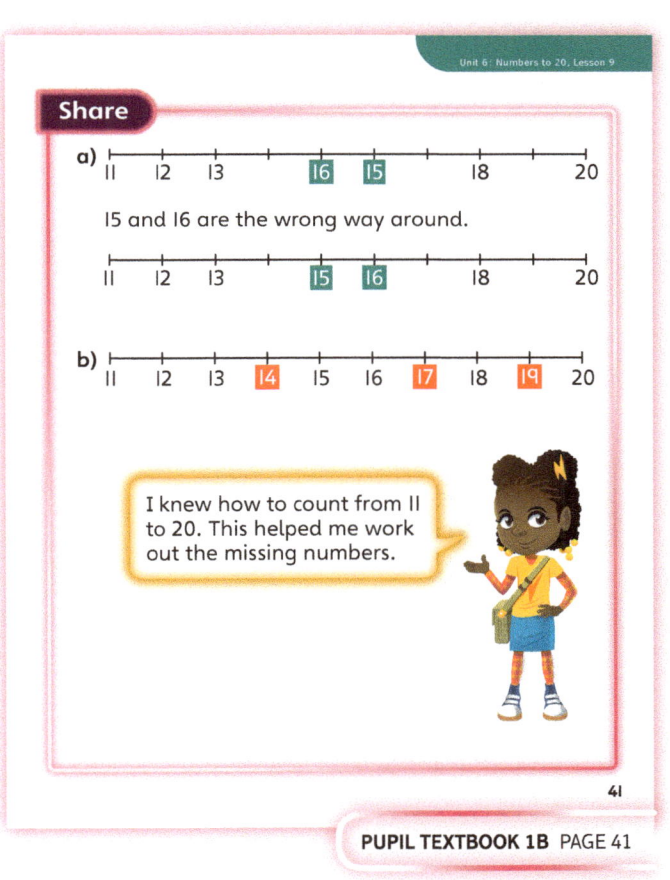

PUPIL TEXTBOOK 1B PAGE 41

Think together

WAYS OF WORKING Whole class teacher led (I do, We do, You do)

ASK

- Question ❶: *What is the same and what is different between the number lines in part a) and part b)? Do both number lines go up in 1s?*
- Question ❷: *How will you work out where each number goes on the number line? Do you need to fill each number in or can you just imagine them?*

IN FOCUS In questions ❶, ❷ and ❸, children are exposed to a variety of number lines with different missing numbers. It is important that children see number lines of the same lengths with different-sized intervals in question ❸ and also number lines of different lengths in question ❶.

STRENGTHEN Create a number line outside with chalk and number cards to recreate the questions and allow children to experiment with them.

DEEPEN Ask children to create some questions for a partner to answer. Check that they have left the right amount of markers on their number lines for their missing numbers.

ASSESSMENT CHECKPOINT In question ❷, check if children complete the whole number line or if they are able to visualise the numbers and only place the required numbers on the number line.

ANSWERS

Question ❶ a): 12, 15 and 18 are the missing numbers.

Question ❶ b): 5, 8, 11 and 14 are the missing numbers.

Question ❷:

Question ❸:

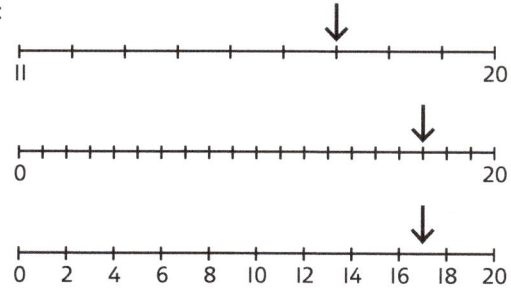

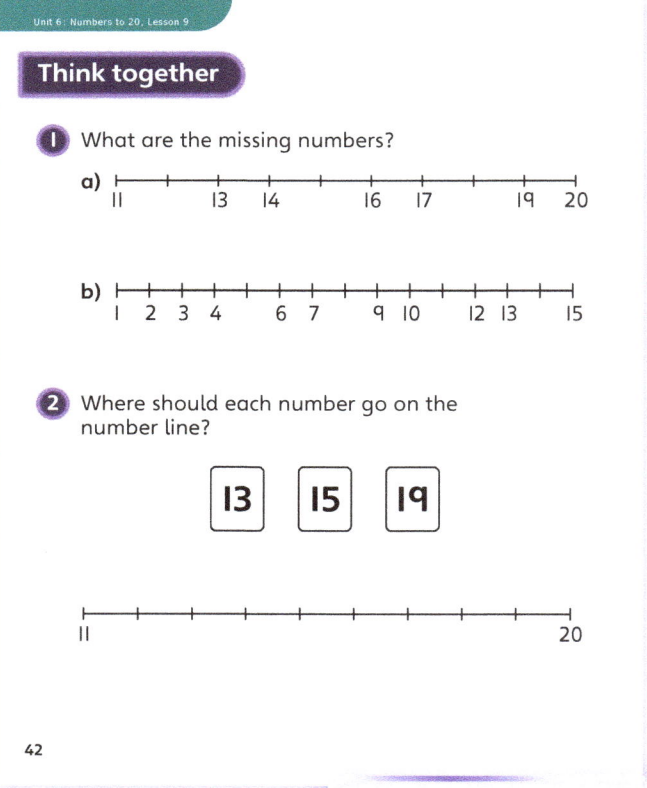

PUPIL TEXTBOOK 1B PAGE 42

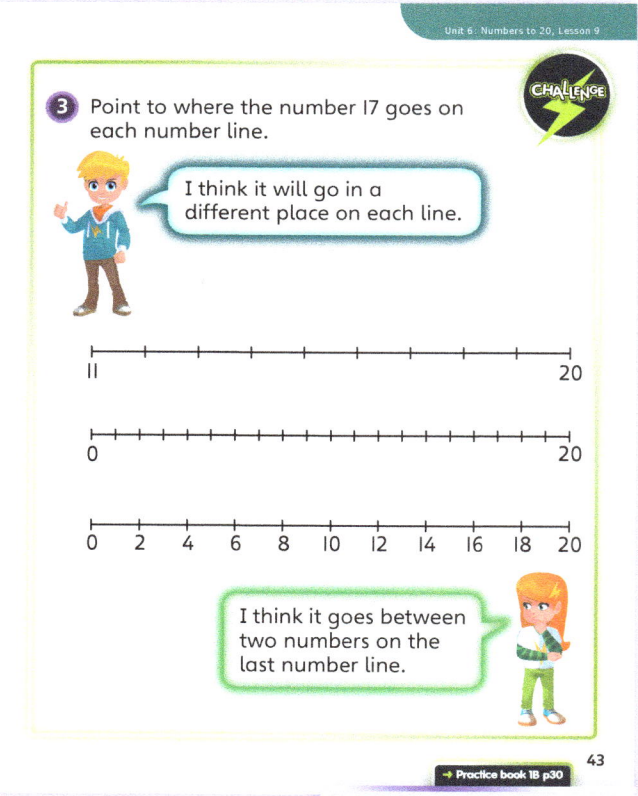

PUPIL TEXTBOOK 1B PAGE 43

Unit 6: Numbers to 20, Lesson 9

Practice

WAYS OF WORKING Independent thinking

IN FOCUS Question ❶ asks children to complete number lines of varying lengths and with varying starting points. It is important that children are exposed to this type of variety. Question ❷ is a little more difficult as children are faced with blank number lines and are asked to draw arrows from number cards to their correct positions on a number line. Encourage children to count and visualise a completed number line before deciding where each number goes. Some children may try to guess instead of count. Use question ❸ as an opportunity to discuss the concept of half-way and emphasise that 15 is half-way between 10 and 20.

STRENGTHEN Ensure children have access to completed number lines to refer to and check their answers. Continue to use an outdoor number line with chalk and number cards to support children with finding missing numbers.

DEEPEN Question ❹ provides an opportunity for children to see that not all number lines go up in 1s and creates a further opportunity to discuss the concept of half-way. To deepen children's understanding further, create some questions for children using a vertical number line.

ASSESSMENT CHECKPOINT Question ❶ is a good opportunity to check if children have grasped the basics of the lesson.

ANSWERS Answers for the **Practice** part of the lesson can be found in the *Power Maths* online subscription.

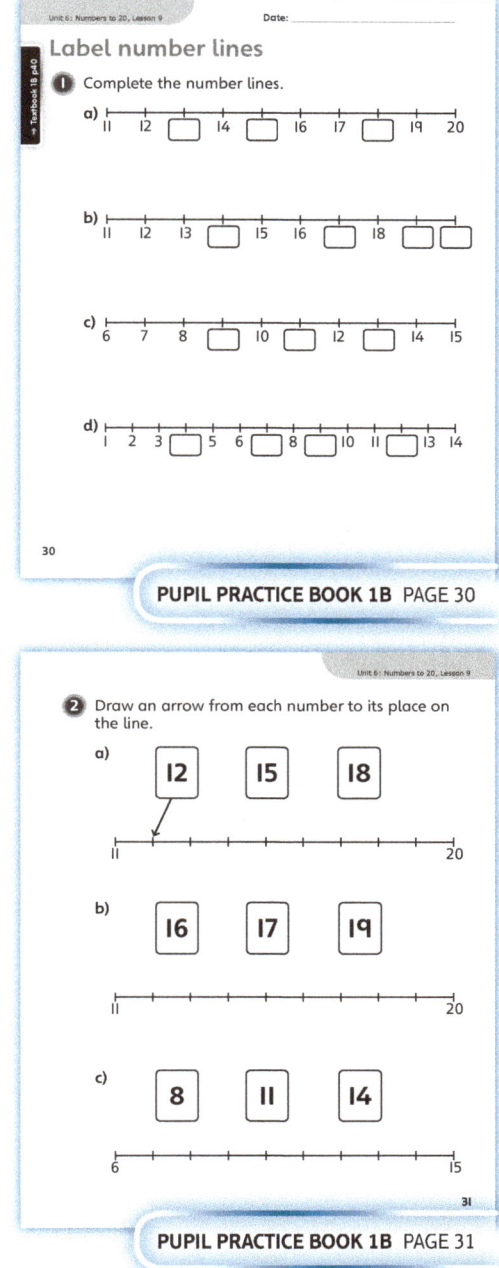

PUPIL PRACTICE BOOK 1B PAGE 30

PUPIL PRACTICE BOOK 1B PAGE 31

Reflect

WAYS OF WORKING Independent thinking

IN FOCUS Children are provided with a final opportunity to complete the numbers from 10 to 20 on a number line.

ASSESSMENT CHECKPOINT Check that children count on in 1s and place all numbers in the correct order.

ANSWERS Answers for the **Reflect** part of the lesson can be found in the *Power Maths* online subscription.

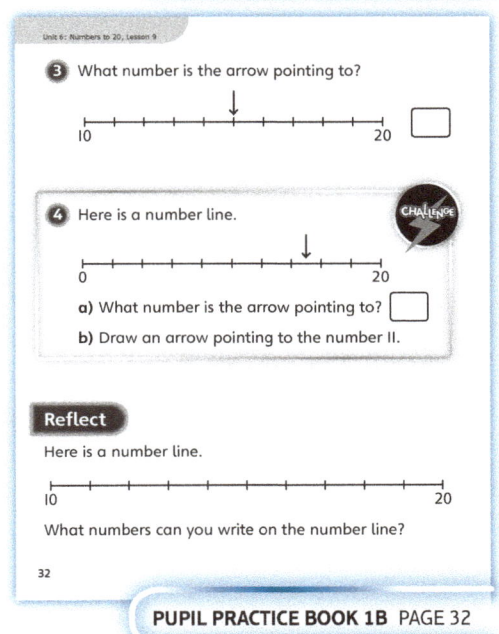

PUPIL PRACTICE BOOK 1B PAGE 32

After the lesson ⏸

- Continue to use the number line to 20 with other learning.
- Ensure you have a number line from 10 to 20 up in your classroom and refer to it regularly.
- Continue to show number lines in different orientations and number lines that do not start at 0.

Unit 6: Numbers to 20, Lesson 10

Estimate on a number line

Learning focus

In this lesson, children are asked to estimate for the first time. This is a new word for children to learn. Previously, they may have been asked to 'guess' and make predictions. To estimate is to roughly calculate or judge a value, number or quantity. When beginning to estimate on a number line with children, take time to explore the half-way point. Where do they think half-way is? How do they know? What informal measurements could they use to check, such as steps in the playground?

Before you teach

- Do children have a good sense of what half-way means?
- Can children complete a number line from 0 to 20?
- Have children had opportunities to guess? For example, *I am thinking of a number that is less than 5. What number could I be thinking of?*

NATIONAL CURRICULUM LINKS

Year 1 Number – number and place value

Identify and represent numbers using objects and pictorial representations including the number line, and use the language of: equal to, more than, less than (fewer), most, least.

ASSESSING MASTERY

Children can estimate where numbers lie on a number line. Children should be able to make an educated guess and reason why they have put a number in a certain place.

COMMON MISCONCEPTIONS

Some children may be reluctant to estimate in case they get it wrong. Introduce estimation in a fun, game-like way so that children feel comfortable having a go and discussing their reasoning. Give children two numbers. Ask:
- *What number is half-way? Can you use a number line to help you?*

STRENGTHENING UNDERSTANDING

Initially, some children may struggle to estimate. Conversation with other children is vital so that they can hear their proportional reasoning and get an insight into their thinking. Some children may find not having an exact answer difficult and need time to grasp the idea of estimating. Children need a good sense of the number line before being able to estimate. For example, if they are estimating where 4 is on a blank number line from 0 to 10, they should be able to reason that it is less than half-way.

GOING DEEPER

Give children a number line from 0 to 20 and ask them to estimate numbers on a larger number line.

KEY LANGUAGE

In lesson: estimate

STRUCTURES AND REPRESENTATIONS

Number lines

RESOURCES

Optional: chalk, blank number lines

 In the eTextbook of this lesson, you will find interactive links to a selection of teaching tools.

Quick recap

Play Guess my number. Tell children that you are thinking of a number between 1 and 20 and they have to ask you questions to guess what your number is. Encourage them to ask questions such as *Is your number less than 10?*

Discover

WAYS OF WORKING Pair work

ASK

- Question 1 a): *Where should Danny stand? Where is 15 compared to 10 and 20? How do you know?*
- Question 1 b): *Is 19 closer to 10 or 20? Is it over half-way? How close to 20 is it? How do you know?*

IN FOCUS In question 1 a), children need to realise that the number 15 is in the middle of the number line. They may need a copy of a number line to help them with this or they could draw their own. In question 1 b), children start to see that 19 is closer to 20. They could count up in 1s and use a blank number line to help them see how close to 20 it is.

PRACTICAL TIPS Create a number line outside with chalk and recreate the scenario.

ANSWERS

Question 1 a):

Question 1 b):

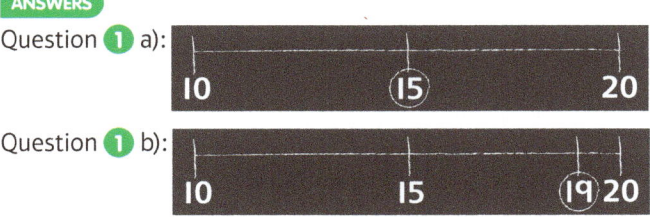

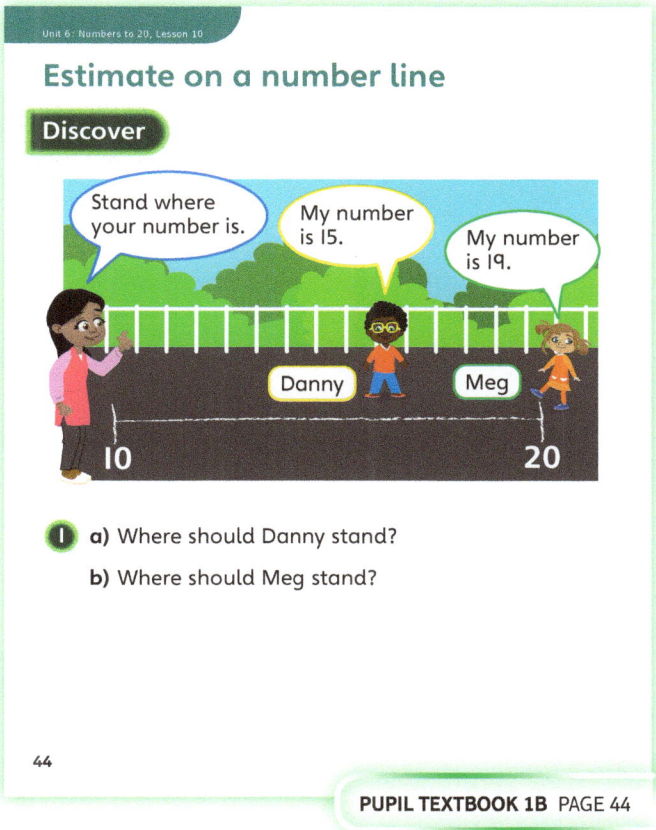

PUPIL TEXTBOOK 1B PAGE 44

Share

WAYS OF WORKING Whole class teacher led

ASK

- Question 1 a): *Is the middle of the number line 15? Why do you think 15 is in the middle of 10 and 20?*
- Question 1 b): *Is 19 closer to 10 or 20? How do you know? Is it more than half-way? How do you know?*

IN FOCUS In question 1 a), children should be encouraged to use a number line to show that 15 is in the middle of the line and so Danny should stand in the middle of the line. In question 1 b), they see from a number line that Meg should stand closer to 20 than to 10. Encourage children to use number lines previously met and compare them to a blank number line. The main skill that children are developing is the ability to estimate where a number lies, but they need to see this on number lines to help them.

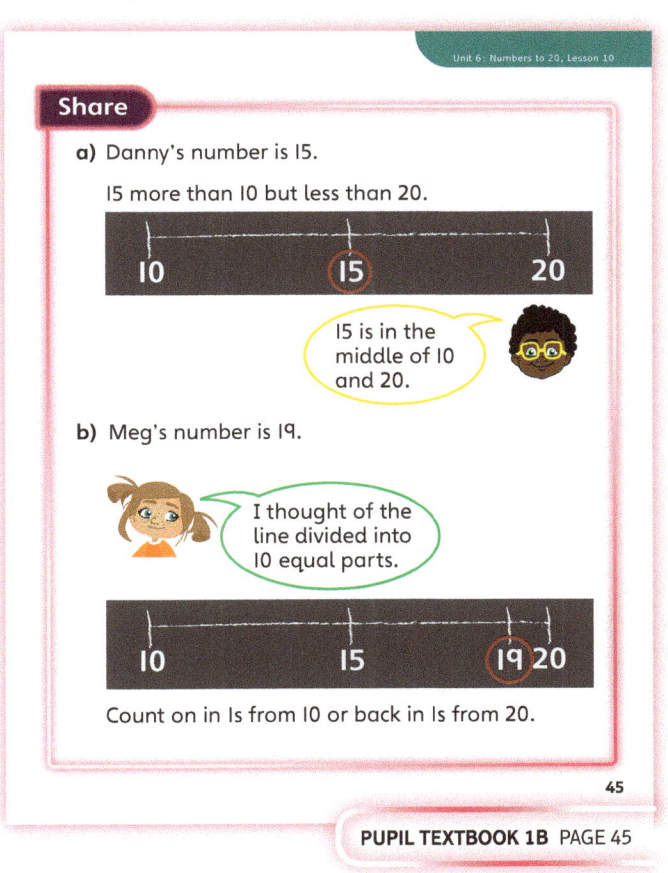

PUPIL TEXTBOOK 1B PAGE 45

83

Think together

WAYS OF WORKING Whole class teacher led (I do, We do, You do)

ASK
- Question ❶ a): *What number is in the middle of the number line? Does 16 come after 15 or before? Where do you think it might go on the number line?*
- Question ❶ b): *Where do you think 11 will go on the number line? How do you know?*
- Question ❷: *What is different about this number line? Which numbers are less than 10?*
- Question ❸: *What is the same and what is different about these number lines? How close is 19 to 20? Will it be in the same place on both number lines? Why or why not?*

IN FOCUS Question ❶ asks children to estimate the position of numbers on a number line from 10 to 20. They should be able to start reasoning about their estimations. Avoid using words such as smaller and larger at this stage as this comes in the next lesson, but consider numbers in relation to the count, for example, 16 comes after 15. Question ❷ asks children to estimate the position of points on a number line from 0 to 20. The half-way point of the number line (10) is marked and so children can consider whether numbers come before or after 10 in the counting sequence to help them.

STRENGTHEN Estimating the position of numbers on a blank number line is quite a hard concept. Some children may need to use a marked number line to help them. Encourage children to divide the number line up themselves or label the half-way point of the number line as a starting point. The whole purpose of this concept is to start to get a sense of location and so for some children seeing a marked number line next to an unmarked line will help significantly.

DEEPEN Question ❸ extends children's understanding by exploring whether the number 19 lies in the same position on different number lines. They explore the reason why 19 might be closer to 20 on one number line than on the other.

ASSESSMENT CHECKPOINT Questions ❶ and ❷ check whether children can estimate the position of points on different number lines. They need to be confident in these types of question before moving on.

ANSWERS

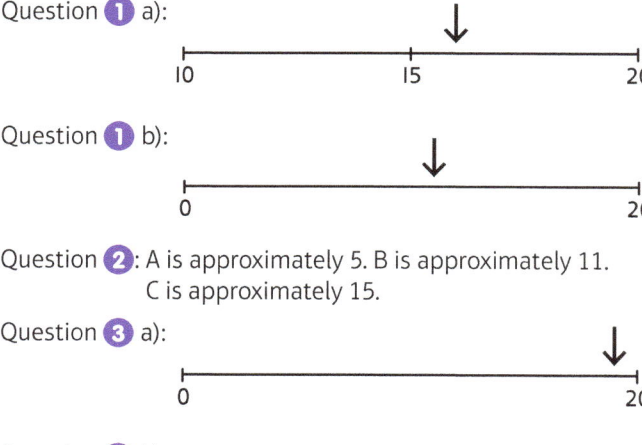

Question ❷: A is approximately 5. B is approximately 11. C is approximately 15.

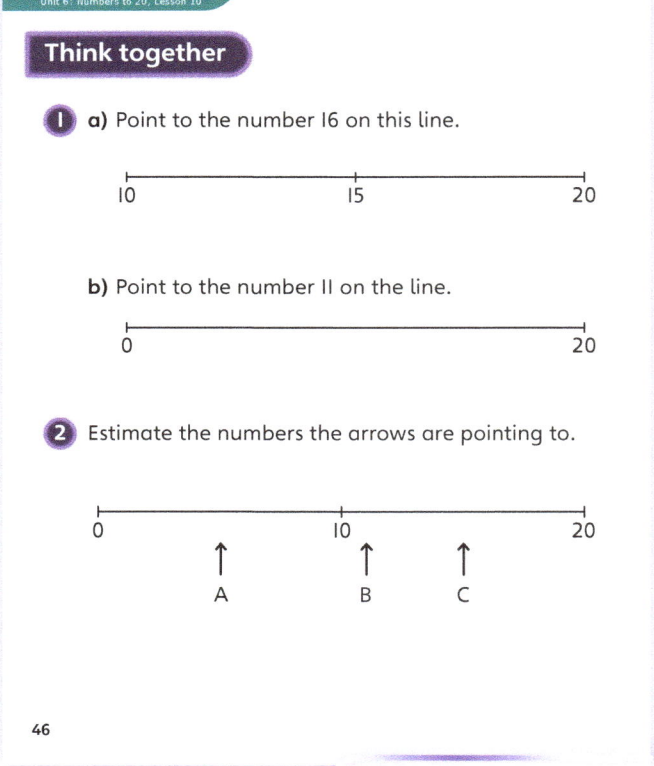

PUPIL TEXTBOOK 1B PAGE 46

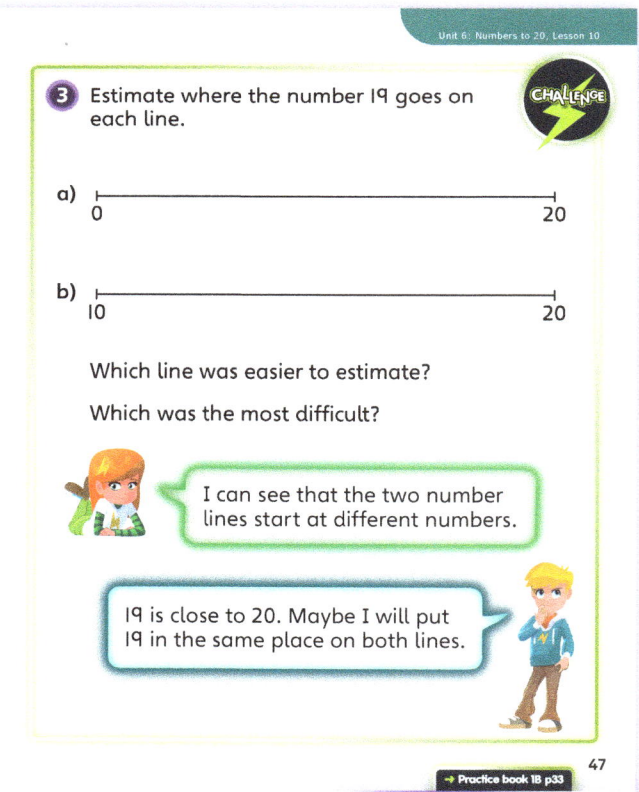

PUPIL TEXTBOOK 1B PAGE 47

Unit 6: Numbers to 20, Lesson 10

Practice

WAYS OF WORKING Independent thinking

IN FOCUS Question ❶ reminds children of the position of numbers on a number line, which they covered in the previous lesson. In question ❷, the marks on the number line have been removed, meaning that children have to work out what numbers lie in between the start and end numbers. Question ❹ is an example of where children have to estimate on a number line between 0 and 20. Encourage children to consider whether one of the arrows is easier to estimate than the others.

STRENGTHEN Some children may need to use a marked number line to help them estimate the position of each number. Further build their confidence by giving some examples similar to questions ❶ and ❷. Children ideally need to see a marked number line next to an unmarked number line, which will help them gain an understanding of the position of each number. When marking on a number line, children may find it easier to count to the number they want to mark.

DEEPEN Question ❺ a) provides children with an example of an unusually marked number line, where the start and end points are non-standard numbers (5 and 15). Ask children to explain their reasoning about the position of the number 12 on the two number lines.

THINK DIFFERENTLY Question ❸ asks children to spot the pattern between the number lines. They should notice that the ones are the same on both number lines; only the number of tens has changed.

ASSESSMENT CHECKPOINT Questions ❸ and ❹ are examples of where children should be able to display an understanding of the position of numbers on a number line.

ANSWERS Answers for the **Practice** part of the lesson can be found in the *Power Maths* online subscription.

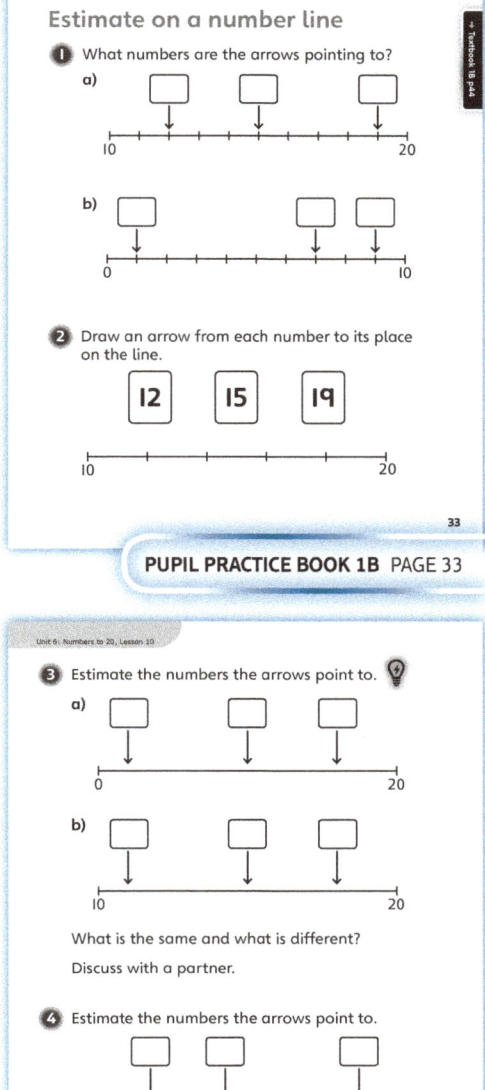

PUPIL PRACTICE BOOK 1B PAGE 33

PUPIL PRACTICE BOOK 1B PAGE 34

Reflect

WAYS OF WORKING Pair work

IN FOCUS Children draw their own number line or use one preprepared without markings. Ask children to estimate where a particular number is and ask a partner to check. You could ask children to challenge each other, but directing as a teacher might be a good starter. Ask children to explain their reasoning of the position of their numbers. If any estimates are incorrect, ask children to explain why.

ASSESSMENT CHECKPOINT Check children can estimate the position of a number.

ANSWERS Answers for the **Reflect** part of the lesson can be found in the *Power Maths* online subscription.

After the lesson

- Can children estimate the position of a number between 10 and 20 on a number line?

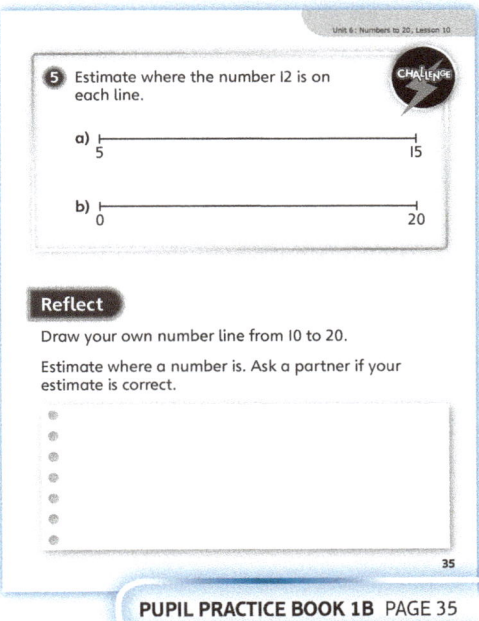

PUPIL PRACTICE BOOK 1B PAGE 35

Unit 6: Numbers to 20, Lesson 11

Compare numbers to 20

Learning focus
In this lesson, children will look at ways that they can compare numbers from 1 to 20.

Before you teach
- Can children represent numbers from 0 to 20?
- Can children use the language 'more than' and 'fewer than' accurately?
- Can children compare numbers between 0 and 10?

NATIONAL CURRICULUM LINKS

Year 1 Number – number and place value

Identify and represent numbers using objects and pictorial representations including the number line, and use the language of: equal to, more than, less than (fewer), most, least.

ASSESSING MASTERY

Children can compare numbers from 0 to 20 using concrete manipulatives to support their understanding. They will use language such as more than, fewer than, greater than and less than accurately.

COMMON MISCONCEPTIONS

When comparing two groups of objects, children may not line up the objects accurately. It is important that children use their knowledge of one-to-one correspondence and that objects are lined up carefully. Ask:
- *Would lining up the objects be easier on a number track?*

When comparing large objects such as footballs, to smaller objects such as golf balls, even though there may be more golf balls, because the footballs are bigger, children may think that there are more of them. Ask:
- *How many footballs are there? How many golf balls are there? Which are there more of? Does the size of a football mean there are more footballs than golf balls?*
- *If you use red and yellow counters in place of the footballs and golf balls, how many red counters are there? How many yellow counters are there?*

STRENGTHENING UNDERSTANDING

Use number tracks to help children line up objects correctly to work out whether there are more or fewer objects. When comparing abstract numbers, ask children to make each number on ten frames to help them. These quantities should help them see which number is greater or smaller.

GOING DEEPER

Ask children to use a number line to compare numbers. Encourage those children who want to go deeper to use the < and > signs throughout and check their accuracy. Ask children to solve missing number problems. For example, can children work out what the possible missing digits are for missing digit problems like 1☐ < 14? Can they use the < and > signs accurately?

KEY LANGUAGE

In lesson: compare, more, fewer, smaller, larger

Other language to be used by the teacher: greater than, less than

STRUCTURES AND REPRESENTATIONS

Counters and cubes arranged on ten frames or number tracks

RESOURCES

Mandatory: counters or cubes

Optional: ten frames, number tracks

In the eTextbook of this lesson, you will find interactive links to a selection of teaching tools.

Quick recap
Check that children know which numbers are less than or greater than a given number. For example, write the number 5 on the board. Ask children to write down all the numbers less than 5.

Unit 6: Numbers to 20, Lesson 11

Discover

WAYS OF WORKING Pair work

ASK
- Question 1 a): *How many marbles does each child have? How did you count them?*
- Question 1 b): *Can you see who has more marbles without counting them? How can you compare the numbers to show who has more marbles?*

IN FOCUS Questions 1 a) and 1 b) require children to use one-to-one correspondence and their knowledge of size of numbers to compare and order numbers. Some children may need to represent the marbles with counters or cubes or use a bead string to help them. Some children may notice which child has more marbles straight away and they should be encouraged to check this using concrete materials.

PRACTICAL TIPS You can use different objects around the classroom to recreate the scenario, comparing these instead of the marbles.

ANSWERS
Question 1 a): Jack has 14 marbles.
Kendi has 11 marbles.

Question 1 b): Jack has more marbles than Kendi.

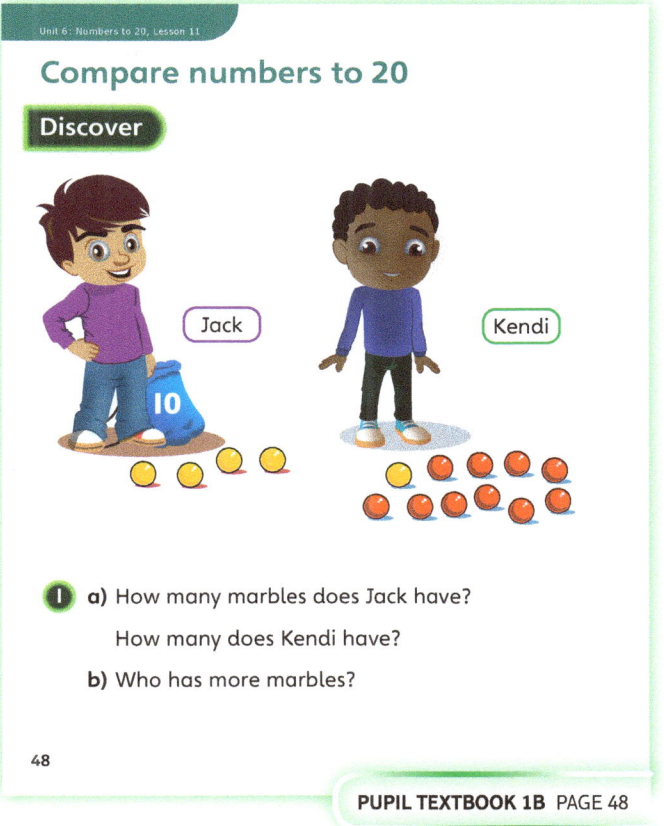

PUPIL TEXTBOOK 1B PAGE 48

Share

WAYS OF WORKING Whole class teacher led

ASK
- Question 1 a): *How did you count the marbles? Was it easy to count them? How could you represent the marbles?*
- Question 1 b): *Can you see who has more marbles straight away? How can you check it with the numbers you have found in part a)?*

IN FOCUS In question 1 b), subitising larger groups can be quite difficult for younger children and so some of them may not be able to see which child has more marbles straight away. For question 1 a), remind children of counting strategies that they may use and how they can represent the objects with counters or cubes. Children may count in 2s or make 10s.

For question 1 b), children start to line up the objects to show the one-to-one correspondence. They line up one of Jack's marbles with one of Kendi's marbles. This will help them to see that Jack has more in his line, so he has more marbles than Kendi. Check to make sure that the marbles or counters are lined up correctly. Explain there are different ways we can write the comparison; 14 is more than or greater than 11. We can say that 14 >11. You may want to bring in the language of 'fewer' at this point, saying that Kendi has fewer marbles than Jack.

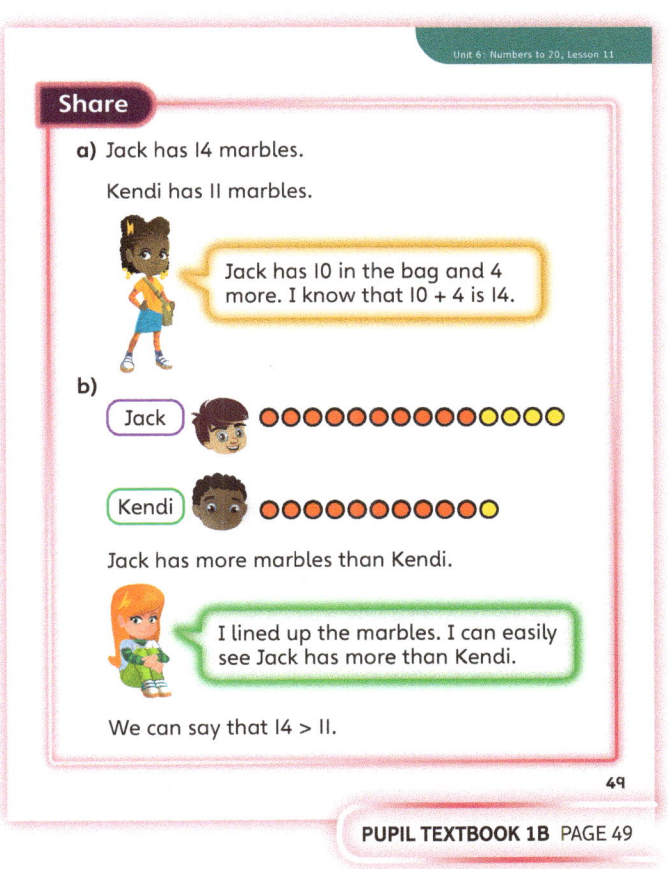

PUPIL TEXTBOOK 1B PAGE 49

Unit 6: Numbers to 20, Lesson 11

Think together

WAYS OF WORKING Whole class teacher led (I do, We do, You do)

ASK
- Question 1 a): *What does the word 'fewer' mean?*
- Question 1 b): *Can you compare Meg and Abdi's cubes easily? What do you need to do with Abdi's cubes?*
- Question 2: *How many votes did each film get? Which film got more votes? How do you know? Can you use the table to compare the numbers 15 and 18?*
- Question 3: *How can you compare the numbers? Do you need to use equipment? What methods do you know?*

IN FOCUS In question 1, children are reminded of the language of fewer. In question 1 b), ensure children realise that they need to show one-to-one correspondence, as it may look like Meg has more cubes because of the arrangement. Question 2 starts to help children compare numbers from 11 to 20 by showing the abstract representation within the concrete counters. Children should be able to see the 15 and 18 counters and see that 18 is more than 15. This will help as they move to question 3, in which they compare abstract numbers. In question 3, some children may need to make the numbers from counters, others may explore using number lines, and some may use inequality signs to compare the numbers straight away.

STRENGTHEN For all questions, use concrete materials to help children compare the numbers. These could be done in lines or on ten frames.

DEEPEN Ask children to use a number line to compare numbers. Encourage those children who want to go deeper to use the < and > signs throughout and check their accuracy. Ask children to solve missing number problems, such as 1☐ < 14. Can they use the < and > signs accurately?

ASSESSMENT CHECKPOINT As children work through the questions, they should be gaining in confidence in comparing numbers. At the end of this section, check that children can compare any two numbers less than 20 using a preferred method.

ANSWERS

Question 1 a): Roz has fewer.

Question 1 b): Meg has fewer.

Question 2: Space Fun has more votes.

Question 3 a): 15 is the smaller number: 15 < 19.

Question 3 b): 20 is the larger number: 20 > 9.
Children should use a variety of ways to prove their answers.

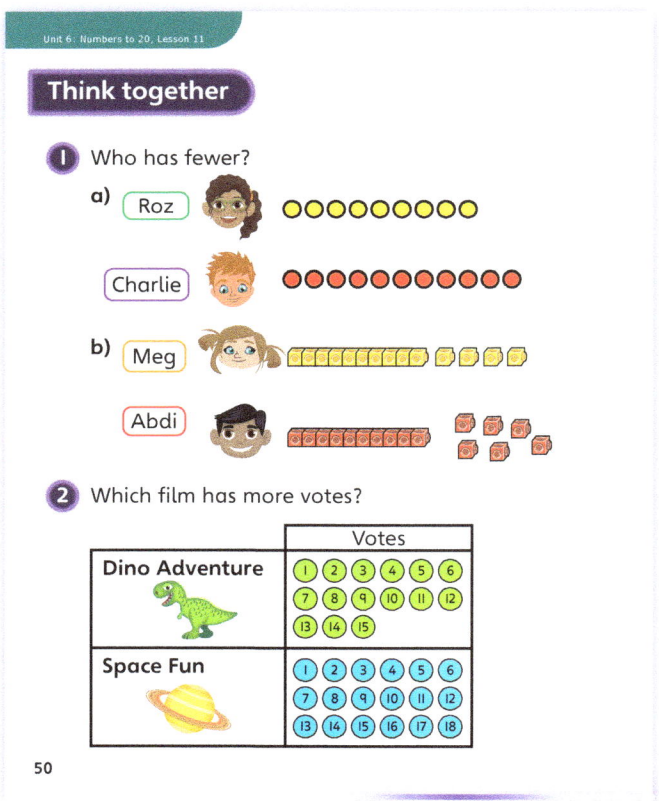

PUPIL TEXTBOOK 1B PAGE 50

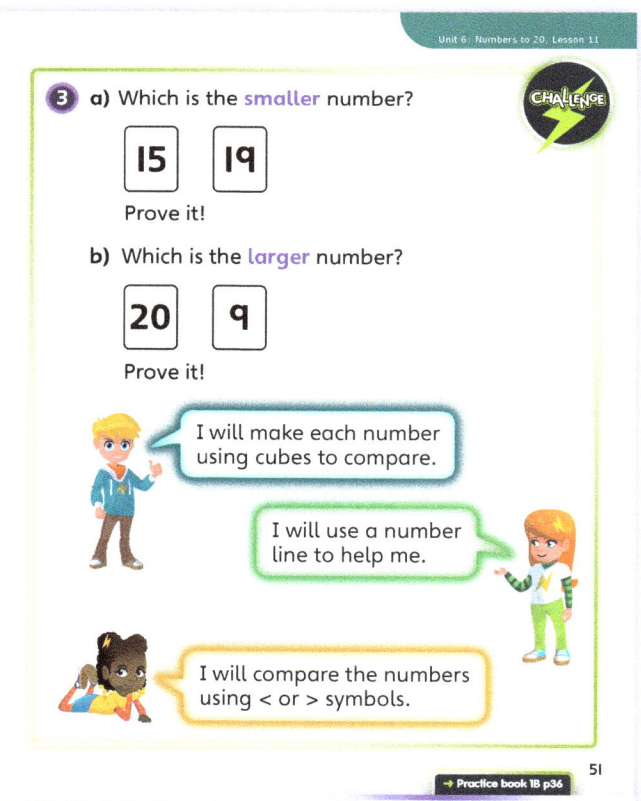

PUPIL TEXTBOOK 1B PAGE 51

Practice

WAYS OF WORKING Independent thinking

IN FOCUS Question ❶ provides children with an opportunity to see who has more or fewer objects by comparing two lines of cubes and counters. Children should be able to see that the longer the line, the more there are. They should notice that the counters and cubes have a one-to-one correspondence. Question ❷ addresses this misconception and, although it may look like they have the same because the lines are the same length, there is a not a one-to-one correspondence. Look for children who get this wrong as it identifies a misunderstanding. Questions ❸ and ❹ allow children to demonstrate more abstract understanding of comparing numbers. Children should call upon the methods they used in the Textbook to help them determine which number is larger and smaller. Question ❺ c) asks children to think about the partition of a number to help them determine which is greater and which is smaller.

STRENGTHEN Throughout these questions, encourage those children who may still be struggling to grasp the concept to use concrete materials to help them compare numbers. This could be done using number lines or ten frames.

DEEPEN Use questions ❻ and ❼ to deepen understanding. Look for children using the inequality signs confidently and being able to identify numbers that could be missing.

ASSESSMENT CHECKPOINT By the end of these questions, children should be able to confidently and accurately compare two numbers. Questions ❸ to ❻ are the key questions to look at in order to check this understanding.

ANSWERS Answers for the **Practice** part of the lesson can be found in the *Power Maths* online subscription.

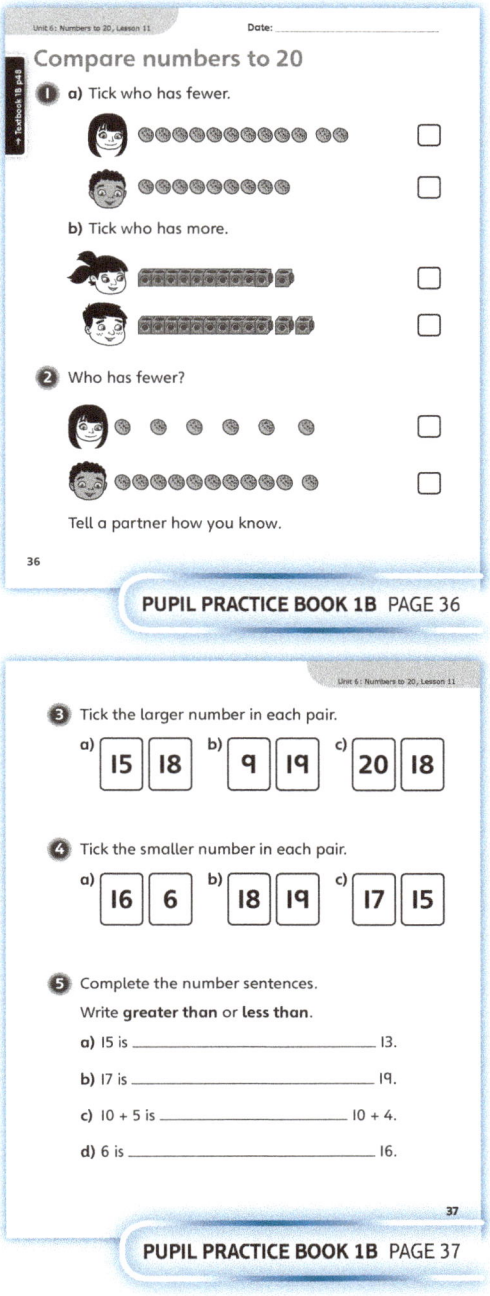

PUPIL PRACTICE BOOK 1B PAGE 36

PUPIL PRACTICE BOOK 1B PAGE 37

Reflect

WAYS OF WORKING Pair Work

IN FOCUS Children should complete the two inequalities using numbers of their own choosing. If some children find the signs too complicated, ask them to complete a sentence such as '☐ is less than ☐'. Initially, children should complete these independently and then ask a partner to check them.

ASSESSMENT CHECKPOINT Check that children's number sentences are correct.

ANSWERS Answers for the **Reflect** part of the lesson can be found in the *Power Maths* online subscription.

After the lesson

- Can children compare two numbers from 0 to 20?
- Can children say which numbers are greater or smaller than a given number?

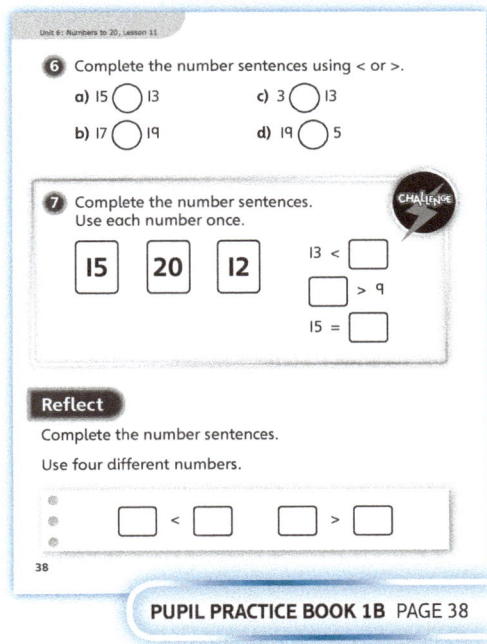

PUPIL PRACTICE BOOK 1B PAGE 38

Unit 6: Numbers to 20, Lesson 12

Order numbers to 20

Learning focus

In this lesson, children order numbers and objects using vocabulary learned in previous lessons and the < and > signs.

Before you teach

- Are children able to write numbers in different ways? For example, can they write 11 as eleven, 1 ten and 1 one, and 10 + 1?
- Are children confident comparing fewer than ten objects using the <, > and = signs?

NATIONAL CURRICULUM LINKS

Year 1 Number – number and place value

Count to and across 100, count on and count back, beginning with 0 or 1, or from any given number (to 20).

Read and write numbers from 1 to 20 in numerals and words.

ASSESSING MASTERY

Children can correctly compare and order numbers and objects using <, > and = signs, and use vocabulary learned in previous lessons, such as more than, less than, fewer than, greater than, etc.

COMMON MISCONCEPTIONS

When ordering numbers of objects, children may count them all rather than set them out in a way that makes them easy to compare. Ask:

- *Do you have to count all the objects to put them in order? How else could you order them?*

STRENGTHENING UNDERSTANDING

Use practical and familiar examples to reinforce words such as 'fewest', 'least', 'most', 'greatest' and 'smallest'. For example, some children could come to the front holding some cubes or a number card. Children can work together to arrange themselves from the least to the most, then from the most to the least.

GOING DEEPER

Some children may be able to explore questions where there is more than one answer or children could create their own questions. For example, children could work systematically to find all of the different ways of filling in ☐ < 10 + 2.

Children can link to prior learning. For example, they could write a number sentence such as '1 ten and 9 ones is fewer than 2 tens but greater than 1 ten and 6 ones because 16 < 19 < 20'.

KEY LANGUAGE

In lesson: order, fewest, least, most, less than, **greater than**, smallest

Other language to be used by the teacher: more than, fewer than, tens, ones, compare, largest

STRUCTURES AND REPRESENTATIONS

Cubes, number line

RESOURCES

Mandatory: cubes

Optional: number cards, ten frame, sweets, other toys or classroom objects for counting, counters

 In the eTextbook of this lesson, you will find interactive links to a selection of teaching tools.

Quick recap

Put two number cards on the board or around the classroom. Ask children to point or move to the smallest or greatest number. Increase the number of cards used and repeat the activity.

Unit 6: Numbers to 20, Lesson 12

Discover

WAYS OF WORKING Pair work

ASK
- Question 1 a): *Without counting, can you guess who has the fewest sweets? Can you guess who has the most sweets?*
- Question 1 a): *Who do you think has fewer than 10 sweets? Who do you think has more than 10 sweets?*
- Question 1 b): *Is it difficult to see who has the most sweets between Anya and Cal?*

IN FOCUS In question 1 a), children order the number of sweets from fewest to most. This helps to reinforce the vocabulary as they can see and compare the size of groups of objects. Using familiar objects like sweets draws them into the problem and makes it more interesting.

PRACTICAL TIPS Recreate the scenario in the classroom, using counters instead of sweets. Choose three children and give them the same number of counters as the number of sweets in the scenario.

ANSWERS
Question 1 a): 5 < 15 < 16
Question 1 b): Cal has the most sweets.

PUPIL TEXTBOOK 1B PAGE 52

Share

WAYS OF WORKING Whole class teacher led

ASK
- Question 1 a): *Has anyone used a different colour to show 10? How does this help you see which tower is bigger?*
- Question 1 a): *Can you think of a picture you could draw to help you get the signs the correct way around?*
- Question 1 b): *By how much is 16 greater than 15?*

IN FOCUS Children use cubes and a number line to compare and order the numbers. Children should try to use language they have learned so far to describe the situation in a variety of ways. For example:
- 5 is less than 15 and less than 16
- 5 sweets are fewer than 15 sweets
- Bo has the fewest sweets
- Cal has the greatest number of sweets.

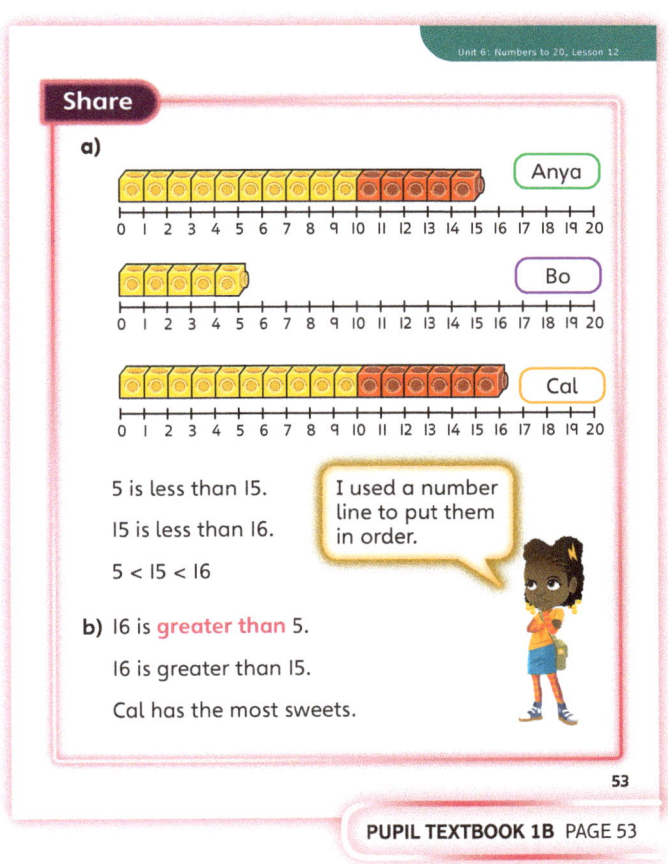

PUPIL TEXTBOOK 1B PAGE 53

Unit 6: Numbers to 20, Lesson 12

Think together

WAYS OF WORKING Whole class teacher led (I do, We do, You do)

ASK
- Question ❶: *Do you need to count on from 1?*
- Question ❶: *Can you write the answer in a different way?*
- Question ❷: *How does the number line help you see which number is smallest?*
- Question ❸: *How many different answers are there for each number sentence?*

IN FOCUS In question ❶, it is important that children make the link between the size of numbers and their place on a number line. Children can line cubes up on a number line and explain why this shows whether a number is greater than or smaller than another. For example, 16 cubes are greater than 15 cubes because they go further up the number line. The further up the number line, the greater the number.

STRENGTHEN Children can draw pictures to help them decide which way around the signs go and use different-coloured cubes to emphasise that they can count on from 10 rather than starting at 1 each time. For question ❷, encourage children to make each number to prove which order they go in. They should be able to talk about the height of each tower to justify their answer.

DEEPEN Question ❸ provides a good opportunity to deepen understanding. There is more than one possible answer for each example given. Interestingly, two will have the same possible answers, whereas the third example looks the same but is different. Encourage children to write their number sentences in different ways, for example:
- 13 < 14 < 17
- 13 is less than 14, which is less than 17
- 1 ten and 3 ones < 1 ten and 4 ones < 1 ten and 7 ones.

ASSESSMENT CHECKPOINT In question ❶, check whether children are still starting to count from 1 rather than starting at 10. Question ❷ checks that they understand how the placement on a number line tells them about the size of the numbers. Ask children to explain their answers either verbally or as a written sentence.

ANSWERS

Question ❶: 12 is less than 16.
16 is less than 19.
12 < 16 < 19

Question ❷: 13 < 14 < 16

Question ❸: Children could put 14, 15 or 16 in the top two boxes. Children could put 18, 19 or 20 in the bottom box.

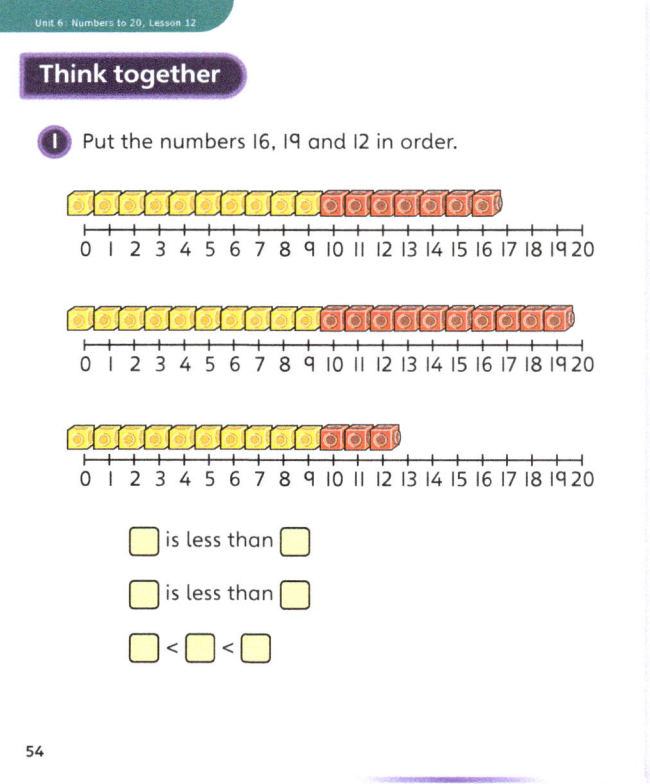

PUPIL TEXTBOOK 1B PAGE 54

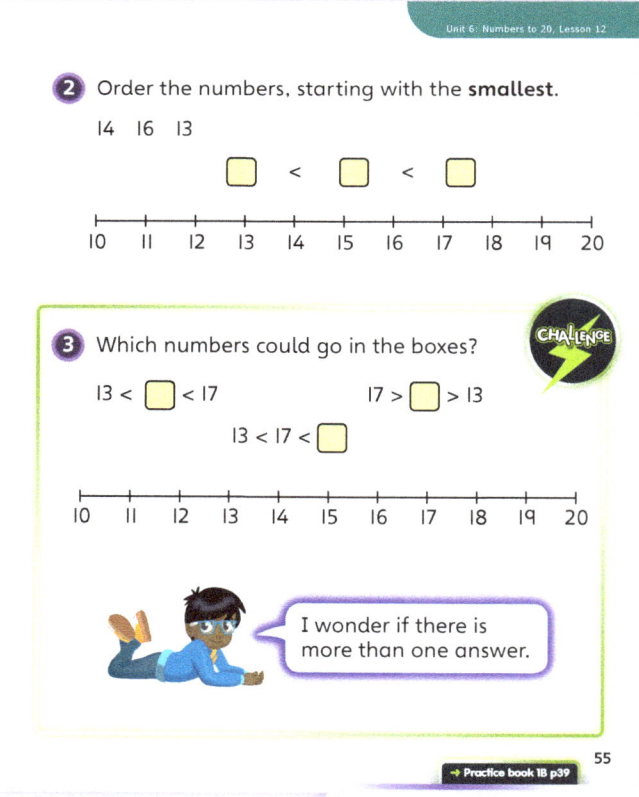

PUPIL TEXTBOOK 1B PAGE 55

Unit 6: Numbers to 20, Lesson 12

Practice

WAYS OF WORKING Independent thinking

IN FOCUS In questions ① to ⑥, children order three numbers from smallest to largest or largest to smallest. It is important that children look at questions in a variety of contexts. The questions use different objects and more abstract contexts such as questions with just numbers and inequality signs. In question ⑦, answers will vary. Bilal has 15 or more sweets. Sam has at least one more than Bilal. Children should be encouraged to discuss the fact that there is more than one correct answer to this question.

STRENGTHEN Ensure children are writing the abstract number sentence alongside their representation. It is important that children gradually manage to use the < and > signs through visualisation. Children should also be encouraged to say their number sentences out loud. This will help embed the key vocabulary in the lesson.

DEEPEN Questions ⑥ and ⑦ both require a deep understanding of numbers 11–20. To deepen this understanding even further, ask children to write their numbers in a variety of ways using both numerals and words. They might order the numbers by partitioning them into 10s and 1s. You might want to make links with previous learning of the part-whole model.

ASSESSMENT CHECKPOINT Question ① assesses that children understand the words 'largest' and 'smallest' when comparing three amounts.

Questions ② and ③ check whether children are ready to answer an ordering question abstractly (without a pictorial representation).

Questions ⑥ and ⑦ are more complex, giving children opportunities to reason and problem solve.

ANSWERS Answers for the **Practice** part of the lesson can be found in the *Power Maths* online subscription.

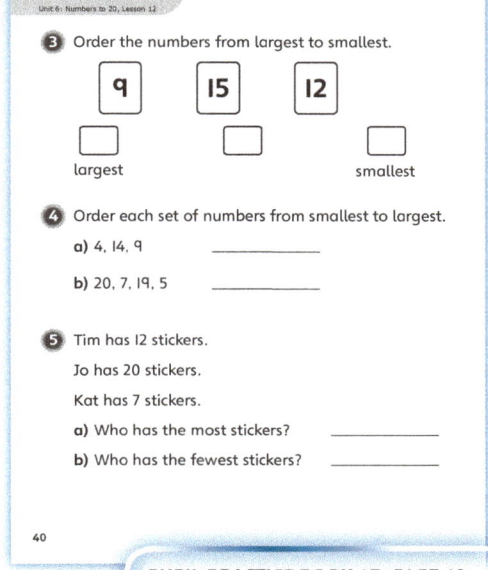

PUPIL PRACTICE BOOK 1B PAGE 39

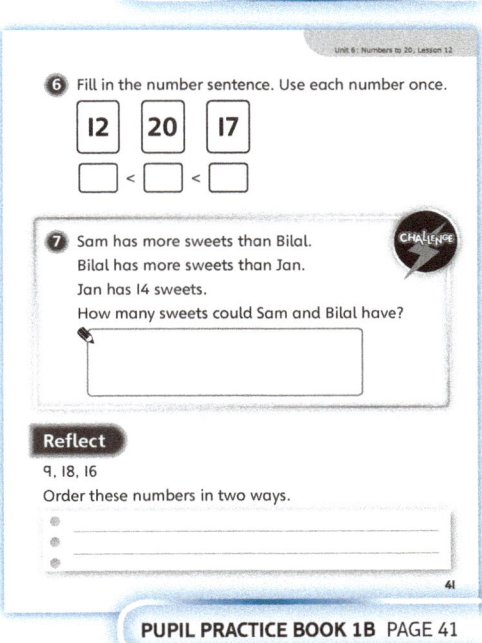

PUPIL PRACTICE BOOK 1B PAGE 40

Reflect

WAYS OF WORKING Independent thinking

IN FOCUS Children think about the different ways they can order a set of numbers.

ASSESSMENT CHECKPOINT Children should be able to recognise that they can order numbers from smallest to largest and largest to smallest.

ANSWERS Answers for the **Reflect** part of the lesson can be found in the *Power Maths* online subscription.

After the lesson

- Compare numbers in different situations throughout the day at school. For example, who has the most carrots on their plate?
- What opportunities can you identify to reinforce and apply this lesson's learning?

PUPIL PRACTICE BOOK 1B PAGE 41

End of unit check

> Don't forget the unit assessment grid in your *Power Maths* online subscription.

WAYS OF WORKING Group work adult led

IN FOCUS
- Question ❷ is designed to assess children's understanding of the structure of 2-digit numbers.
- Question ❺ requires children to be able to compare two numbers and use the language of comparison.

Think!

WAYS OF WORKING Pair work or small groups

IN FOCUS
- This question presents numbers using a variety of representations. Children need to interpret each representation to read the number, identify which is the odd one out and justify their decision with an explanation.
- Encourage children to use the vocabulary at the bottom of the **My journal** page. Can they describe each number using 'tens' and 'ones'?
- Encourage children to think through or discuss what is the same and what is different about the five numbers before writing their answer in **My journal**.

ANSWERS AND COMMENTARY Children who have mastered the concepts of this unit will be able to work confidently with numbers within 20. They will be able to count on and back from any number and count one more and one less. They will know that a number between 10 and 20 is made up of 1 whole ten and some ones and use this knowledge to order and compare numbers within 20.

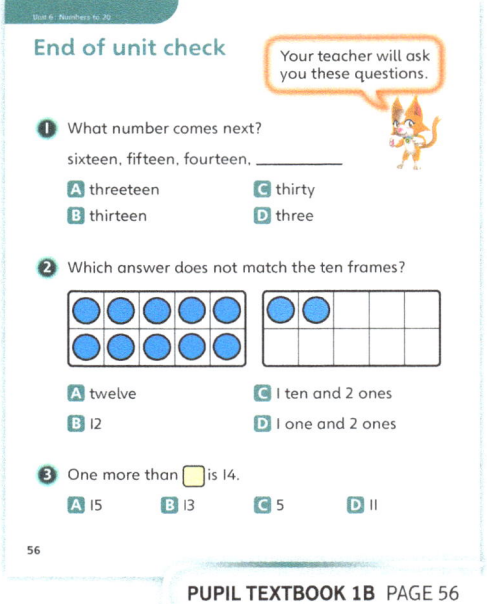

PUPIL TEXTBOOK 1B PAGE 56

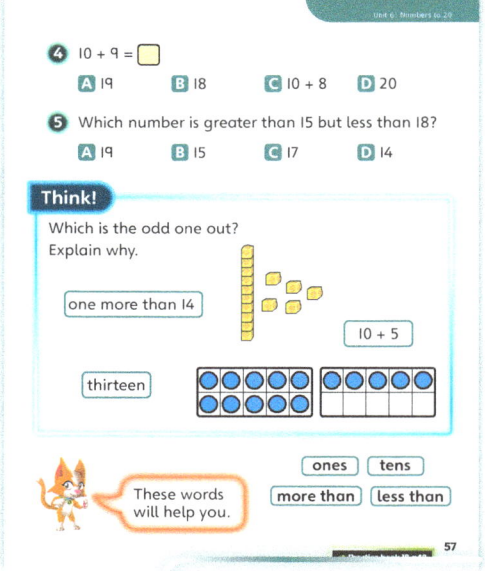

PUPIL TEXTBOOK 1B PAGE 57

Q	A	WRONG ANSWERS AND MISCONCEPTIONS	STRENGTHENING UNDERSTANDING
1	B	Any wrong answer indicates that children are not confident with the number names; spoken or written.	Allow children to use cubes to check and prove their answers as they work through the questions..
2	D	Any wrong answer suggests that children do not have a secure understanding of the structure of 2-digit numbers. They may think that the '1' in 12 stands for one, not ten.	
3	B	A suggests that children have not carefully read and understood the question. C or D indicates that children have only looked at part of the number sentence; they may not fully understand the concept of partitioning numbers.	
4	A	Any wrong answers suggests that children are not confident at partitioning a 2-digit number.	
5	C	Any wrong answer suggests that children do not have a thorough understanding of the language of comparison.	

My journal

WAYS OF WORKING Independent thinking

ANSWERS AND COMMENTARY

The card that is the odd one out is 13, because the other cards and the ten frame show 15.

Children are likely to give a simple explanation for their choice. Encourage them to expand upon it, using some of the key words. For example: 'The card that does not match is 13, because it is made up of 1 ten and 3 ones and the others are made up of 1 ten and 5 ones. 13 is less than 15.'

Observe how children identify the number represented on the ten frame. Did children count all the counters in the full ten frame? If so, they may need further support in partitioning numbers into tens and ones, and recognising 10 on a full ten frame, before moving on to addition within 20 in the next unit.

Power check

WAYS OF WORKING Independent thinking

ASK
- *What part of the unit did you find the most challenging?*
- *What helped you understand this part the most?*

Power puzzle

WAYS OF WORKING Pair work or small groups

IN FOCUS Use this **Power puzzle** to see if children can find one more or one less than a number and identify numbers between two given numbers. They could either complete the puzzle together, or work individually and then compare their answers. It is unlikely that they will all complete it in the same way, so this activity will promote a lot of discussion. Incorrect answers are most likely to be the result of using the same number twice, or approaching the puzzle from top row to bottom row. It is more efficient to fill in the boxes that cannot change (the 'one more' and 'one less' columns) first, and then address the middle column.

ANSWERS AND COMMENTARY Here is a possible solution.

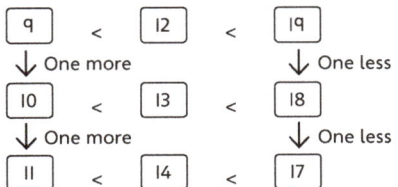

After the unit

- How confident are children using the greater than (>) and less than (<) signs to compare numbers? What strategies could you give those who are not confident?
- Are children willing to approach unfamiliar problems like the **Power puzzle**?

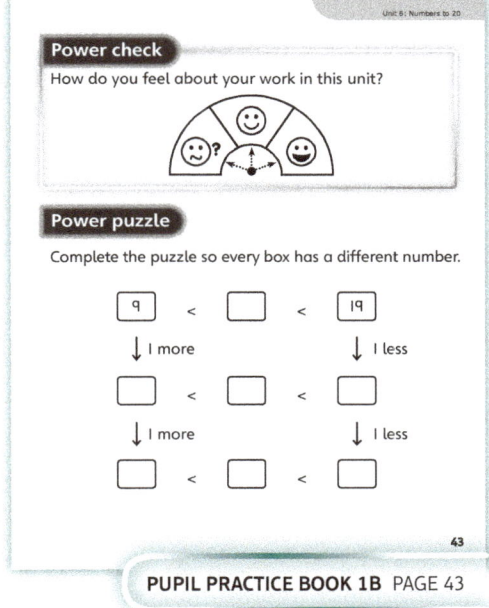

PUPIL PRACTICE BOOK 1B PAGE 42

PUPIL PRACTICE BOOK 1B PAGE 43

Strengthen and **Deepen** activities for this unit can be found in the *Power Maths* online subscription.

Unit 7
Addition and subtraction within 20

Mastery Expert tip! 'Teaching this unit really helped my children deepen their understanding of place value. Using two ten frames gave my children a clear visual representation of how to move beyond using counting strategies to using number facts.'

Don't forget to watch the Unit 7 video!

WHY THIS UNIT IS IMPORTANT

In this unit, children choose the most appropriate addition and subtraction strategies by thinking about the numbers involved in the calculations. It is a vital unit, as understanding how to add and subtract by crossing a 10 is very important for later addition and subtraction strategies, including the formal methods introduced in Key Stage 2.

Children will have learned about subtraction before, but only within 10. This unit is a good chance to revise previously taught methods and move on from them to a range of efficient subtraction strategies.

Within this unit, children progress from using a counting strategy to using known number bonds to derive answers to additions, including adding the 1s separately from the 10, and also when to add by crossing 10.

This unit forms the foundation for understanding efficient and effective calculation strategies throughout the rest of Key Stage 1 and into Key Stage 2, and children will also be making decisions about when to apply different approaches, which is a very important mathematical thinking skill.

WHERE THIS UNIT FITS

→ Unit 6: Numbers to 20
→ **Unit 7: Addition and subtraction within 20**
→ Unit 8: Numbers to 50

This unit builds on children's understanding of addition and subtraction from Units 3 and 4, as well as on their knowledge of numbers to 20 from Unit 6. It requires children to understand how numbers can be split apart into number bonds and fact families, and how to represent numbers using manipulatives, as well as on number lines and number tracks.

Before they start this unit, it is expected that children:
- know how to count accurately up to 20
- understand how to represent numbers up to 20 on ten frames and on a bead string
- know number bonds to 10 and how to split numbers up to 10 into two parts using a part-whole model
- are confident with subtraction skills within 10
- can subtract more than one number mentally or by using a representation such as a number line.

ASSESSING MASTERY

Children who have mastered this unit will be able to use different strategies for adding and subtracting numbers within 20, choosing an appropriate method dependent on the numbers in the calculation. They will be able to use representations of ten frames, bead strings and number lines to explain different methods for adding and subtracting numbers up to 20.

Children will use their knowledge of addition and subtraction to correctly interpret word problems and explain their solutions using the correct mathematical language.

COMMON MISCONCEPTIONS	STRENGTHENING UNDERSTANDING	GOING DEEPER
Children may resort to counting strategies, including 'count all', instead of using known number bonds.	Encourage children to derive the answer by hiding a complete 10. Use part-whole models to display number bonds that can be used in a calculation.	Challenge children to find families of related facts from a known number bond.
Children may not know which strategy to apply in a given context.	Prompt children to look for whether or not the addition or subtraction would involve crossing the 10.	Ask children to create their own story problems for different kinds of additions and subtractions.
Children may misinterpret word problems and use the incorrect operation.	Make a classroom display featuring the key mathematical language of the unit.	If a problem has a range of answers, see if children can find all of them. Encourage them to work methodically.

Unit 7: Addition and subtraction within 20

UNIT STARTER PAGES

Use these pages to introduce the main learning of this unit, and to remind children of the key models you use to represent numbers to 20. Use the characters to introduce the concepts and different ways of trying to solve a problem.

STRUCTURES AND REPRESENTATIONS

Ten frame: The ten frame helps give children a sense of 10 and of number bonds to 10.

It is especially powerful when two ten frames are placed side by side, to be used when adding and subtracting numbers to 20.

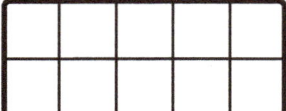

Bead string: The bead string helps children represent numbers and split numbers into parts. It can also be used to show the effect of adding two numbers together.

Part-whole model: This model helps children understand that two or more parts combine to make a whole. It also helps strengthen children's understanding of number bonds within 20.

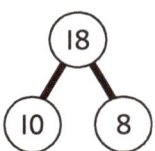

Number line and number track: These help children identify whether an addition or subtraction will require 'crossing the 10' and can also be a useful support when children use 'count on' to check an addition or 'count back' to check a subtraction.

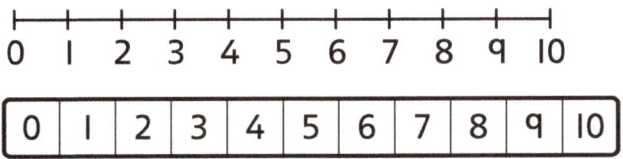

KEY LANGUAGE

There is some key language that children will need to know as part of the learning in this unit:

- count, count on, count back
- add, addition, additions, plus (+)
- subtract, difference, subtraction, take away, minus (−)
- altogether, in total
- number bonds, fact family
- tens, ones
- number stories, represent
- part, whole, part-whole
- compare, greater, less, how many more? how many are left? how many fewer?
- predict

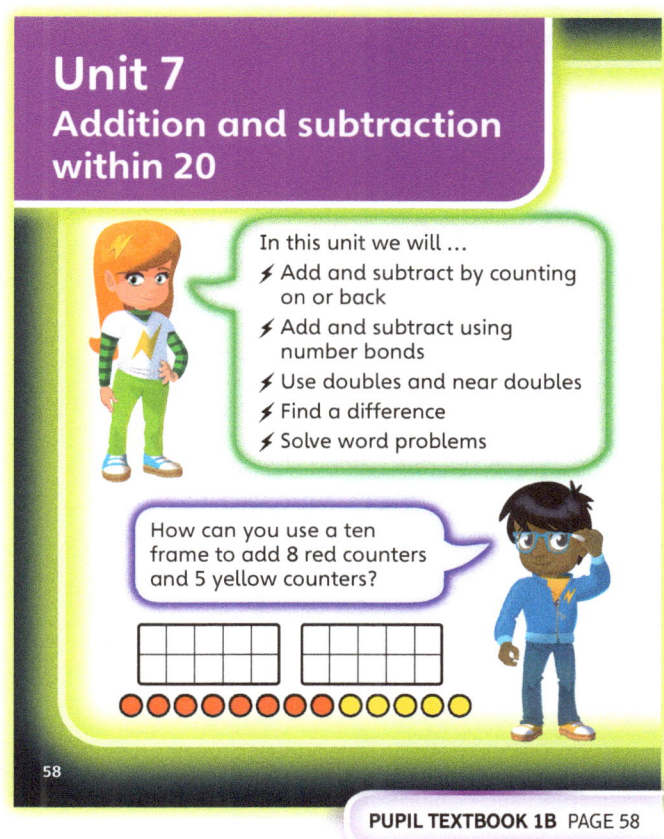

PUPIL TEXTBOOK 1B PAGE 58

PUPIL TEXTBOOK 1B PAGE 59

Unit 7: Addition and subtraction within 20, Lesson 1

Add by counting on within 20

Learning focus

In this lesson, children will add numbers by counting on from one number. Children will know where to start the count and when to stop the count.

Before you teach

- Are children secure counting to 20?
- Are children confident counting to 20 from different starting numbers?
- Can children identify the missing number in a sequence such as 8, 9, 10, 12, 13?

NATIONAL CURRICULUM LINKS

Year 1 Number – addition and subtraction

Add and subtract 1-digit and 2-digit numbers to 20, including zero.

ASSESSING MASTERY

Children can solve an addition with a total of up to 20 by counting on, including where additions cross 10. Children can decide which number to count on from.

COMMON MISCONCEPTIONS

When counting on, children may include the starting number in the count. To challenge this misconception, ask:
- *Show how to count 1 more than 8 on a number track. Do you count the 8?*

When counting on, children may not recognise when to stop the count. Ask:
- *How will you know when to stop counting on?*

STRENGTHENING UNDERSTANDING

Use a number track on the playground, or draw one with chalk. Children can physically hop along the number track. Give instructions such as: *You are on 8, jump on 3 more. Count the numbers as you land. Where did you finish?* Embed the learning by showing how this activity links to the calculation 8 + 3 = ☐ on a number track and by representing it on ten frames.

GOING DEEPER

Discuss the effect of adding 0, or adding to 0, in additions such as 13 + 0 or 0 + 13. To deepen understanding further, you could prompt children to explain what they notice when they compare additions such as 5 + 8 and 8 + 5. What is the same and what is different about the additions and about how they find the total?

KEY LANGUAGE

In lesson: count on, ten, ones (1s), number track, add, addition, in total, more, altogether

Other language to be used by the teacher: represent, calculation, method, solve

STRUCTURES AND REPRESENTATIONS

Number track, number line, ten frame

RESOURCES

Mandatory: counters or cubes, ten frames, number tracks, number lines

Optional: large number track on playground

In the eTextbook of this lesson, you will find interactive links to a selection of teaching tools.

Quick recap

Draw a number track on the board or give children individual ones. Ask children to show simple additions, such as 5 + 3. Ask them where they start, then how they show their sum. Repeat for similar additions.

Unit 7: Addition and subtraction within 20, Lesson 1

Discover

WAYS OF WORKING Pair work

ASK

- Question 1 a): *How can you count the children? How can you check you have counted them all?*
- Question 1 b): *What do you need to do to solve this problem? Do you need to count all the children on the bus again? What calculation does this show? Did you work it out in the same way as a partner?*

IN FOCUS In question 1 a), children practise counting 8 children. They may want to double check to ensure they have counted all the children or not counted one child twice. In question 1 b), provide children with a number track, if needed, to help them. Look for children not needing to count the 8 children on the bus again and using 8 as the starting number. Determine if children revert to addition by a 'count all' strategy, or if they recognise that they can use a different addition strategy such as 'count on'.

PRACTICAL TIPS You could act out the situation with the children in the classroom.

ANSWERS

Question 1 a): There are 8 children on the bus.

Question 1 b): There are 11 children in total.

PUPIL TEXTBOOK 1B PAGE 60

Share

WAYS OF WORKING Whole class teacher led

ASK

- Question 1 a): *Did we all get the same answer? How can we make sure we have counted correctly?*
- Question 1 b): *Why can we count on from 8? How do you know when to stop counting? Will the total be greater than or less than 10? Can you tell before finding the answer?*

IN FOCUS In question 1 a), count the number of children as a class. Ask them all to point and count to check that there are 8 children on the bus. Listen to some children who count too fast. Some children may want to represent the children on the bus with counters on a ten frame. In question 1 b), children count on from 8. Explain they do not need to count the 8 children again and can just count on from 8. Ask them to point to each child and say the next number as they count. Again, they may want to use counters on a ten frame to help them. Highlight the misconception of including the starting number in the count.

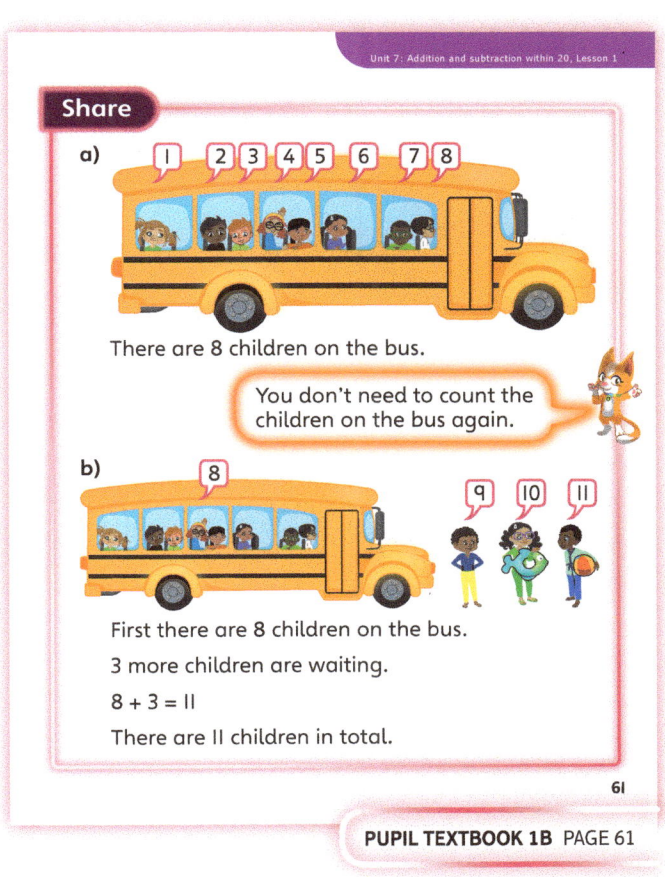

PUPIL TEXTBOOK 1B PAGE 61

Unit 7: Addition and subtraction within 20, Lesson 1

Think together

WAYS OF WORKING Whole class teacher led (I do, We do, You do)

ASK
- Question ①: *Where do we need to start the count? How do you know? What can you do to help you count?*
- Question ②: *What number do we need to start at? How can we use the number line to help? How many jumps do we need to make? Is there a quicker way we can count?*
- Question ③: *Can you explain the method that Hassan has used? What number would you start from in each example?*

IN FOCUS In question ①, children are forced to count on from the number in the box as the number of items are not countable. Discuss different methods that children may use. They are most likely going to point and count. Children should be confident in counting these numbers. In question ②, children use a number line to support them in working out simple additions that cross 10. In question ③, Hassan counts on from the larger number in the addition, even though the smaller number is given first. Children may use this method to work out the addition. In the examples, the smaller number comes first. Discuss with children how starting with the larger number might be easier.

STRENGTHEN Support children with a number line and draw on jumps to help children find the answers. Ensure that children are clear where they start and how they know how many to add on. Count with children as they count on.

DEEPEN Ask children to consider the most efficient method to solve 4 + 9. Ask them to work out 4 + 9 and 9 + 4 by counting on. What do they notice? Children should start to generalise that it might be easier to start with the largest number.

ASSESSMENT CHECKPOINT Children's responses to questions ① and ② will demonstrate whether they include the starting number in their count or know how to count on accurately. Question ③ will show whether children can generalise the counting on method and whether they use the larger number to start the count or just the first number in the calculation.

ANSWERS

Question ① a): 6 + 2 = 8
Question ① b): 9 + 2 = 11
Question ① c): 12 + 2 = 14
Question ② a): 6 + 3 = 9
Question ② b): 7 + 7 = 14
Question ② c): 9 + 4 = 13
Question ② d): 12 + 3 = 15
Question ③ a): 3 + 9 = 12
Question ③ b): 2 + 9 = 11
3 + 14 = 17
2 + 16 = 18

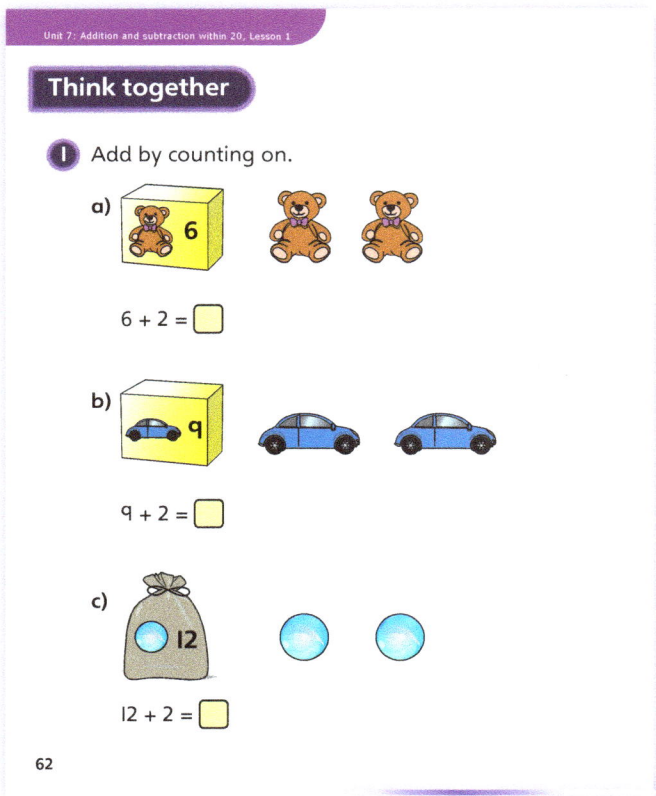

PUPIL TEXTBOOK 1B PAGE 62

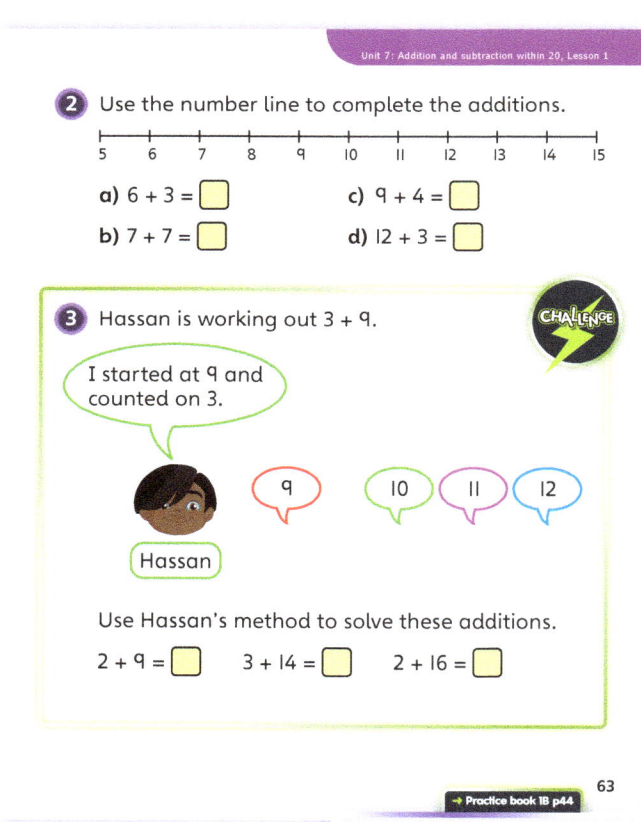

PUPIL TEXTBOOK 1B PAGE 63

100

Unit 7: Addition and subtraction within 20, Lesson 1

Practice

WAYS OF WORKING Independent thinking

IN FOCUS In questions ①and ②, the starting number is not countable, so children are channelled into using a count on method. They are most likely to point and count or use a number track/number line to help them. Questions ③ and ④ use a number line for pictorial support. Check carefully that children are starting on the correct number and that they know how to count on the correct number of steps to get to the answer. Look for children who count the starting number and challenge this misconception. Question ⑤ prompts children to think about how the facts 8 + 5 and 5 + 8 are related.

STRENGTHEN Children can use counters or cubes to support their counting, especially when deciding where to stop the count. Having a number track or number line on display or on the table will help children use the counting on method.

DEEPEN Encourage children to discuss what is the same and what is different about the two additions in question ⑤. Can they use the answer to one of the additions to predict the answer to the other, then check both using number lines? Ask: *What do you notice? Was it easier to work one out than the other?*

ASSESSMENT CHECKPOINT Responses to questions ①, ② and ③ will show if children understand where the count starts and whether they are including the starting number in the count. Question ④ will show whether children can add confidently by counting on using a number line to support them. Question ⑥ is an opportunity to assess understanding of the = sign in different positions.

ANSWERS Answers for the **Practice** part of the lesson can be found in the *Power Maths* online subscription.

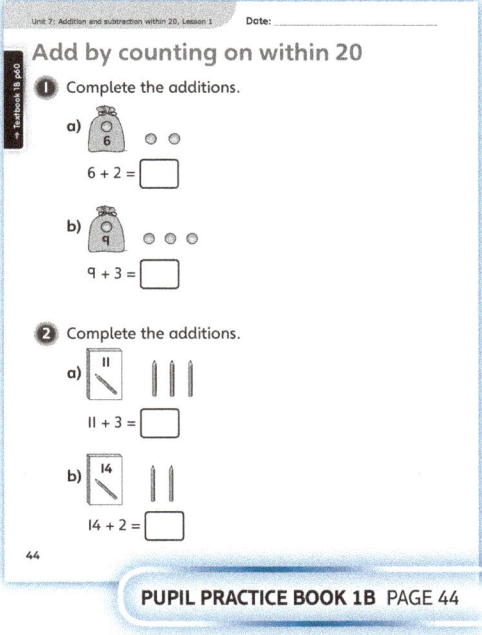

PUPIL PRACTICE BOOK 1B PAGE 44

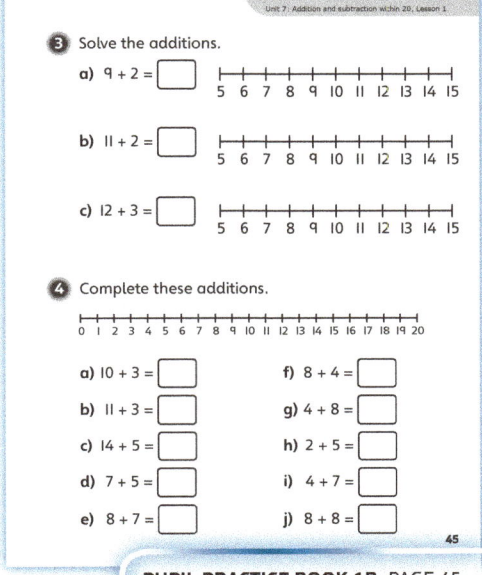

PUPIL PRACTICE BOOK 1B PAGE 45

Reflect

WAYS OF WORKING Pair work

IN FOCUS In this part of the lesson, children explain how they would find 5 + 9. Children may count on 5 from 9 or 9 from 5.

ASSESSMENT CHECKPOINT Assess whether children are able to explain the key points: where to start the count, when to stop the count and whether to include the starting number.

ANSWERS Answers for the **Reflect** part of the lesson can be found in the *Power Maths* online subscription.

After the lesson

- Were children able to use count on rather than the count all strategy?
- Did children understand how the representation on a number line showed the count on method in relation to the context of the problem?
- Could children accurately stop the count to find the correct total?

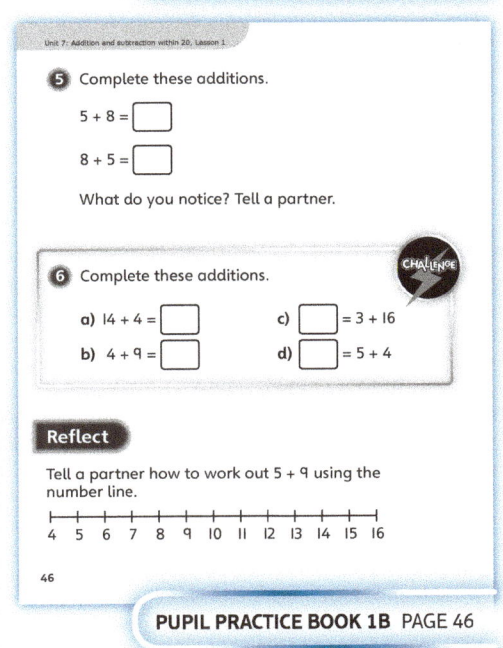

PUPIL PRACTICE BOOK 1B PAGE 46

101

Unit 7: Addition and subtraction within 20, Lesson 2

Add ones using number bonds

Learning focus
In this lesson, children will add a 2-digit number to a 1-digit number by adding the ones.

Before you teach
- Can children add two 1-digit numbers?
- Are children confident identifying which digit is the tens and which is the ones in a 2-digit number?
- Are children secure in ordering numbers?

NATIONAL CURRICULUM LINKS

Year 1 Number – addition and subtraction

Represent and use number bonds and related subtraction facts within 20.

Add and subtract 1-digit and 2-digit numbers to 20, including zero.

ASSESSING MASTERY

Children can explain how knowing a number fact such as 2 + 3 helps find the related fact 12 + 3. Children can illustrate this using representations on a ten frame or a bead string.

COMMON MISCONCEPTIONS

Children may not understand that the tens digit represents 1 ten and may think it just represents 1. Ask:
- *What does the '1' represent in the numbers 13 and 15? What is the 1 worth in 13 and 15?*

Children may find the language of 'ones' confusing. For example, the 'ones' in 13 refers to 3 ones, not to the digit '1'. Ask:
- *What is the ones digit in the number 13, 15 or 18 worth? Which digits are in the ones place?*

STRENGTHENING UNDERSTANDING

Use manipulatives in different colours, such as different coloured counters, to represent the two numbers to be added. Arranging the counters on ten frames will support understanding of the 10 and the ones. Children can also use bead strings to find the 10 and the 1s parts of 2-digit numbers.

GOING DEEPER

Ask children to explain the link between additions such as 13 + 4 and 3 + 14, using place value to support their reasoning. Encourage children to use their knowledge of number bonds within 10 to generate other additions. For example, if they know 1 + 4 = 5, can they find 11 + 4, 14 + 1, 4 + 11 and 1 + 14?

KEY LANGUAGE

In lesson: ten (10), ones (1s), add, addition, in total, altogether, digit

Other language to be used by the teacher: number fact, number bond

STRUCTURES AND REPRESENTATIONS

Ten frame, bead string, part-whole model, rekenreks

RESOURCES

Mandatory: ten frames, counters, bead strings, scissor boxes or boxes of pencils

Optional: rekenreks

 In the eTextbook of this lesson, you will find interactive links to a selection of teaching tools.

Quick recap

Check that children know simple number bonds. Ask them to write down as many number bonds to 5 as they can. Look at the different strategies they use. Repeat, asking children to write down as many number bonds to 8 that they know.

Unit 7: Addition and subtraction within 20, Lesson 2

Discover

WAYS OF WORKING Pair work

ASK

- Question 1 a): *How many pairs of scissors are there in the full scissor box?*
- Question 1 b): *Do you have to count all the scissors to work out the total? Where will the 3 extra pairs of scissors go? What calculation will find the total number of pairs of scissors? How can you use 2 + 3 to help you?*

IN FOCUS In question 1 a), the scissor boxes are designed to have 2 rows of 5 in each box. Encourage children to notice the similarity with ten frames. Children should be able to use their understanding to recognise that there must be 12 scissors in the two boxes by identifying the 1 complete ten and the 2 ones. In question 1 b), as one of the boxes is full, children can think about adding the 3 scissors to the right-hand box. For support, children can use counters and ten frames to represent the scissors. Ask children if they can use their knowledge of 2 + 3 to work out the total number of scissors.

PRACTICAL TIPS Use scissor boxes or boxes of pencils to recreate the scenario in the classroom.

ANSWERS

Question 1 a): There are 12 pairs of scissors.

Question 1 b): There are now 15 pairs of scissors.

PUPIL TEXTBOOK 1B PAGE 64

Share

WAYS OF WORKING Whole class teacher led

ASK

- Question 1 a): *How do you know the ten frames show 12?*
- Question 1 b): *How can we show the 3 extra pairs of scissors? Do we need to count all of the counters to work out the total? Why are three counters a different colour?*

IN FOCUS The important learning in this part of the lesson is recognising how 2 + 3 helps us to solve 12 + 3 in question 1 b). This links with children's understanding of numbers to 20 from their work in the previous unit. The ten frame representations show the link very clearly, and children should discuss how the full ten frame links to the place value of the number 12, using the word 'ones' accurately. You may want to use Astrid's comment to help further the explanation of how children can use number bonds within 10 to help us work out the addition 12 + 3.

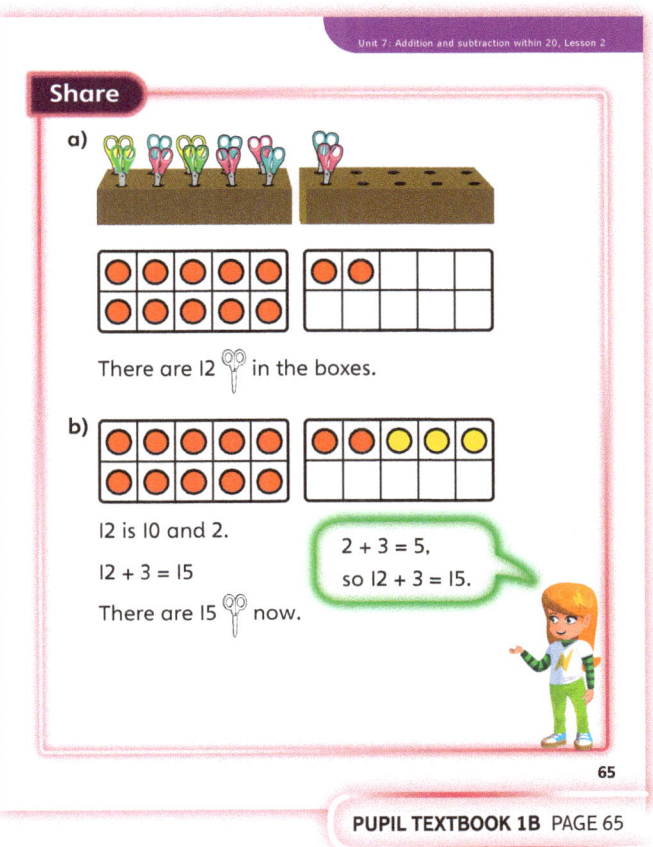

PUPIL TEXTBOOK 1B PAGE 65

Think together

WAYS OF WORKING Whole class teacher led (I do, We do, You do)

ASK
- Question ❶: *How do the ten frames show the calculation 15 + 4? What number bond can help us work out the answer?*
- Question ❷: *What number is shown on the ten frames? What number do you need to add on? What is the answer? How can you work it out using number bonds?*
- Question ❸: *How many answers can you find? How can you use ten frames, a bead string or a rekenrek to help you?*

IN FOCUS In question ❶, children work through how to work out an addition using knowledge of number bonds within 10. Children see the 15 red counters and 4 yellow counters and realise they just need to work out 5 + 4, as the 10 has not changed. They should realise they are just adding the 1s in this example. In question ❷ a), the second number is not represented in the ten frames and children will need to add more counters. Question ❷ b) requires children to set up their own ten frames. Some children may be able to work out the answers without representing the addition using ten frames. Question ❸ requires children to focus on place value.

STRENGTHEN Use ten frames and bead strings to support understanding, but ensure children are not relying on a count all method. They should 'see' the complete 10 and use that to deduce the total number. If children continue to count all, cover the ten frame with a piece of paper once it is complete, so that they have to think of it as a complete 10, rather than as 10 ones.

DEEPEN Encourage children to convince a partner that they have found all possible solutions to question ❸. Ask them to give their reasons for thinking there cannot be any more options. Many children may not spot 18 + 0. Children will start to work out the answers using knowledge of number bonds rather than making numbers on ten frames each time.

ASSESSMENT CHECKPOINT Children's responses to questions ❶ and ❷ will reveal if they understand how they can just add the 1s then work out the total by using place value. Their responses to question ❸ will reveal if they can work systematically, and whether they have a deep understanding of place value for the 10s and 1s digits.

ANSWERS

Question ❶: 5 + 4 = 9
15 + 4 = 19

Question ❷ a): 3 + 2 = 5
13 + 2 = 15

Question ❷ b): 4 + 3 = 7
14 + 3 = 17

Question ❸ a): Possible solutions are: 10 + 8, 11 + 7, 12 + 6, 13 + 5, 14 + 4, 15 + 3, 16 + 2, 17 + 1, 18 + 0.

Question ❸ b): Possible solutions are: 10 + 5, 11 + 4, 12 + 3, 13 + 2, 14 + 1, 15 + 0.

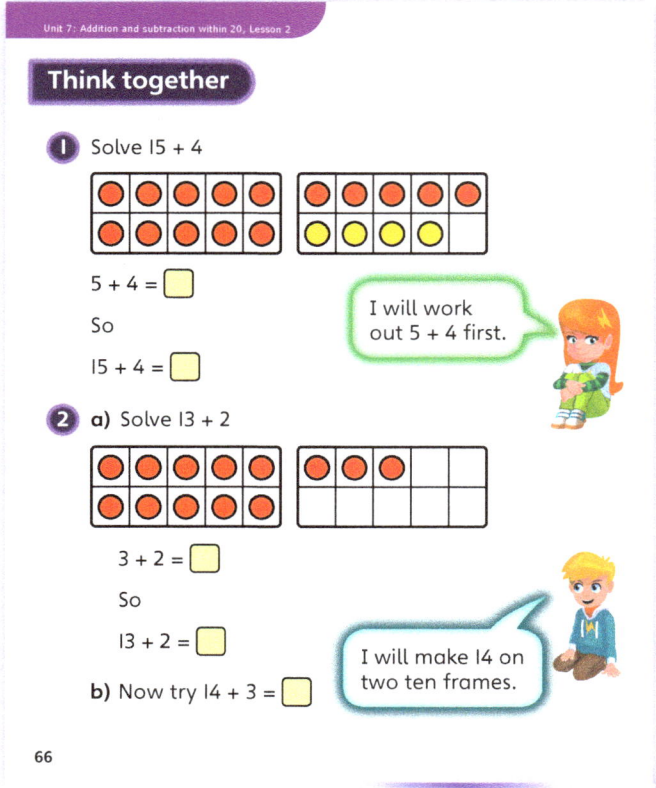

PUPIL TEXTBOOK 1B PAGE 66

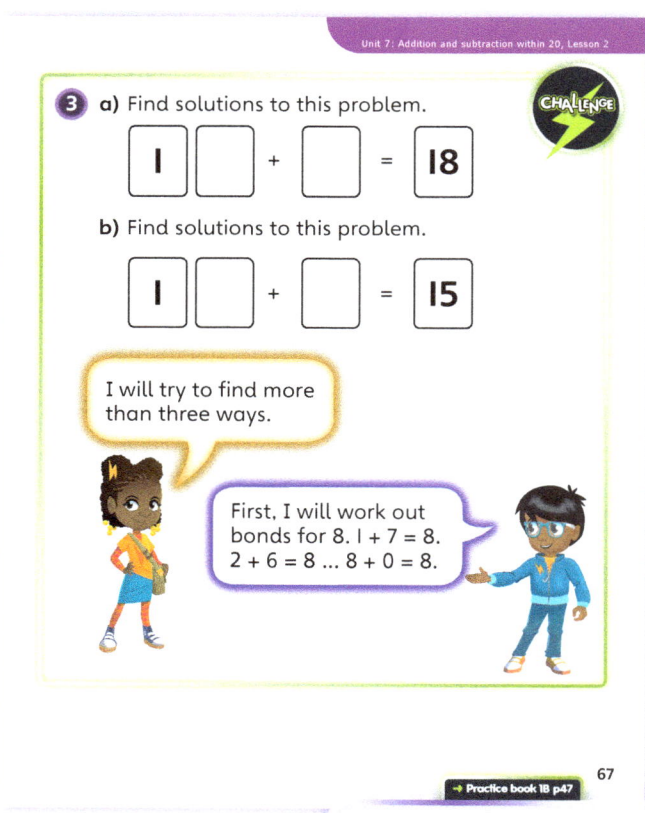

PUPIL TEXTBOOK 1B PAGE 67

Unit 7: Addition and subtraction within 20, Lesson 2

Practice

WAYS OF WORKING Independent thinking

IN FOCUS These questions focus on adding a 1-digit number to a 2-digit number, where the 1s digits add to less than 10. The method encouraged is to add the 1s digits, adding that to the 10, rather than counting on from the 2-digit number. Question ❻ focuses on children's understanding of place value for the 10s and 1s digits.

STRENGTHEN Strengthen understanding by asking children to represent their addition of 1s on a ten frame. Children can then use that to help them combine the number fact with the complete 10. Children can check their answers using bead strings, but should be encouraged to use this as a checking strategy, rather than a first resort.

DEEPEN Question ❹ is a simple word problem. Ask children to make up similar word problems involving adding a 2-digit number to a 1-digit number. You may need to specify pairs of numbers to ensure the 1s digits add to less than 10.

ASSESSMENT CHECKPOINT Assess children's approaches to questions ❶, ❷ and ❸ to see if they revert to a count on strategy. Questions ❺ and ❻ will be an opportunity to see if children can solve additions by using knowledge of number facts and additions within 10.

ANSWERS Answers for the **Practice** part of the lesson can be found in the *Power Maths* online subscription.

Reflect

WAYS OF WORKING Pair work

IN FOCUS In this question, children will use their knowledge of number bonds to find related additions within 20. This will require some discussion around place value, so children should agree with their partners about their decisions.

ASSESSMENT CHECKPOINT Children's answers will demonstrate if they have understood how the place value of the 10s digit allows them to find addition facts by using knowledge of number bonds.

ANSWERS Answers for the **Reflect** part of the lesson can be found in the *Power Maths* online subscription.

After the lesson

- Can children explain why they should add the 1s and then combine the answer with the complete 10?
- Can children use a known addition within 10 to complete a related addition within 20?

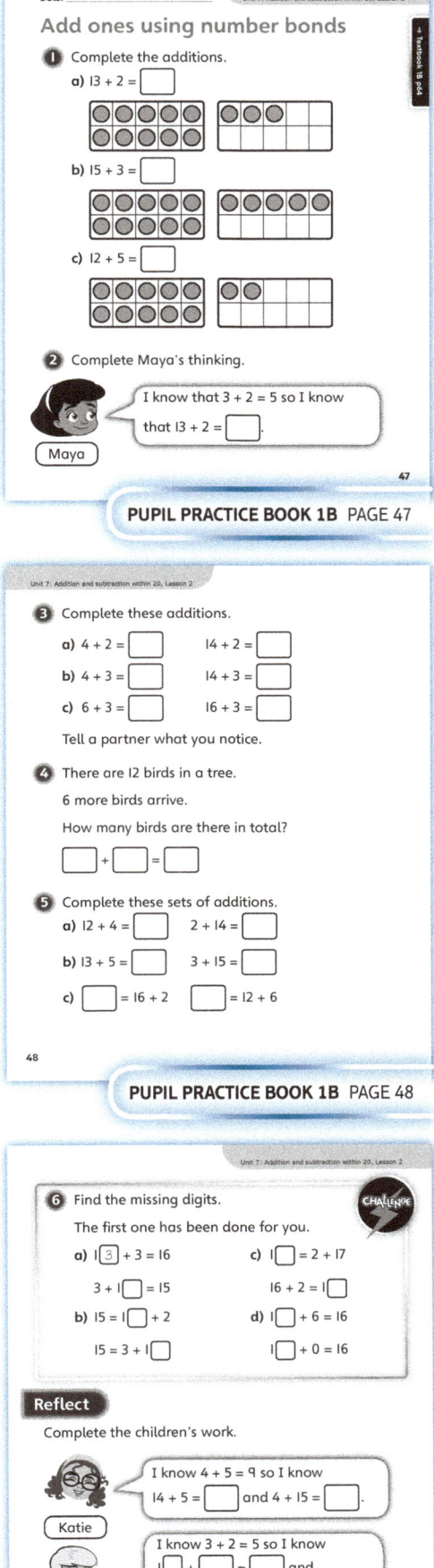

PUPIL PRACTICE BOOK 1B PAGE 47

PUPIL PRACTICE BOOK 1B PAGE 48

PUPIL PRACTICE BOOK 1B PAGE 49

105

Unit 7: Addition and subtraction within 20, Lesson 3

Find and make number bonds to 20

Learning focus
In this lesson, children will find number bonds to 20. They will learn the link between number bonds to 10 and number bonds to 20.

Before you teach
- How many counters fill two ten frames?
- What does the '2' represent in '20'?
- How many number bonds to 10 can you remember?

NATIONAL CURRICULUM LINKS

Year 1 Number – addition and subtraction

Represent and use number bonds and related subtraction facts within 20.

ASSESSING MASTERY

Children can deduce number bonds to 20 and explain the link with number bonds to 10. Children can recall common number bonds to 20, such as 10 + 10, 1 + 19, 5 + 15.

COMMON MISCONCEPTIONS

Children may use an inefficient method, such as counting on or back, rather than using known facts and developing instant recall. Ask:
- What do you add to 10 to make 20?
- What do you add to 6 to make 10? So, what do you add to 16 to make 20?

STRENGTHENING UNDERSTANDING

Use two ten frames and counters to give a visual representation of how the number bonds to 20 are related to number bonds to 10. Make a list of calculations with a total of 20, including a 10, and highlight the bond to 10: 10 + **1 + 9** = 20; 10 + **2 + 8**; 10 + **3 + 7**, and so on.

GOING DEEPER

Ask children to explain how the number bonds to 20 are linked to the number bonds to 10 and ask them to create families of related facts. For example, from 3 + 7, can they produce 13 + 7, 3 + 17, 17 + 3, and 7 + 13?

KEY LANGUAGE

In lesson: number bond, ten frame, part-whole model, ten (10), ones (1s)

Other language to be used by the teacher: represent, related facts, bond to 20

STRUCTURES AND REPRESENTATIONS

Ten frames, bead string, part-whole model

RESOURCES

Mandatory: counters or cubes, ten frames, double-sided counters

Optional: bead strings

 In the eTextbook of this lesson, you will find interactive links to a selection of teaching tools.

Quick recap

Revise number bonds to 10. Give children 10 double-sided counters and a ten frame and ask them to use their counters to make and write down all the number bonds to 10.

Unit 7: Addition and subtraction within 20, Lesson 3

Discover

WAYS OF WORKING Pair work

ASK
- Question 1 a): *How many red cubes are there? How many yellow cubes are there? What number bond to 20 does it show?*
- Question 1 b): *Can you use cubes or counters and ten frames to show another number bond to 20?*

IN FOCUS Throughout this section, children are using either counters or cubes to make 20. In question 1 a), children can see that 10 + 10 = 20. In question 1 b), provide children with 20 double sided counters and two ten frames to help them write down a different number bond to 20. They may make other number bonds by turning over counters.

PRACTICAL TIPS Encourage children to use double-sided counters and ten frames to make number bonds to 20.

ANSWERS

Question 1 a): There are 10 red cubes.

There are 10 yellow cubes:

10 + 10 = 20

There are 20 cubes altogether.

Question 1 b): Various answers are possible. For example, **Share** shows 17 + 3 = 20.

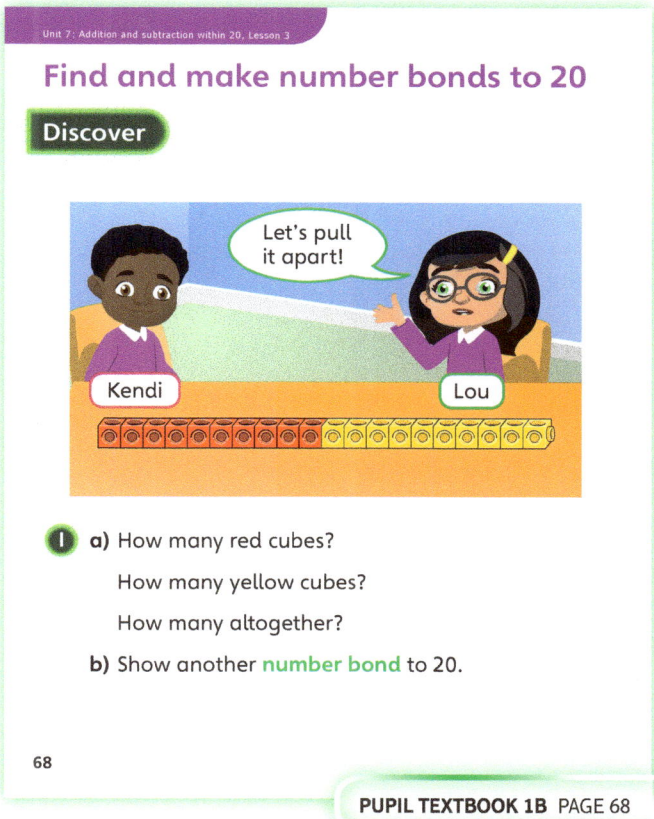

PUPIL TEXTBOOK 1B PAGE 68

Share

WAYS OF WORKING Whole class teacher led

ASK
- Question 1 a): *How many red cubes are there? How have these been shown on the ten frames? How many yellow cubes are there? How have these been shown? How do we know we have made 20?*
- Question 1 b): *How many counters can you see in total? How do you know? How many are red? How many are yellow? What number bond to 20 can you see?*

IN FOCUS In question 1 a), explain to children how the cubes have been represented by counters on the ten frames. You may want to count the cubes out loud with children to check there are 10. Explain we have used counters and ten frames so that we can see how many there are straight away. Discuss how 2 tens are the same as 20. In question 1 b), explain to children how they can use the ten frames they have made to make other number bonds to 20. They can do this by changing the colour of some of the counters on just one of the ten frames. Explain why these ten frames show that 17 + 3 = 20. They may want to use knowledge of number bonds to 10 to help them too.

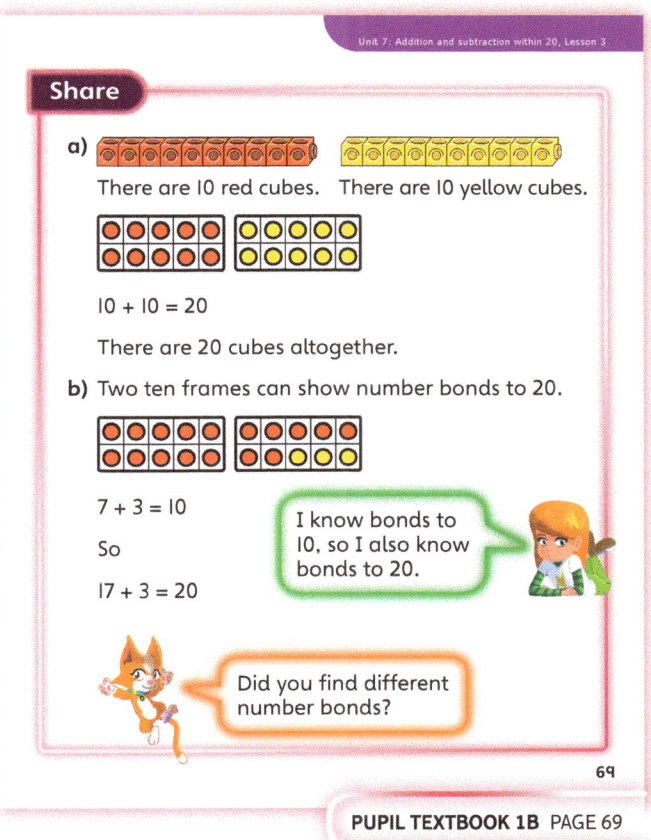

PUPIL TEXTBOOK 1B PAGE 69

107

Unit 7: Addition and subtraction within 20, Lesson 3

Think together

WAYS OF WORKING Whole class teacher led (I do, We do, You do)

ASK

- Question ①: *How many red counters are there? How many yellow counters are there? How do you know? What number bond to 20 is shown? How can you use 4 + 6 to help you?*
- Question ②: *What number bond to 10 is shown? How can you use this to make a number bond to 20? Can you make another number bond to 20 using this same ten frame?*
- Question ③: *Which of these number bonds to 20 do you know straight away? How do you know them?*

IN FOCUS In question ①, children are practising what they have just done in the **Share** section by using counters to help them create a number bond to 20. They can see there are 20 counters because the 2 ten frames are full. The first part of the question gets them to focus on the second ten frame by writing down a number bond to 10. Show the link between the number bond to 10 and the number bond to 20. Question ① should set children up for question ② as they use the number bonds to 10 to help them write two number bonds to 20. Ask children to have a go at the abstract number bonds to 20 in question ③ and discuss their methods. Encourage them to use their number bonds to 10 to help them instead of needing to use ten frames and counters.

STRENGTHEN The ten frames and counters in questions ① and ② should help children see the number bonds to 20. It is important for children to see how a number bond to 10 can help them work out a number bond to 20. Ask children to make the number bond to 10 first and then add another full ten frame of one of the counters, then ask them if they can see at least one number bond to 20. Write the number bonds down next to the ten frames so that children can clearly see the number bonds.

DEEPEN In question ③ a), ask children to find all the number bonds to 20 systematically. Look at the different ways they approach this, so that they do not miss any out.

ASSESSMENT CHECKPOINT Questions ① and ② will help determine if children can form a number bond to 20 using counters on ten frames. They will help you assess if they can also use their number bonds to 10 to help them. Question ③ checks whether children can work out their number bonds to 20 abstractly, without necessarily needing to use counters and ten frames.

ANSWERS

Question ①: 14 + 6 = 20

Question ②: 2 + 8 = 10
12 + 8 = 20
2 + 18 = 20

Question ③ a): 1 + 19 = 20   2 + 18 = 20
3 + 17 = 20   11 + 9 = 20
12 + 8 = 20   13 + 7 = 20
11 + 9 = 20   1 + 19 = 20

Question ③ b): The bonds to 20 are: 1 + 19, 2 + 18, 3 + 17, 4 + 16, 5 + 15, 6 + 14, 7 + 13, 8 + 12, 9 + 11, 10 + 10, 11 + 9, 12 + 8, 13 + 7, 14 + 6, 15 + 5, 16 + 4, 17 + 3, 18 + 2 and 19 + 1.

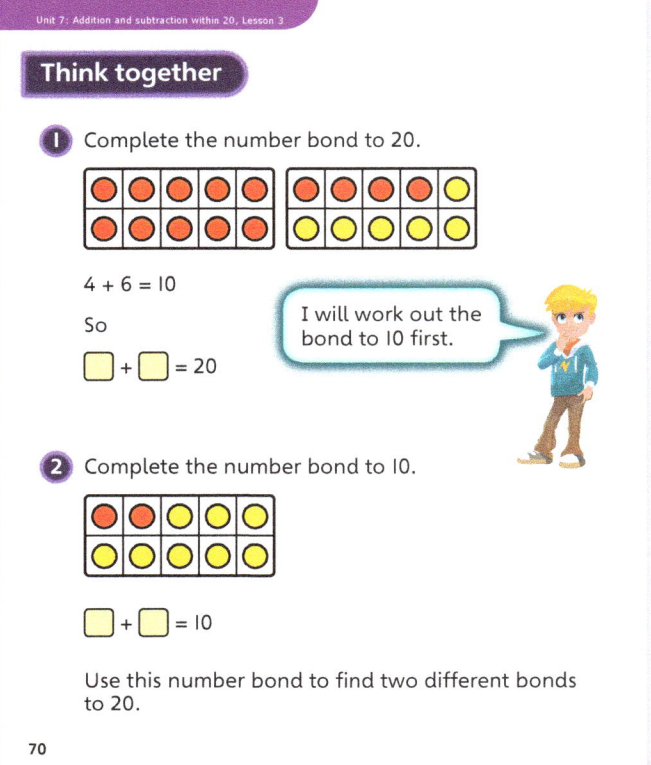

PUPIL TEXTBOOK 1B PAGE 70

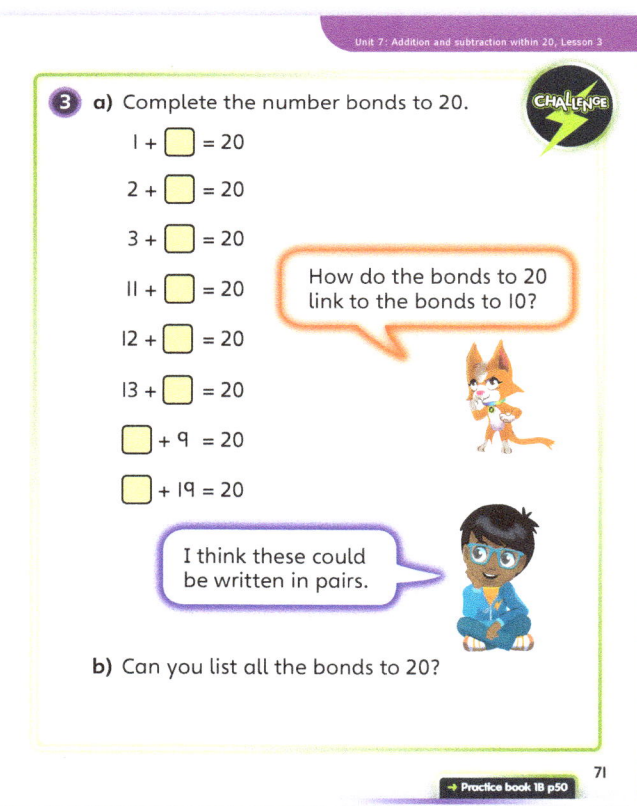

PUPIL TEXTBOOK 1B PAGE 71

Unit 7: Addition and subtraction within 20, Lesson 3

Practice

WAYS OF WORKING Independent thinking

IN FOCUS Questions ❶ and ❷ use concrete contexts to introduce some number bonds to 20. In question ❷, there are two strategies children could use: two rows of 10, giving 10 + 10, or dark grey and light grey, giving 5 + 15. The focus in question ❸ is to understand the different kinds of apparatus that can be used to represent numbers, whereas question ❹ prompts children to reason explicitly from number bonds to 10 to find number bonds to 20.

STRENGTHEN Children should have access to manipulatives to support their thinking, although they should be using these to support, not replace, the method of knowing number bonds. Once a complete 10 has been formed, children could cover it or discard it to make sure they are using place value rather than a count all strategy. Children could then use the counting approach as a checking strategy.

DEEPEN Challenge children to list number bonds to 20. Can they find four related bonds to 20 from the fact 8 + 2 = 10?

THINK DIFFERENTLY Question ❺ encourages children to use logic to match abstract pairs of numbers that make a number bond to 20.

ASSESSMENT CHECKPOINT Children's responses to question ❹ will demonstrate if they have understood the link between number bonds to 10 and number bonds to 20. Answers to question ❺ will illustrate if children have understood place value correctly. For example, some may join 16 and 14 as a bond to 20, reasoning that the 1s add up to 10.

ANSWERS Answers for the **Practice** part of the lesson can be found in the *Power Maths* online subscription.

Reflect

WAYS OF WORKING Pair work

IN FOCUS Children are encouraged to find as many number bonds to 20 as they can. Refer them back to the previous questions to find more.

ASSESSMENT CHECKPOINT Assess which strategies, if any, children are using to generate the number bonds to 20. Which children are using number bonds to 10? Are any children using the commutative law of addition to generate pairs of number bonds?

ANSWERS Answers for the **Reflect** part of the lesson can be found in the *Power Maths* online subscription.

After the lesson

- Can children find number bonds to 20 without using a count all strategy?
- Can children explain the link between number bonds to 10 and number bonds to 20?
- Are children able to see that 13 + 7 gives the same result as 17 + 3, 3 + 17 and 7 + 13, for example?

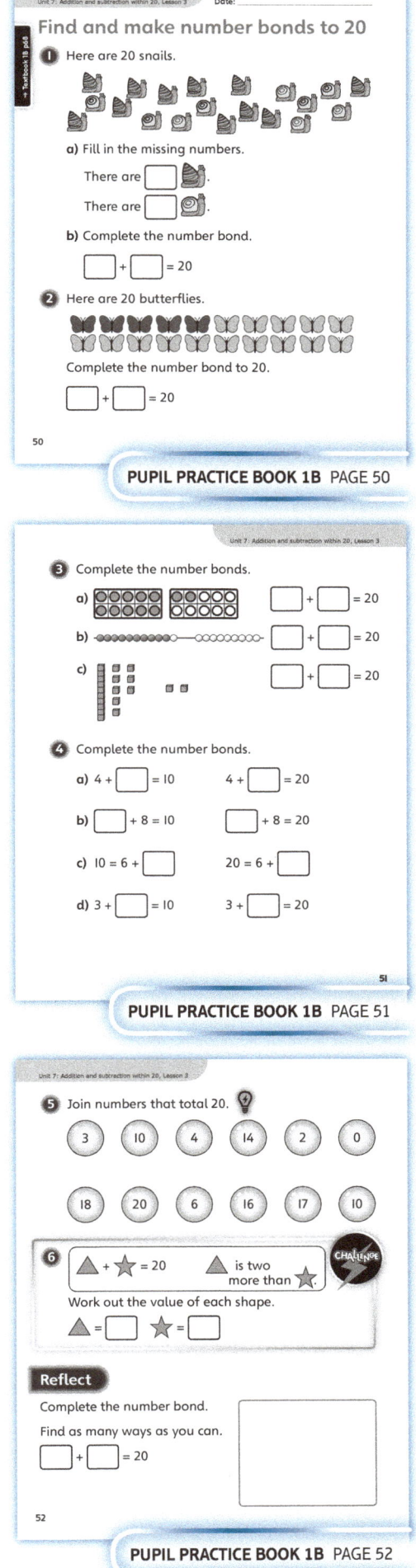

PUPIL PRACTICE BOOK 1B PAGE 50

PUPIL PRACTICE BOOK 1B PAGE 51

PUPIL PRACTICE BOOK 1B PAGE 52

Unit 7: Addition and subtraction within 20, Lesson 4

Doubles

Learning focus

In this lesson, children will understand what the word 'double' means and understand different ways doubles can be represented. Children will begin to learn their doubles up to 10 + 10.

Before you teach

- Can children represent numbers up to 20 on ten frames?
- Can children add two 1-digit numbers using at least one method?

NATIONAL CURRICULUM LINKS

Year 1 Number – addition and subtraction

Represent and use number bonds and related subtraction facts within 20 (within 10).

ASSESSING MASTERY

Children can find doubles from double 1 through to double 10. They know many of the doubles off by heart and can work out others using different strategies.

COMMON MISCONCEPTIONS

Children think that double 1 is 11 and not 2. They simply put the numbers together rather than realising they need to add them. Explain to children that when they find a double they need to add the same number twice. Ask:
- *What does the word 'double' mean? What is 1 + 1?*

STRENGTHENING UNDERSTANDING

Use ten frames and counters or a rekenrek to show doubles. You should put counters on the ten frames in a pair-wise fashion as opposed to completing columns or rows. This will allow children to easily see the same number next to each other. Once children can see the double, help children work out the total. They use methods that they have learnt earlier in the year to work out the total. They may subitise smaller doubles, but for larger doubles they may need to use a 'count on' method.

GOING DEEPER

Children should know their doubles off by heart and be able to work out double what number makes a given number. Ask children what they notice about doubles. They may say they notice numbers that have been doubled all end in 2, 4, 6, 8 or 0.

KEY LANGUAGE

In lesson: double

Other language to be used by the teacher: pair

STRUCTURES AND REPRESENTATIONS

Dice, ten frames

RESOURCES

Mandatory: ten frames, counters

Optional: dice, rekenreks

 In the eTextbook of this lesson, you will find interactive links to a selection of teaching tools.

Quick recap

Check that children can represent different numbers using their fingers or using counters on ten frames. They should work together, for example, to show the number 15 on their fingers. Ask: *Can you show 15 in different ways?*

Discover

WAYS OF WORKING Pair work

ASK

- Question 1 a): *What game are the children playing? Why do you think it is called the mirror game? How many fingers is each child showing? How many fingers are being shown in total?*
- Question 1 b): *Do you think you can play the game with a partner?*

IN FOCUS Questions 1 a) and 1 b) focus on a game that children can play in pairs. Each child shows 1 to 5 fingers on their hand and the other child has to show the same. They could use counters instead of fingers. They then need to work out how many fingers or counters in total. Check which doubles children know straight away as they play a few times. Some children may want to represent the numbers on ten frames to work out the total. Some children may just count the total number of fingers or counters. If so, try and encourage a count on method.

PRACTICAL TIPS Play the game with children, ensuring they all know the rules before they begin.

ANSWERS

Question 1 a): There are 2 fingers in total.

Question 1 b): Children should find the following doubles:
1 + 1 = 2, 2 + 2 = 4, 3 + 3 = 6, 4 + 4 = 8 and 5 + 5 = 10.

Share

WAYS OF WORKING Whole class teacher led

ASK

- Question 1 a): *What do you think the word 'double' means? How many fingers are shown in total? How did you work out the answer?*
- Question 1 b): *Did you show any of these fingers when you played the game? How did you work out the total? Can you say the doubles?*

IN FOCUS Question 1 b) shows all the possible answers that children could have made in the game they have just played. Discuss different methods that children may have used to work out the doubles. Some may be able to see by subitising the total number of fingers. Some may need to count them by counting on. You may want to present double 5 as counters on a ten frame to show a different representation. Use the key word 'double' throughout, explaining that double means when you add two numbers that are the same.

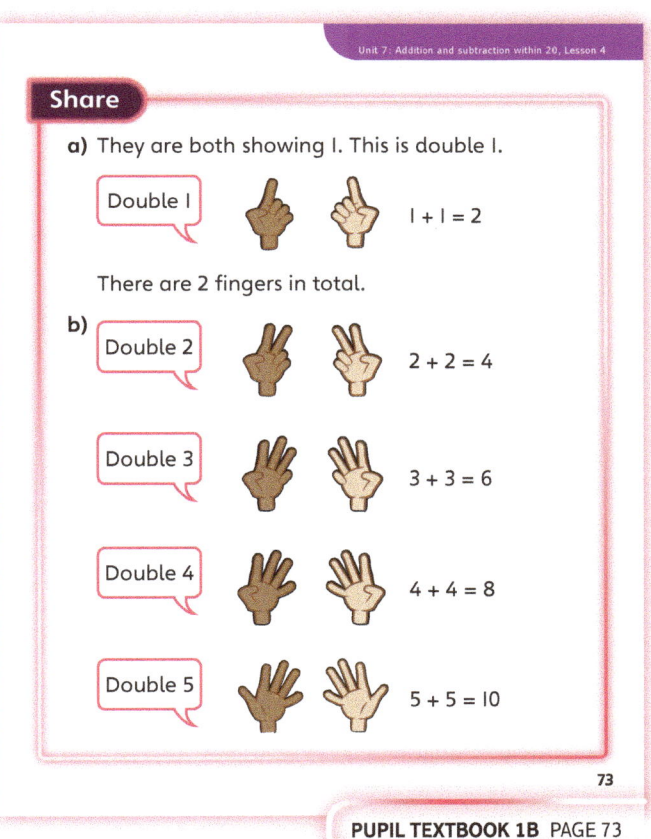

111

Think together

WAYS OF WORKING Whole class teacher led (I do, We do, You do)

ASK
- Question ❶: *Which pairs of dice show doubles? How do you know they show doubles?*
- Question ❷: *Which double is shown? How can you work out what double 6 is?*
- Question ❸: *What doubles can you see? How can you work out the doubles?*

IN FOCUS In question ❶, children recognise which pairs of dice show doubles and which do not. Explain that a double is when both dice show the same number. Encourage them to say the doubles without working out the answer. Explain that it is important to remember their doubles. They understand that a double is an addition of the same number. In question ❷, children work out their first double with an answer above 10. Ask them to make the doubles using counters on two ten frames to help them. Can they see why the answer to double 6 is 12? Ensure they make the doubles on ten frames in a pair-wise way so they can see the double 6 clearly. Question ❸ asks children to work out further doubles. They may not know these off by heart, but should be able to work them out using counters and ten frames or using a 'count on' method.

STRENGTHEN Throughout these questions, encourage children to use counters on ten frames or a rekenrek to show doubles. You should put counters on to ten frames in a pair-wise fashion as opposed to completing columns or rows, so that children can easily see the same number next to each other. Once children can see the double, help children work out the total. They use methods that they have learnt earlier in the year to work out the total. They may subitise smaller doubles, but for larger doubles they may need to use a count on method.

DEEPEN Children should be able to answer questions such as double ☐ = 12 to show that they know their doubles. Ask children to write out their doubles and ask what they notice about their answers as the doubles increase. What patterns do they see?

ASSESSMENT CHECKPOINT Question ❶ assesses that children know their doubles to 10. Questions ❷ and ❸ move to doubles within 20. Ensure children have a method to work out their doubles before moving on. They may know the answers or use a count on method.

ANSWERS

Question ❶ a):

Question ❶ b): 2 + 2 = 4   4 + 4 = 8
              5 + 5 = 10   3 + 3 = 6

Question ❷: 6 + 6 = 12

Question ❸: 6 + 6 = 12   7 + 7 = 14
             8 + 8 = 16   9 + 9 = 18

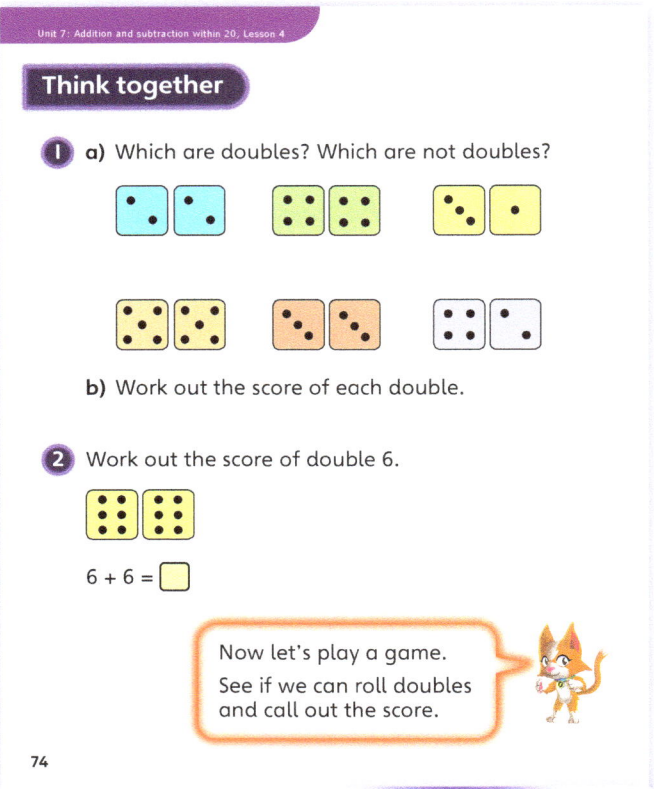

PUPIL TEXTBOOK 1B PAGE 74

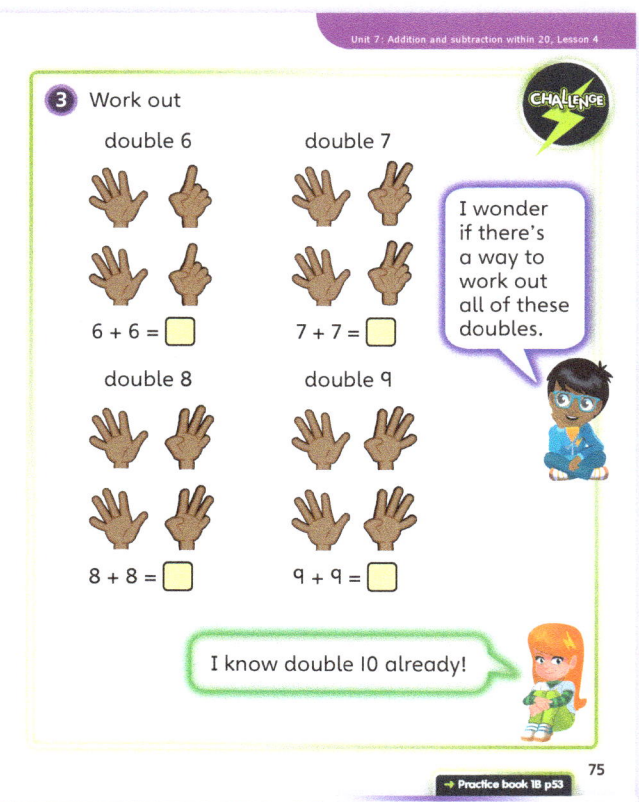

PUPIL TEXTBOOK 1B PAGE 75

112

Unit 7: Addition and subtraction within 20, Lesson 4

Practice

WAYS OF WORKING Independent thinking

IN FOCUS Question ❶ asks children to recognise doubles presented on fingers and write these as abstract additions. Encourage them to write down the answers without having to count. Questions ❷ and ❸ represent doubles in different ways, using dots on ladybirds and dice. They recognise that doubles show the same number twice. Question ❹ checks children's understanding of doubles by asking them to make or draw double 7. They may use some of the representations already shown or may use their own, such as counters on ten frames. In question ❺, children work out the doubles in order. Are there any that children already know? Ask them how they can work out any they do not know. What methods will they use? Throughout these questions, children should be using the correct terminology of 'double'.

STRENGTHEN Throughout these questions, encourage children to use counters on ten frames or a rekenrek to show doubles. You should put counters on ten frames in a pair-wise fashion, so that children can easily see the same numbers next to each other. Once children can see the double, help them to work out the total. They use methods that they have learnt earlier in the year to work out the total. They may subitise smaller doubles, but for larger doubles they may need to use a count on method.

DEEPEN As children work through question ❺, ask them what they notice about their answers. What patterns can they see?

ASSESSMENT CHECKPOINT Questions ❶, ❷ and ❸ check that children know their doubles up to double 6. These are the most common doubles that children need to know as they are seen most often. Question ❺ assesses children's understanding of larger doubles. Check that children know their doubles off by heart.

ANSWERS Answers for the **Practice** part of the lesson can be found in the *Power Maths* online subscription.

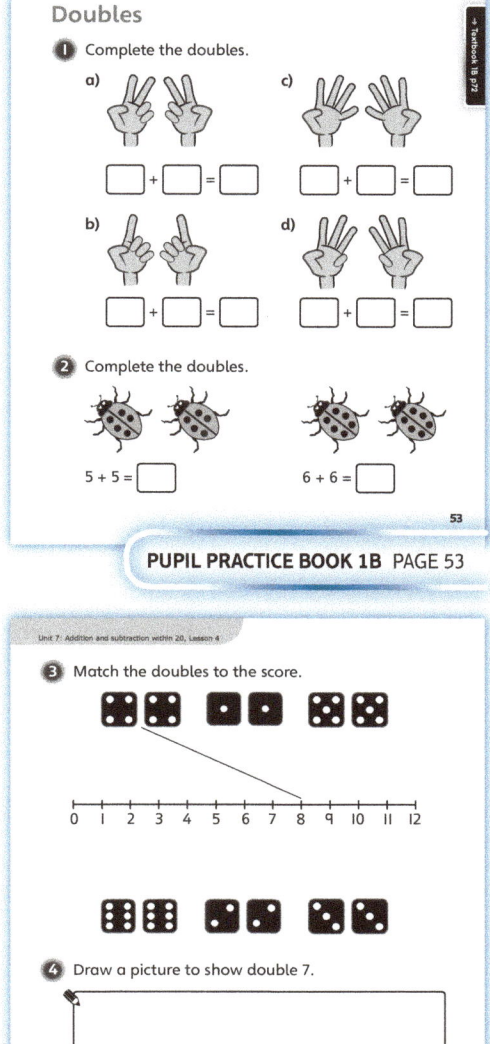

PUPIL PRACTICE BOOK 1B PAGE 53

PUPIL PRACTICE BOOK 1B PAGE 54

Reflect

WAYS OF WORKING Pair work

IN FOCUS This question gives children an opportunity for them to tell a partner and their class the doubles that they know off by heart. Ask them if they can draw or make their doubles too. Children should take it in turns with their partner. Challenge them to say all their doubles to double 10.

ASSESSMENT CHECKPOINT Check which doubles children know confidently. Check which ones they still need to work out by using a method of counting on.

ANSWERS Answers for the **Reflect** part of the lesson can be found in the *Power Maths* online subscription.

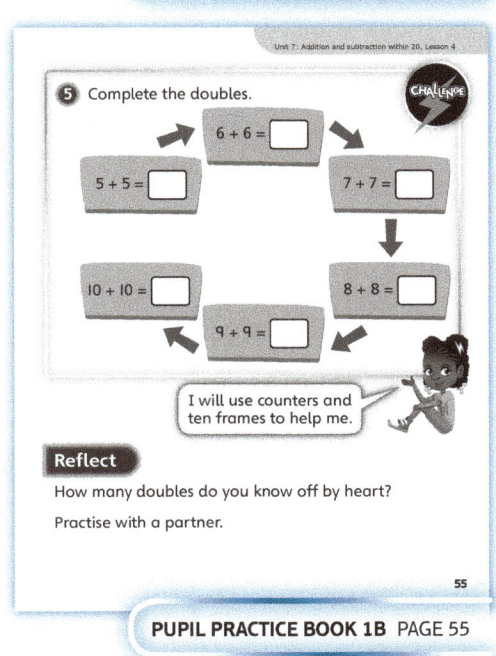

PUPIL PRACTICE BOOK 1B PAGE 55

After the lesson

- Do children know different representations of doubles?
- Do children know double 1 to double 6 off by heart? Do they have a method to work out the other doubles if they do not know them?

Unit 7: Addition and subtraction within 20, Lesson 5

Near doubles

Learning focus
In this lesson, children will find near double additions by considering doubles that they know. For example, they will use 5 + 5 to work out 5 + 6.

Before you teach
- Do children know their doubles to at least 6 + 6?
- Do children know common number bonds within 10?
- Do children know how to find one more and one less than a number within 20?

NATIONAL CURRICULUM LINKS

Year 1 Number – addition and subtraction

Represent and use number bonds and related subtraction facts within 20 (within 10).

ASSESSING MASTERY

Children should be able to confidently work out near doubles using their knowledge of doubles.

COMMON MISCONCEPTIONS

Children may not know their doubles, so they are unable to work out near doubles quickly and efficiently. Knowing their doubles is essential for children to be able to find near doubles. Children should avoid using count all or count on methods to work out the answers. Ask:
- *What does the word 'double' mean? Can you use counters on ten frames to help you show doubles?*

STRENGTHENING UNDERSTANDING

Children need to know their doubles to at least double 6 before being able to confidently find near doubles. To work out 7 + 8, first show children double 7 on a pair of ten frames or on a rekenrek. Ask children what they notice about 7 + 8. They should see that they just need to add one more counter or bead. Work through several examples so that children understand the concept of finding near doubles.

GOING DEEPER

Children explain how they can use both double 6 and double 7 to work our 6 + 7. Discuss with children which method they prefer (if any) and why.

KEY LANGUAGE

In lesson: double, near double

Other language to be used by the teacher: one more, one less

STRUCTURES AND REPRESENTATIONS

Multilink cubes

RESOURCES

Mandatory: cubes, counters, ten frames, dice

Optional: rekenreks

 In the eTextbook of this lesson, you will find interactive links to a selection of teaching tools.

Quick recap

In pairs, ask children to roll two dice. Then ask them to work out the total of the numbers on the two dice.

Unit 7: Addition and subtraction within 20, Lesson 5

Discover

WAYS OF WORKING Pair work

ASK

- Question 1 a): *How can you show 5? With a partner, how can you show 5 + 5? What is the answer to 5 + 5? How can you change this to show 5 + 6? How does your answer compare to double 5?*
- Question 1 b): *Can you show double 3? How can you change this to show 3 + 4? How can you use double 3 to work out 3 + 4?*

IN FOCUS Questions 1 a) and 1 b) provide two near double questions for children to explore. They should use their known doubles to work out the near doubles. The questions take them through the steps they may consider. The task relies heavily on children quickly knowing their doubles rather than having to count on to find a double. Children should see, as they make the tower of cubes, that to work out 5 + 6 it is just 1 more than 5 + 5.

PRACTICAL TIPS Use cubes throughout to support understanding.

ANSWERS

Question 1 a): Children should make two towers of 5 cubes. Then they should add one more cube to one of the towers to show one tower of 5 and one tower of 6.

5 + 5 = 10

5 + 6 = 11

Question 1 b): 3 + 3 = 6

3 + 4 = 7

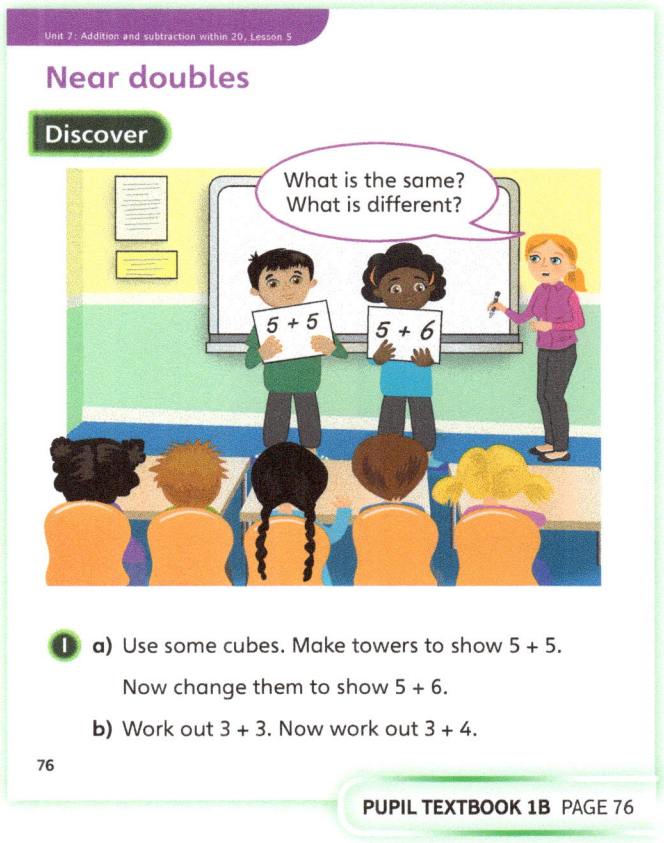

PUPIL TEXTBOOK 1B PAGE 76

Share

WAYS OF WORKING Whole class teacher led

ASK

- Question 1 a): *What does the first set of towers show? How have the towers been changed to show 5 + 6? How can you work out 5 + 6 from knowing 5 + 5?*
- Question 1 b): *How can you work out 3 + 4 using double 3? How do the cubes show this?*

IN FOCUS In question 1 a), make double 5 with the children. Use the same-coloured cubes for each tower. Ask them what double 5 is. Discuss how they can change this to show 5 + 6. Compare this to what Ash has done. Does it matter which cube they add the tower to? Work through question 1 b) using 3 + 3 to work out 3 + 4. Use towers of cubes to clearly show that the answer to the near double is 1 more.

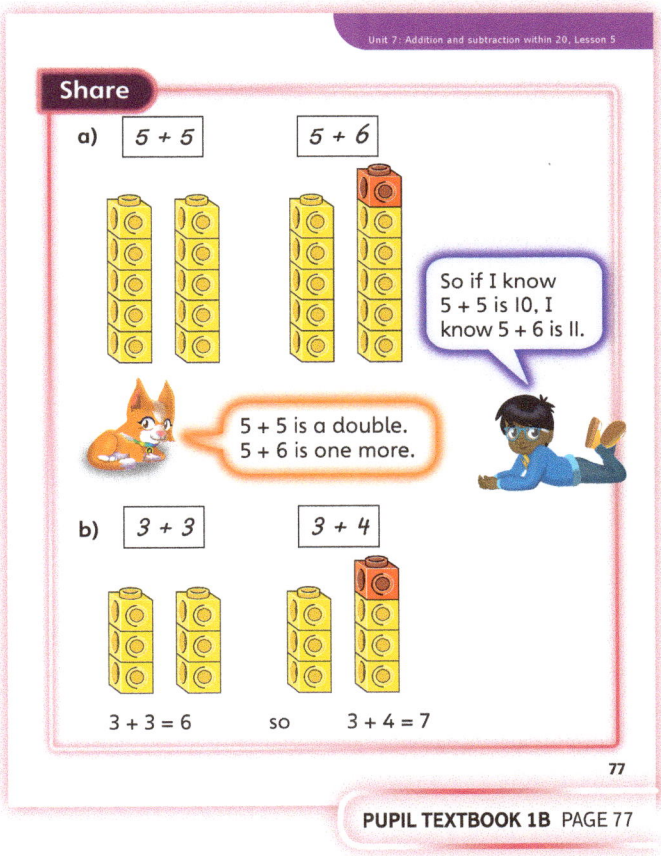

PUPIL TEXTBOOK 1B PAGE 77

115

Unit 7: Addition and subtraction within 20, Lesson 5

Think together

WAYS OF WORKING Whole class teacher led (I do, We do, You do)

ASK

- Question ①: *What doubles are shown? What are the answers? How can you use the double to work out the near double?*
- Question ②: *What are the answers to the doubles? How can we use the answers to work out the other additions? What do you notice about the last two questions in each column?*
- Question ③: *What do you notice about near doubles compared to doubles? Can you work out 5 + 6 from double 5? How can you work out 5 + 6 from double 6?*

IN FOCUS In question ①, children are presented with fingers showing doubles. They should make the doubles themselves on their fingers or using counters and then change them to show the near double. They should show they are just adding 1 more to the double. In question ②, children explore using doubles to work out two near doubles without a pictorial representation. In question ③, children explore the position of near doubles on a number line compared to doubles. Using counters on the number line will help them see that these near doubles are 1 more or less than a double.

STRENGTHEN Children need to ensure they know their doubles rather than having to count on to find a double. Work with children on quickly recapping this. They could use their fingers or counters on ten frames. To work out 7 + 8, first show children double 7 as cubes, counters or on a rekenrek. Ask them what they notice about 7 + 8. They should see that they just need to add 1 more counter or bead to find 7 + 8. Work through several examples so that children understand the concept of finding near doubles.

DEEPEN Question ③ will help children see how they can use both double 5 and double 6 to work our 5 + 6. Discuss which method they prefer (if any) and why.

ASSESSMENT CHECKPOINT Use questions ① and ② to check that children can find near doubles using known doubles. Ensure children are using their knowledge of doubles and not using a count all or count on method.

ANSWERS

Question ①: 2 + 3 = 5
4 + 5 = 9

Question ②: 1 + 1 = 2   6 + 6 = 12   9 + 9 = 18
1 + 2 = 3   6 + 7 = 13   9 + 10 = 19
2 + 1 = 3   7 + 6 = 13   10 + 9 = 19

Question ③ a): Counters should be placed on 10 for double 5 and 12 for double 6.

Children should notice that 5 + 6 = 11 is in between double 5 and double 6 on the number line.

Question ③ b): 2 + 3 = 5   6 + 5 = 11
5 + 4 = 9   9 + 8 = 17
7 + 8 = 15   8 + 9 = 17

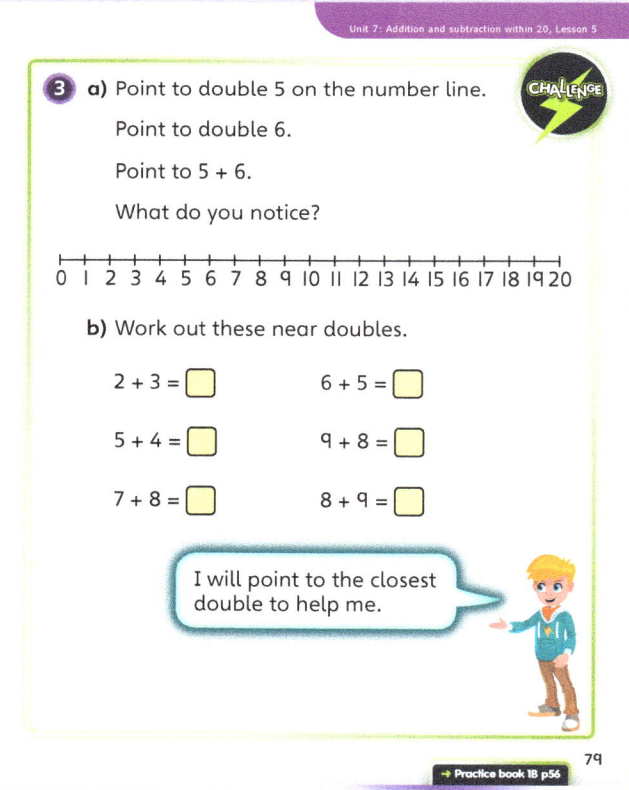

PUPIL TEXTBOOK 1B PAGE 78

PUPIL TEXTBOOK 1B PAGE 79

Unit 7: Addition and subtraction within 20, Lesson 5

Practice

WAYS OF WORKING Independent thinking

IN FOCUS Question ❶ provides practice on the main method children used in the lesson to find near doubles. In each of the examples, the double is shown as towers of cubes and the second set of cubes has been changed by adding 1 more.

In questions ❷ and ❸, children are presented with near doubles on dice. Some children may struggle to see the connection with the double. Ask them which double they can use to work out the near double. Also ask them why it is a near double. All the examples stay within simple doubles that children are likely to know off by heart. You may want to ask children to roll their own dice to see if they can recognise doubles and near doubles.

Question ❹ provides a different representation to find greater near doubles, where children might not know the answer straight away. They have to change the ten frames by adding an additional counter.

In question ❺, children use methods learnt during the lesson to find near doubles in abstract form. Look for children who need support and for those children who can work out the answer in their head, because they are confident with their doubles.

STRENGTHEN Children need to ensure they know their doubles. Work with children on quickly recapping this. They could use their fingers or counters on a ten frame. To work out 7 + 8, first show children double 7 as cubes, counters or on a rekenrek. Ask them what they notice about 7 + 8. They should see that they just need to add 1 more counter or bead to find 7 + 8. Work through several examples so that children understand the concept of finding near doubles.

DEEPEN Children should be working towards finding near doubles using known doubles. Ask them to explain the methods that they use and ask if they can use two different doubles to work out their near doubles. Discuss methods they might use to work out 6 + 8.

ASSESSMENT CHECKPOINT Questions ❶, ❷ and ❸ check whether children can work out near doubles confidently using equipment for support. Check whether they are counting or using their knowledge of doubles.

ANSWERS Answers for the **Practice** part of the lesson can be found in the *Power Maths* online subscription.

Reflect

WAYS OF WORKING Whole class

IN FOCUS Ask children to shade in the doubles that they know. Ask them what they notice about the doubles and where they are on the number track. Ask them to say their doubles as a class as they shade them in. Ensure that all children have found all the doubles. Then ask them how they can use a double to work out a near double. Choose examples from the lesson.

ASSESSMENT CHECKPOINT Check to ensure children are using doubles to work out near doubles and not counting all or counting on.

ANSWERS Answers for the **Reflect** part of the lesson can be found in the *Power Maths* online subscription.

After the lesson

- Can children find near doubles such as 5 + 6 or 9 + 8?

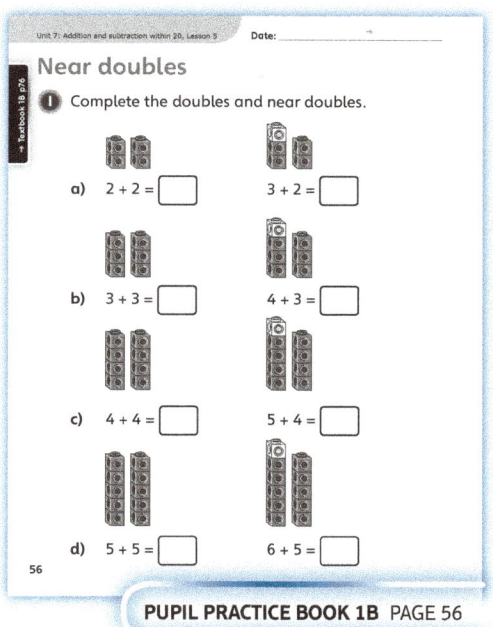

PUPIL PRACTICE BOOK 1B PAGE 56

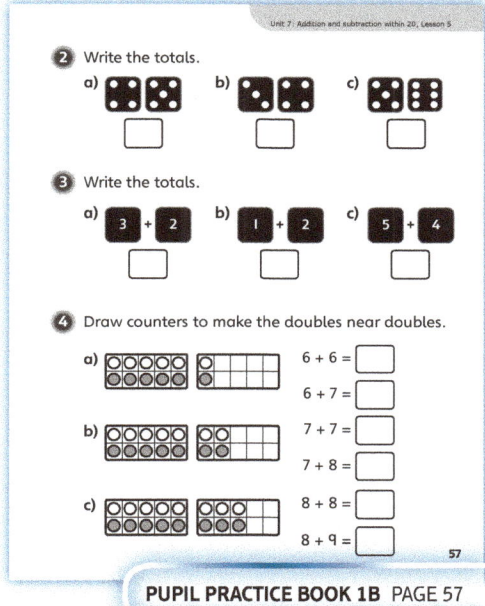

PUPIL PRACTICE BOOK 1B PAGE 57

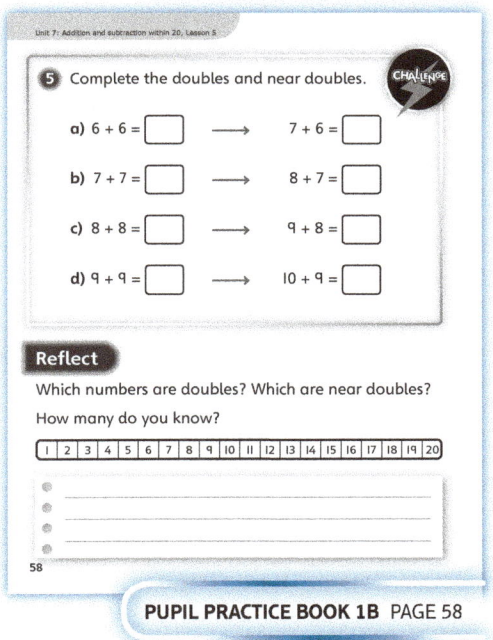

PUPIL PRACTICE BOOK 1B PAGE 58

117

Unit 7: Addition and subtraction within 20, Lesson 6

Subtract ones using number bonds

Learning focus
In this lesson, children will learn how to subtract by counting back. They will then go on to solve number sentences that have missing numbers.

Before you teach
- Will children need number lines to support them in this lesson?
- Would a classroom display help children with methods of subtraction?
- How will you introduce new vocabulary?

NATIONAL CURRICULUM LINKS

Year 1 Number – addition and subtraction

Add and subtract 1-digit and 2-digit numbers to 20, including zero.

Represent and use number bonds and related subtraction facts within 20.

ASSESSING MASTERY

Children can use efficient strategies to subtract 1s, and to find the solution to number sentences with missing numbers. Children can use their knowledge of number bonds within 20 to work out calculations mentally and to make connections between number sentences, such as 16 – 4 and 6 – 4.

COMMON MISCONCEPTIONS

Miscounting will probably happen in this lesson. Sometimes children miss out numbers, count numbers more than once, or start counting on (rather than back). Ask:
- Could you use a number line so you do not miss any numbers?

Children may use the wrong operation when there are missing numbers in a number sentence. For instance, with ☐ – 5 = 2, children may do 5 – 2. Ask:
- Can you check your answer?

STRENGTHENING UNDERSTANDING

Some children will need counting support, especially when counting back. Run some intervention groups, in which children practise both with and without a number line.

GOING DEEPER

To deepen understanding in this lesson, ask children to represent subtraction calculations using a bar model.

KEY LANGUAGE

In lesson: how many are left?, subtraction, subtract, take away, left, count back, ones (1s), number sentence, missing number

Other language to be used by the teacher: find the difference, word problem, picture problem, digits, tens

STRUCTURES AND REPRESENTATIONS

Ten frames, bead strings, number tracks and number lines

RESOURCES

Mandatory: counters, number lines, ten frames

Optional: bead strings, number tracks, footballs

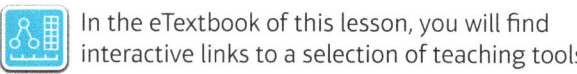 In the eTextbook of this lesson, you will find interactive links to a selection of teaching tools.

Quick recap
Practise subtraction facts within 10. For example, put the numbers from 0 to 10 on cards and select two cards. Ask children to find the difference quickly.

Unit 7: Addition and subtraction within 20, Lesson 6

Discover

WAYS OF WORKING Pair work

ASK

- Question 1 a): *How many footballs are there? How can you count them carefully to make sure you count them all and do not miss any?*
- Question 1 b): *What does the word 'left' mean? If the footballs are being kicked away, do we need to do an addition or a subtraction?*

IN FOCUS Question 1 b) introduces children to a subtraction problem. Draw out the key vocabulary and show that you are taking away (you could use counters to represent the balls). 'Left' is a keyword here: demonstrate what it means with the counters.

PRACTICAL TIPS Recreate this scenario using footballs either outside or in the sports hall.

ANSWERS

Question 1 a): There are 15 footballs.

Question 1 b): There are 12 footballs left.

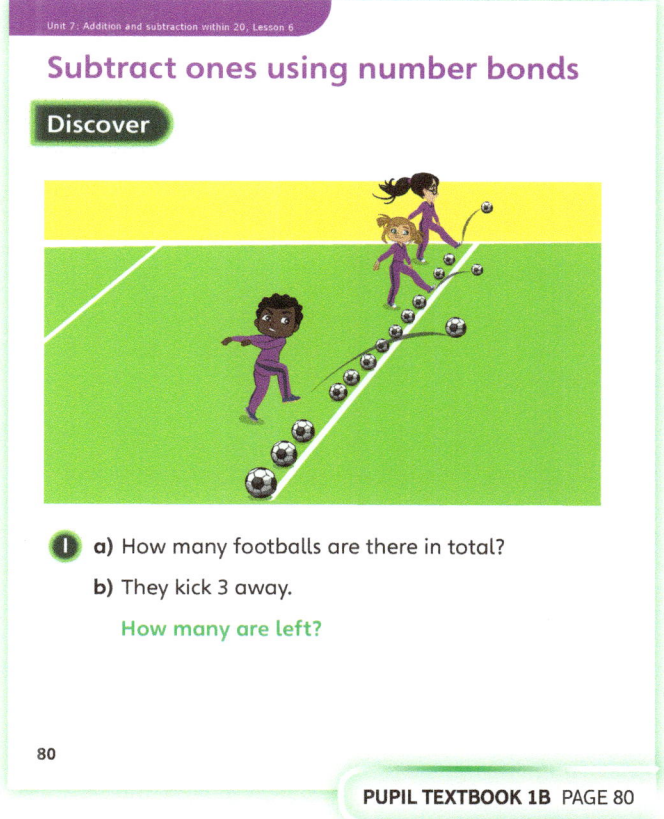

PUPIL TEXTBOOK 1B PAGE 80

Share

WAYS OF WORKING Whole class teacher led

ASK

- Question 1 a): *Do the ten frames help you see how many footballs there are in total?*
- Question 1 b): *How many footballs do we need to cross out? How can you work out how many are left? How can you use 5 – 3 to help you work out 15 – 3?*

IN FOCUS In question 1 a), explain to children how the characters have put the footballs on ten frames to help them count. Children will have likely used counters to represent the footballs. In question 1 b), children cross out or remove 3 counters from the ten frame to represent taking away. Explain to children what Astrid has done in using the fact that 5 – 3 = 2 to work out that 15 – 3 = 12. She has subtracted the 1s as we do not need to cross 10. Ask: *Is Astrid's method a quicker method than Dexter's method? Which do you prefer?*

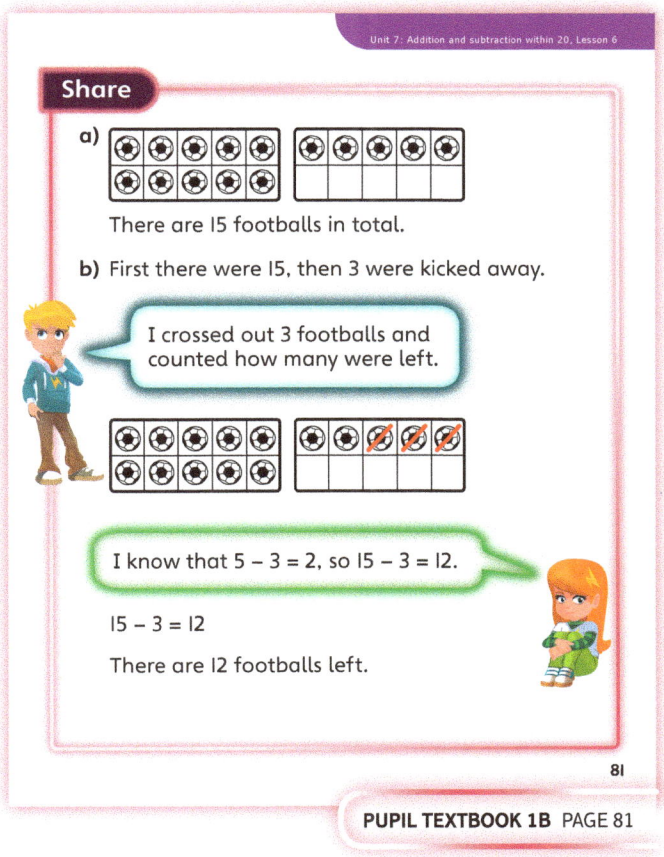

PUPIL TEXTBOOK 1B PAGE 81

Think together

WAYS OF WORKING Whole class teacher led (I do, We do, You do)

ASK
- Question 3 c): *How could you work it out? Can you show me the method you used? Can you show me your method using counters?*

IN FOCUS Question 3 will involve some deeper thinking. Focus on Ash's comment, 'I will think about what would make the answer to a subtraction 0.' Can children explain this and the reason why? Can they provide some more examples to prove it? Children may make decisions about how to represent the calculations using concrete manipulatives.

STRENGTHEN Make the learning more practical for children by asking them to arrange counters into a ten frame and then getting them to physically take away the counters. Children could also do some activities in the hall or the playground. Ask: *Get 10 balls. Now throw away 4. How many have you got left?*

DEEPEN Question 3 gives children an opportunity to work with missing numbers. Deepen learning by giving children number sentences with only one number given. For example, 12 – ☐ = ☐. Then ask them to work out all of the different solutions. Can children work methodically? Question 3 also makes connections between subtraction and place value such as 7 – 5 and 17 – 5. Can children give more examples of this? What patterns can they spot?

ASSESSMENT CHECKPOINT Questions 1 and 2 will allow you to assess children's subtraction knowledge. Look carefully at the strategies they use and note whether they depend on a visual representation such as a ten frame.

ANSWERS

Question 1 a): 19 – 6 = 13
Question 1 b): 16 – 4 = 12
Question 2: 16 – 5 = 11
Question 3 a): 7 – 5 = 2
 17 – 5 = 12
Question 3 b): 8 – 3 = 5
 18 – 3 = 15
Question 3 c): 7 – 7 = 0
 17 – 7 = 10

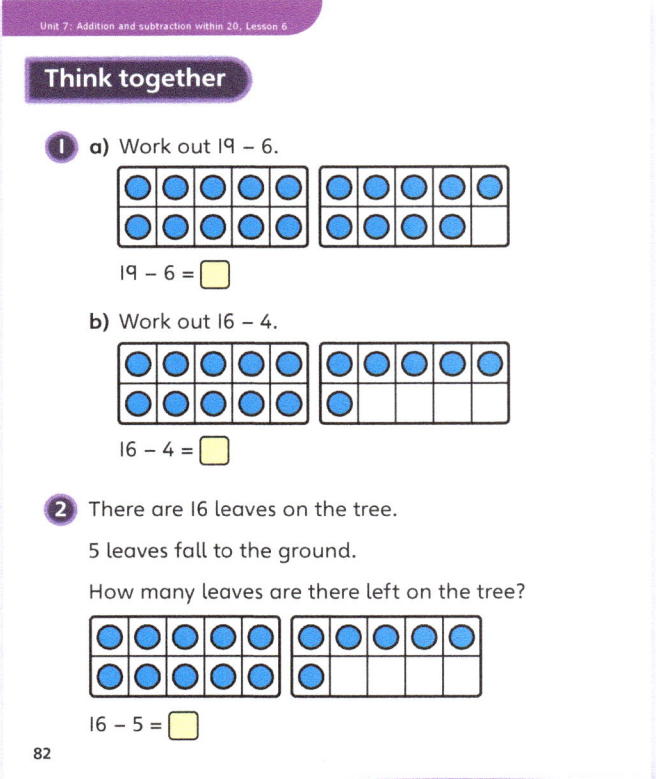

PUPIL TEXTBOOK 1B PAGE 82

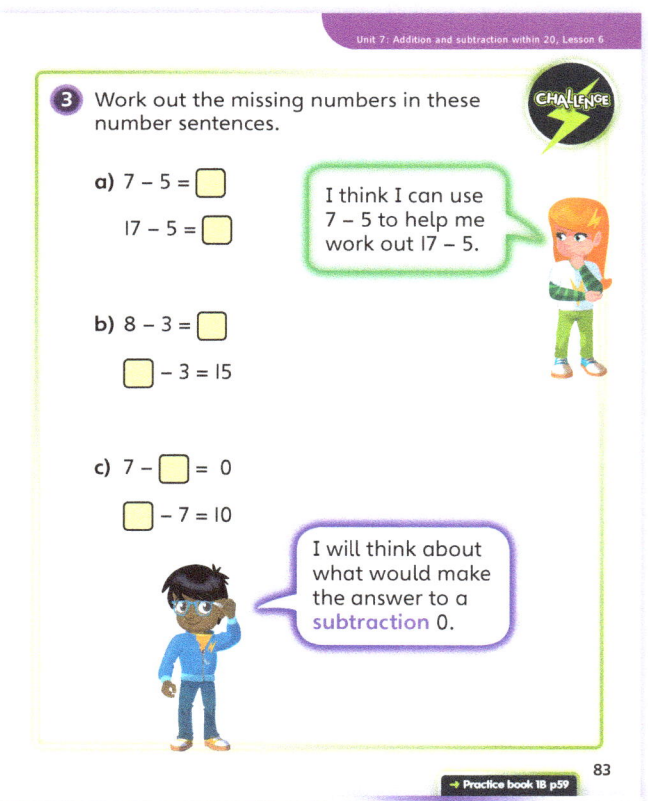

PUPIL TEXTBOOK 1B PAGE 83

Unit 7: Addition and subtraction within 20, Lesson 6

Practice

WAYS OF WORKING Independent thinking

IN FOCUS When children approach question ④, discuss what it means when there is a number missing in a number sentence. When the second number is missing, children should subtract the other two; however, when the first number is missing, they will have to add the other two. Modelling this with apparatus such as counters will help children gain a more secure understanding.

STRENGTHEN For question ⑤, encourage children to discuss it and write down what they know. When they have worked out what a shape represents, ask them to write it inside the shape; this will break down the question.

DEEPEN Challenge children to make their own problem similar to the one in question ⑤. They should use different shapes and numbers (but stick to a subtraction problem).

ASSESSMENT CHECKPOINT Question ④ will tell you which children have mastered the lesson. Look for use of effective strategies to solve the missing number problems.

ANSWERS Answers for the **Practice** part of the lesson can be found in the *Power Maths* online subscription.

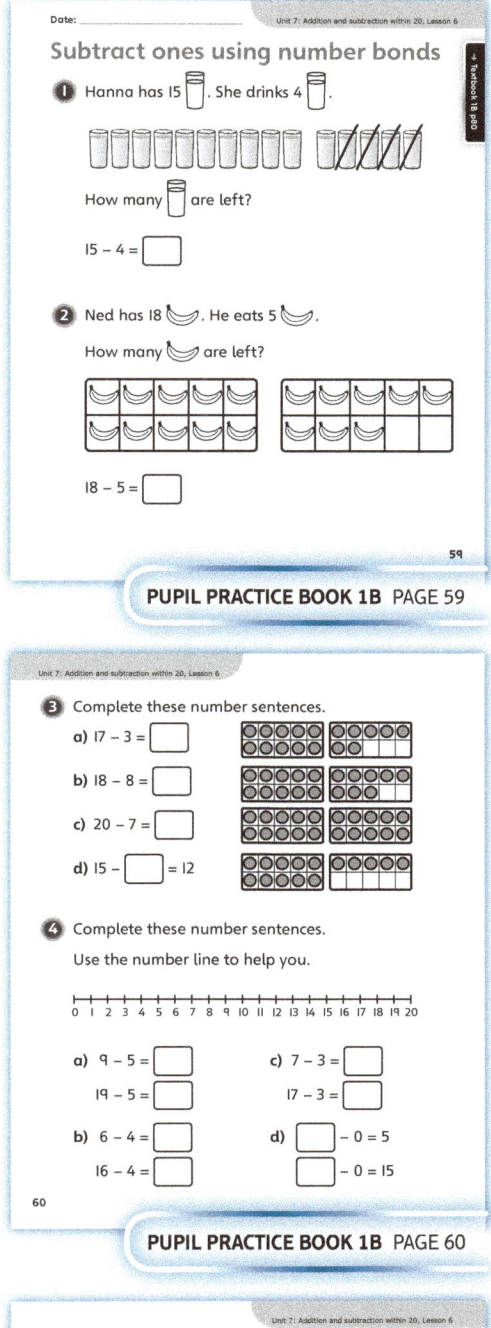

PUPIL PRACTICE BOOK 1B PAGE 59

PUPIL PRACTICE BOOK 1B PAGE 60

Reflect

WAYS OF WORKING Independent thinking

IN FOCUS This question involves some deeper thinking. Children may start with trial and improvement. Ask them what numbers could not go in each box. Some children may want to use number lines or ten frames to support their thinking. To extend learning, ask children if there is only one solution. Can they find all of them?

ASSESSMENT CHECKPOINT This question will help you to assess children's reasoning and knowledge of place value when subtracting.

ANSWERS Answers for the **Reflect** part of the lesson can be found in the *Power Maths* online subscription.

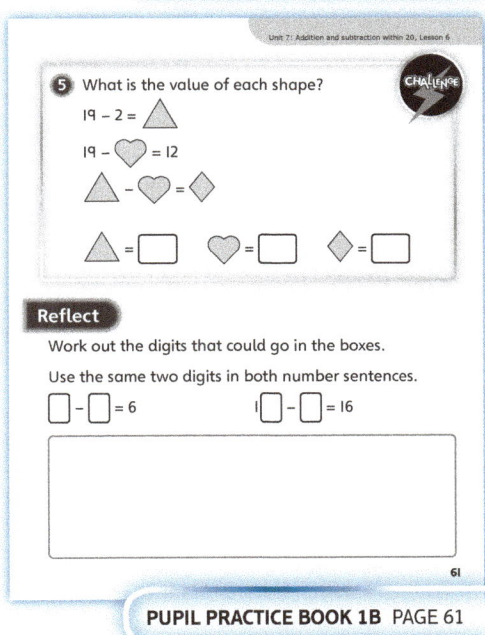

PUPIL PRACTICE BOOK 1B PAGE 61

After the lesson ⏸
- Are children ready to move on to subtracting 10s and 1s?
- Do children need more support with their counting?
- Would children benefit from more subtraction practice with counters or representations such as ten frames?

Unit 7: Addition and subtraction within 20, Lesson 7

Subtraction – count back

Learning focus
In this lesson, children will count back across 10 to work out answers to subtraction problems.

Before you teach
- Do children know common subtraction facts within 10?
- Can children find the answer to a subtraction by counting back on a number track or number line?
- Can children find the answers to subtractions, such as 15 – 2, using number bonds and subtracting 1s?

NATIONAL CURRICULUM LINKS

Year 1 Number – addition and subtraction

Solve one-step problems that involve addition and subtraction, using concrete objects and pictorial representations, and missing number problems such as 7 = ☐ – 9.

Add and subtract one-digit and two-digit numbers to 20, including zero.

ASSESSING MASTERY

Children can confidently find the answer to subtraction problems that cross 10 by counting back on a number line.

COMMON MISCONCEPTIONS

When counting back, children will not make enough jumps or jump back too far. For example, when working out 15 – 7, children may include 15 in the count back. Also, children may not keep track of how many they have counted back by, if they are trying to do this in their head. Ask:
- *Can you use counters on ten frames or a number line to help you count back? What number do you need to start on? Is this number included in the count back?*

STRENGTHENING UNDERSTANDING

Children use a number track or number line to help them find the answers to subtractions. Encourage children to count back slowly, each jump represents subtracting 1 and they should check the correct number of jumps for the subtraction once they think they have arrived at the answer.

GOING DEEPER

When working out problems, such as 15 – 7, instead of counting back 7, some children may realise they can jump back 5 and then jump back 2 more. This is a quicker way of completing a subtraction across 10 compared to counting back in 1s. In this example, children jump back to 10 first and then jump back the remaining amount.

KEY LANGUAGE

In lesson: subtraction, minus, count back

Other language to be used by the teacher: left

STRUCTURES AND REPRESENTATIONS

Number tracks, ten frames, number lines

RESOURCES

Mandatory: straws or counters

Optional: number tracks, number lines

 In the eTextbook of this lesson, you will find interactive links to a selection of teaching tools.

Quick recap

Ask children to work out 7 – 3 by counting back on a number line. Check that children start at the correct number and they count back 3 correctly. Ask them to check each other's number lines. Repeat with different subtractions to ensure children are confident with counting back.

Unit 7: Addition and subtraction within 20, Lesson 7

Discover

WAYS OF WORKING Pair work

ASK

- Question 1 a): *How many packs of 10 pencils does the teacher have? How many single pencils does she have? How many pencils does the teacher have altogether?*
- Question 1 b): *How many children would like a pencil? How can you work out how many pencils will be left over? Will the teacher need to open the box of 10 pencils?*

IN FOCUS In question 1 a), children need to work out the number of pencils the teacher has. They see that the teacher has one pack of 10 pencils and 3 single pencils. In class, you can represent this with pencils, straws or counters. In question 1 b), they realise that 5 children need a pencil and use their equipment to count out 5 pencils from the 13 pencils that the teacher starts with. Then, they may count how many pencils they have left. Another way children may do it is to begin at 13 and count back 1 each time they give out a pencil.

PRACTICAL TIPS Recreate the situation in the classroom and act it out practically.

ANSWERS

Question 1 a): The teacher has 13 pencils.

Question 1 b): $13 - 5 = 8$

The teacher has 8 pencils left.

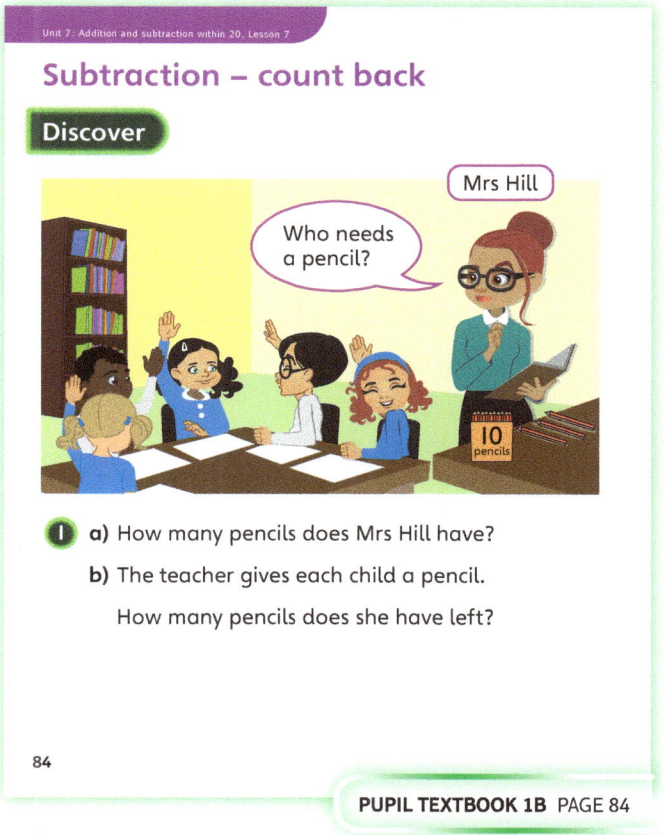

PUPIL TEXTBOOK 1B PAGE 84

Share

WAYS OF WORKING Whole class teacher led

ASK

- Question 1 a): *How may packs of 10 pencils can you see? How many single pencils are there? How many pencils are there altogether?*
- Question 1 b): *How many children need a pencil? How many pencils are left?*

IN FOCUS In question 1 a), children have represented the number of pencils as counters on a pair of ten frames. They will need this for question 1 b), so ask children to replicate this using their own counters on ten frames. In question 1 b), children realise that 5 children in the image need a pencil as they have their hand up. Demonstrate with children what they can do with the counters: they can give 1 away to 5 people. They can work out the number of counters left by counting. Discuss with children a different method where they count back 5 from 13. This means they do not have to count at the end. Explain that 13 is where we start and then, as a class, they will count back 5 with their fingers or using counters, showing the jumps on the number line as they go.

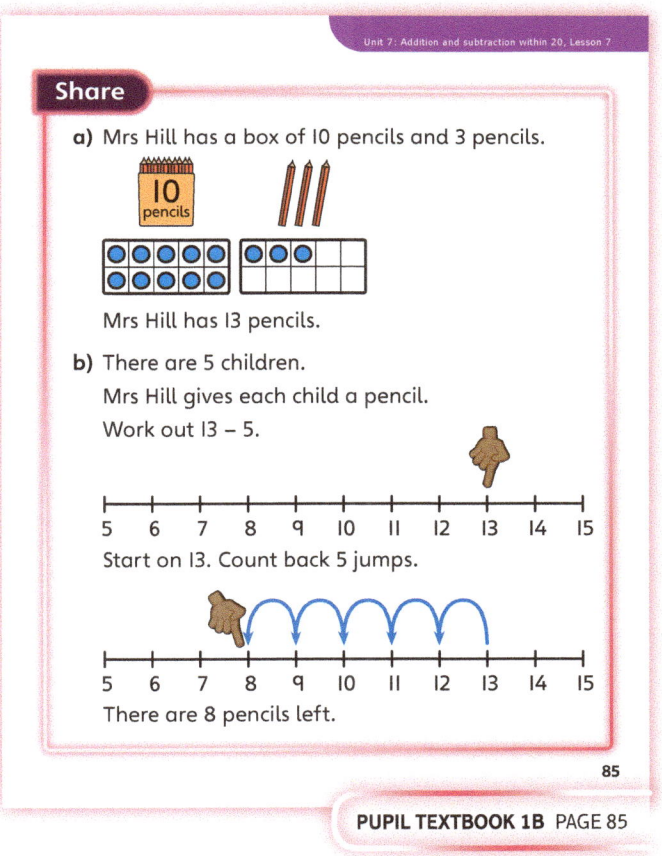

PUPIL TEXTBOOK 1B PAGE 85

123

Think together

WAYS OF WORKING Whole class teacher led (I do, We do, You do)

ASK
- Question ❶: *What number do you need to start on? How many do you need to count back by? What is the answer?*
- Question ❷: *For each question, what number do you need to start on? Do you need to count on or count back? Which number tells you how many you need to count on and back by?*
- Question ❸ a): *How can you subtract 10 quicker than counting back in 1s? How can you subtract 15?*

IN FOCUS Question ❶ provides an example to work through with the class on subtracting using counting back through 10. Children realise they start at 11 and then count back 2 to 9, which gives the answer to the problem. Question ❷ provides more practice on the same concept. Focus on ensuring children understand which number tells them where they need to start and which number tells them how many they need to count back by. Each time, children should use the number line for support. Children may want to use the ten frame and counters alongside the number line.

STRENGTHEN Children use a number track or number line to help them find the answers to the subtractions. Encourage children to count back slowly, each jump representing subtracting 1, and they should check the correct number of jumps for the subtraction once they think they have arrived at the answer. Alongside the number line, some children may find it useful to use counters on ten frames that they remove one at a time. Link the jump back on the number line to removing a counter from the ten frames.

DEEPEN Question ❸ provides the first example of subtracting a 2-digit number from a 2-digit number. Discuss with children the methods that they can use to find the answers. In question ❸ b), ask children if there is a more efficient way than subtracting 15 ones. Question ❸ a) guides children to subtract 10 as 1 ten rather than 10 ones. Children should be able to see that they can cross out the whole ten frame. They can use this in question ❸ b) by first subtracting 10 from 18 and then count back or use their knowledge of number bonds to subtract the 1s. Discuss the most efficient way to get the answers.

ASSESSMENT CHECKPOINT Questions ❶ and ❷ assess whether children can work out subtractions that cross 10 by counting back.

ANSWERS

Question ❶: 11 – 2 = 9

Question ❷ a): 12 – 3 = 9

Question ❷ b): 14 – 1 = 13

Question ❷ c): 15 – 7 = 8

Question ❸ a): 18 – 10 = 8   16 – 10 = 6   15 – 10 = 5
Children should notice that the 1s digit does not change.

Question ❸ b): 18 – 15 = 3

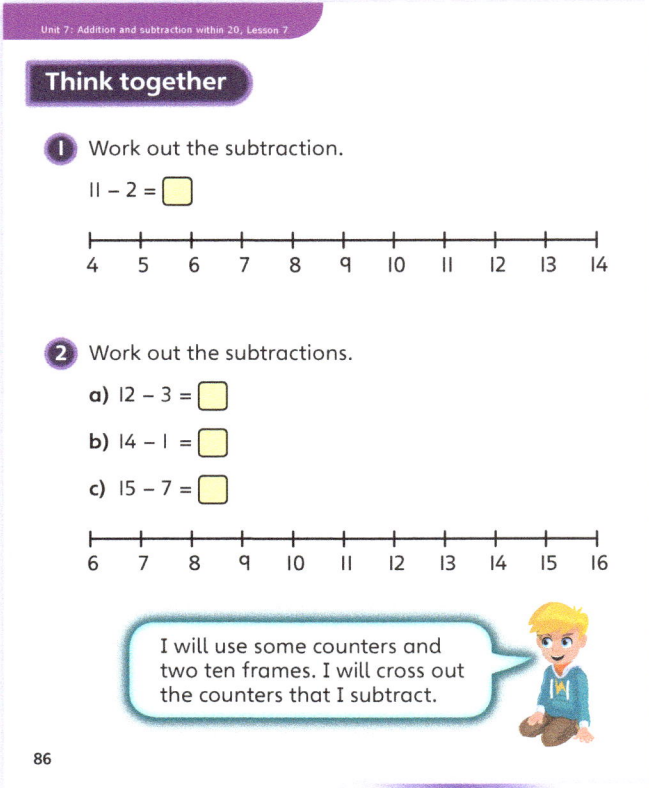

PUPIL TEXTBOOK 1B PAGE 86

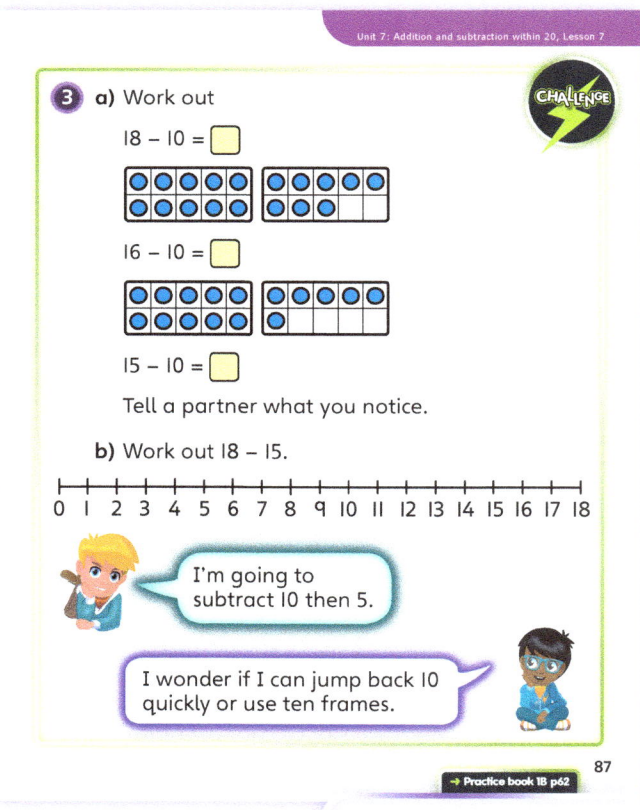

PUPIL TEXTBOOK 1B PAGE 87

Unit 7: Addition and subtraction within 20, Lesson 7

Practice

WAYS OF WORKING Independent thinking

IN FOCUS In question ❶, children use counters and ten frames to work out the subtractions. They cross out the correct number of counters that they are subtracting. Some children may want to use a number line to help them instead.

Question ❸ provides examples where children use a number line to work out a subtraction by counting back across 10. Children should realise the first number tells them where they start and the second number tells them how many they are counting back by.

Question ❹ provides examples where children subtract a 2-digit number from a 2-digit number within 20. Children should be encouraged to use a different method to counting back in 1s. They could subtract a 10 first and then subtract the 1s. Children can jump back 10 on the number line or make the numbers using counters on ten frames.

STRENGTHEN Children use a number track or number line to help them find the answers to the subtractions. Encourage children to count back slowly, each jump representing subtracting 1 and they should check the correct number of jumps for the subtraction once they think they have arrived at the answer. Alongside the number line, some children may find it useful to replicate subtractions as counters on ten frames that they remove one at a time. Link the jump back on the number line to removing a counter from the ten frames.

DEEPEN When working out problems such as 15 – 7, instead of counting back 7, some children may realise they can jump back 5 and then jump back 2 more. This is a quicker way of completing a subtraction that crosses 10 compared to counting back in 1s. Ask children to apply this method throughout the exercise if they understand this.

ASSESSMENT CHECKPOINT Questions ❶ and ❸ assess if children can subtract across 10 by counting back either by using counters on ten frames or by counting back on a number line.

ANSWERS Answers for the **Practice** part of the lesson can be found in the *Power Maths* online subscription.

PUPIL PRACTICE BOOK 1B PAGE 62

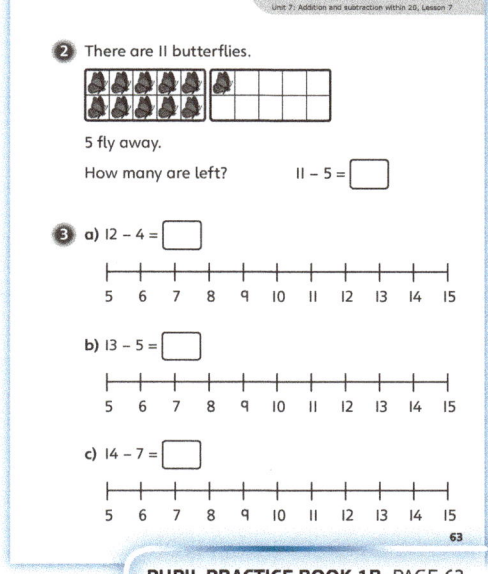

PUPIL PRACTICE BOOK 1B PAGE 63

Reflect

WAYS OF WORKING Independent thinking

IN FOCUS Children explore different methods to work out the subtraction of a 2-digit number from a 2-digit number. They may, for example, count back 13 using a number line or may subtract 13 counters from 17 counters on two ten frames. Some children may subtract 10 first and then 3 ones, possibly even using number bonds.

ASSESSMENT CHECKPOINT Check the method that children use to subtract 13 from 17. Look to see if children can do it more efficiently.

ANSWERS Answers for the **Reflect** part of the lesson can be found in the *Power Maths* online subscription.

After the lesson

- Can children find the answer to a subtraction by counting back across 10?
- Can children subtract a 2-digit number from a 2-digit number less than 20?

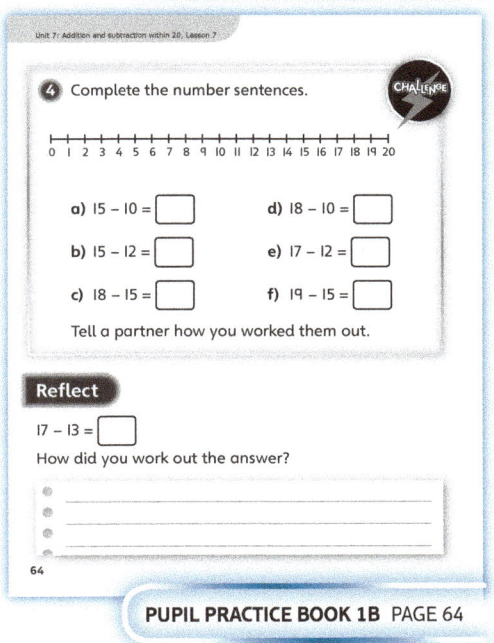

PUPIL PRACTICE BOOK 1B PAGE 64

Unit 7: Addition and subtraction within 20, Lesson 8

Subtraction – find the difference

Learning focus
In this lesson, children will answer questions worded 'how many more' and 'how many fewer'. They will compare quantities of objects to find the difference and represent this on a number line.

Before you teach
- Are all children secure with counting on or back from different starting points?
- How familiar are children with the words 'more' and 'fewer'?
- How could you link 'finding more' and 'finding fewer' to subtraction facts?

NATIONAL CURRICULUM LINKS

Year 1 Number – addition and subtraction

Solve one-step problems that involve addition and subtraction, using concrete objects and pictorial representations, and missing number problems such as 7 = ☐ – 9.

ASSESSING MASTERY

Children can compare different quantities of objects in terms of how many more or fewer there are. Children can understand that the difference is the same when comparing two groups in terms of more and fewer and express this correctly in a sentence (for example, 'there are 2 more in the first group' and 'there are 2 fewer in the second group').

COMMON MISCONCEPTIONS

Children may line up groups of objects inaccurately when comparing them, meaning that objects from one group do not correspond with those from the other. Demonstrate how to line up the objects consistently and explain that anything left over (anything that cannot be matched) is the difference between the groups. Ask:
- How are you going to line up the objects? Which objects are the ones left over? Which group has more? Which group has fewer?

STRENGTHENING UNDERSTANDING

Strengthen understanding by asking two unequal groups of children to sit in rows on the carpet, with the smaller group spread out to take up the same amount of space. Then ask children to pair up with someone from the other group. Explain that the children with no partners make up the difference between these two groups: there are more children in one group and fewer in the other.

GOING DEEPER

Use classroom stationery to deepen understanding by asking children to go on a comparison hunt. Choose a number for comparison and ask them to find out how many more or fewer pens, pencils, rubbers, sharpeners, etc. there are in the room. Provide blank sentence scaffolds and number sentences to help children record their thinking.

KEY LANGUAGE

In lesson: how many more, how many fewer, compare, difference, count on, count back

Other language to be used by the teacher: left

STRUCTURES AND REPRESENTATIONS

Number line

RESOURCES

Mandatory: blank number lines, cubes and/or counters

Optional: classroom stationery, blank sentence scaffolds and number sentences, cuddly toys

 In the eTextbook of this lesson, you will find interactive links to a selection of teaching tools.

Quick recap

Children practise subtractions from the previous lessons. On the board, write down simple subtractions that children should know the answer to and ones where they need to count back across 10.

Unit 7: Addition and subtraction within 20, Lesson 8

Discover

WAYS OF WORKING Pair work

ASK
- Question 1 a): *How many children are in the front row? How many are there in the back row?*
- Questions 1 a) and 1 b): *What do 'how many more' and 'how many fewer' mean?*
- After question 1 b): *Why are both answers the same?*

IN FOCUS Question 1 a) requires children to understand the concept of more and question 1 b) requires children to understand the concept of fewer. Children may count how many children are in each row in the picture and then not know what to do with those numbers, or mistakenly say there are '8 more' because 8 is more than 6.

PRACTICAL TIPS Recreate the scenario in the classroom by positioning six children in the front row and eight children in the back row. If there are not enough children to replicate the scenario, you could use cuddly toys instead.

ANSWERS

Question 1 a): There are 2 more children in the back row.

Question 1 b): There are 2 fewer children in the front row.

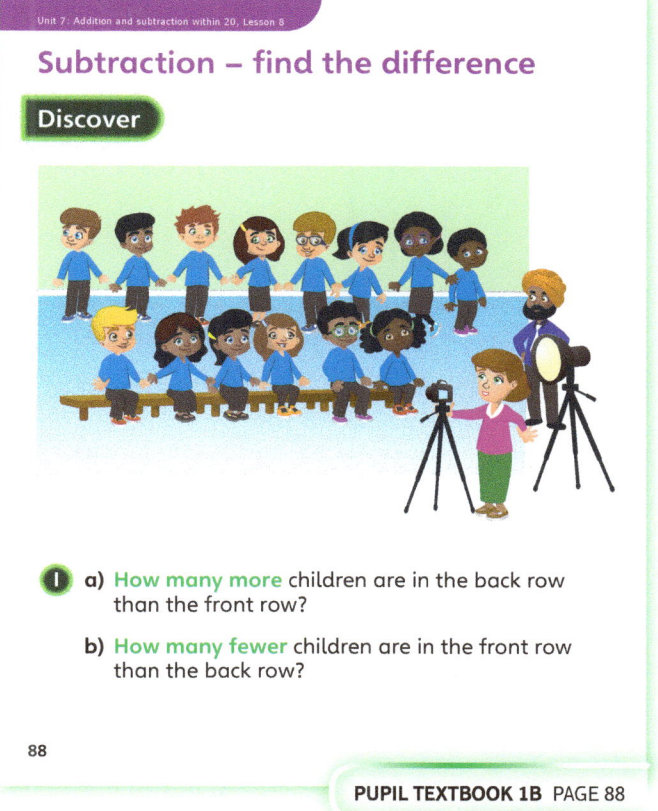

PUPIL TEXTBOOK 1B PAGE 88

Share

WAYS OF WORKING Whole class teacher led

ASK
- Questions 1 a) and 1 b): *Why are the arrows going in different directions on the number lines?*

IN FOCUS In question 1 a), discuss Sparks's use of the word 'difference', and that it explains how many more or fewer things there are when you compare them.

DEEPEN Ask children to arrange counters in place of the children, and then to rearrange them so that there is no difference between the rows. Ask children to explain how this would look in the picture in the **Discover** section: for example, 'The child on the end of the standing line could come and sit on the bench.'

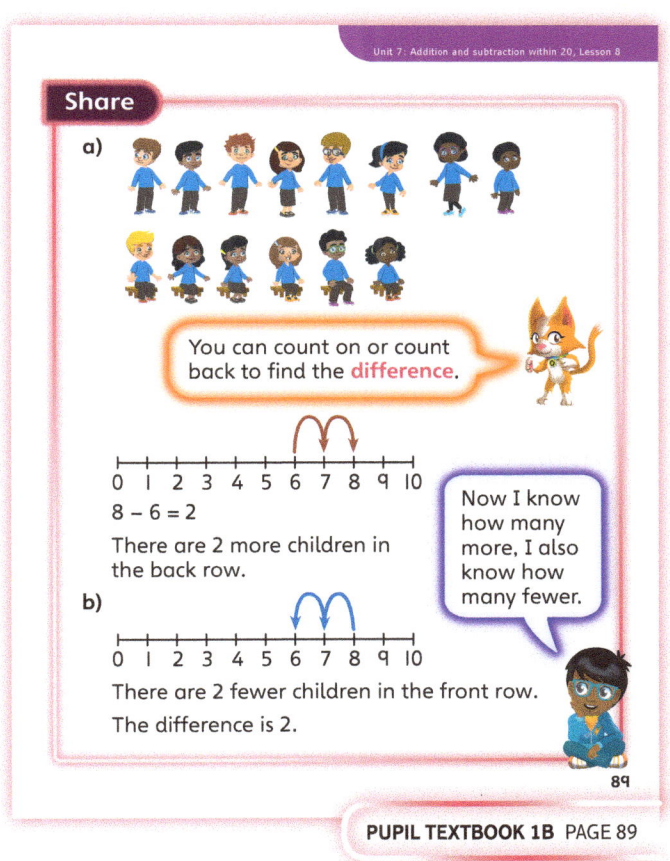

PUPIL TEXTBOOK 1B PAGE 89

127

Unit 7: Addition and subtraction within 20, Lesson 8

Think together

WAYS OF WORKING Whole class teacher led (I do, We do, You do)

ASK
- Question ❶: *How will you represent the children? How will you organise the counters? Can you see what the difference is? How can you find the difference? What is the subtraction you are going to complete?*
- Question ❷: *How many red counters are there? How many yellow counters are there? How many more red counters than yellow counters are there? What subtraction do you need to do to work out the answer? What methods can you use to find the answer?*

IN FOCUS In question ❶, encourage children to represent the children in the image by lining up counters. They can then see how many more are in the front row. Point out to children where the difference is and link the subtraction 6 – 5 = 1 to the image (for example, the 6 represents the children in the front row). In question ❷, children look at finding the difference between two numbers. Explore with children the different methods they could use such as counting on or counting back. Ensure children know what each number represents in the subtraction and that they can see from the image where the difference is. Use a number line to support children through both questions.

STRENGTHEN In question ❶, guide children to an understanding that the number of children in the longer row (6) should be their starting point on a number line, and can be circled. The number in the shorter row (5) is the number of jumps back they need to make. The number they land on after making the jumps back (1) is the difference between the two groups.

DEEPEN Question ❸ brings together concepts from earlier in this unit and children explore the different methods of working out the abstract calculation 15 – 14. For example, they may use their knowledge of number bonds, counting back or finding the difference by counting all. They may even count on from 14 to 15 if they recognise the link between addition and subtraction here. Explore the different methods with children and show that all the methods give the same answer, but some are more efficient than others.

ASSESSMENT CHECKPOINT Assess whether children recognise that the difference between two points on the number line is the number of jumps back. Ensure that they understand that the number that is more in one row is the same as the number that is fewer in the other row.

ANSWERS

Question ❶: 6 – 5 = 1. There is 1 more child in the front row.

Question ❷: 10 – 2 = 8. There are 8 fewer counters.

Question ❸ a): Amy and Danny's methods work but are time consuming.
Hassan's method is the quickest and easiest.
15 – 14 = 1

Question ❸ b): 13 – 11 = 2
17 – 16 = 1
20 – 17 = 3

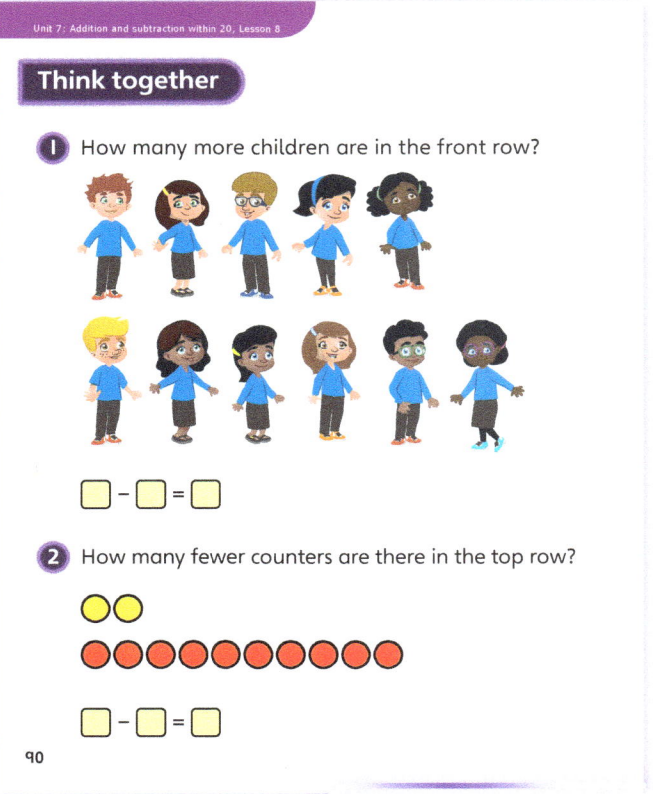

PUPIL TEXTBOOK 1B PAGE 90

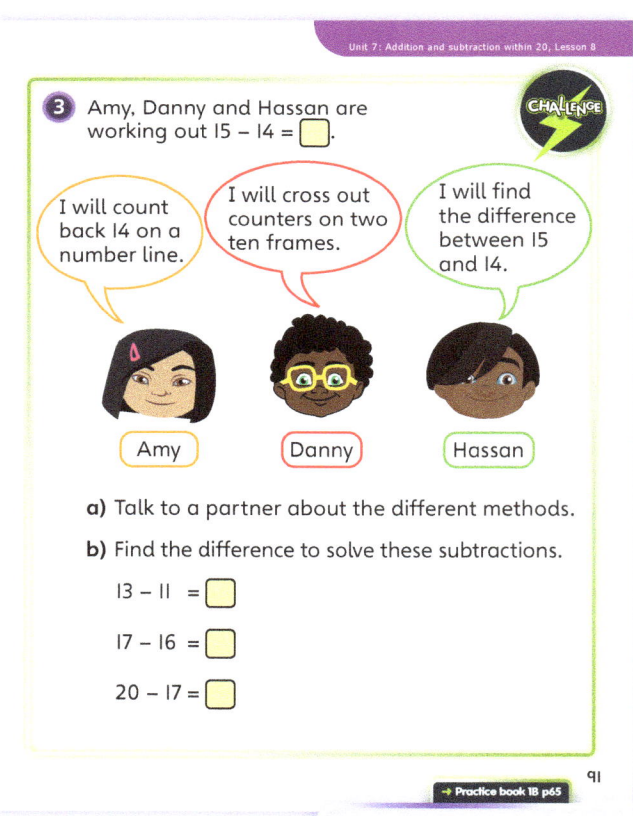

PUPIL TEXTBOOK 1B PAGE 91

128

Unit 7: Addition and subtraction within 20, Lesson 8

Practice

WAYS OF WORKING Independent thinking

IN FOCUS Questions ❶ and ❷ present finding the difference questions with pictorial support. Question ❸ helps children make the link between counting on and finding the difference. Question ❺ provides no pictorial representations to support thinking. (Children could draw the problem or represent it on a number line to help them work it out.)

STRENGTHEN Strengthen understanding in question ❹ by asking children to write some number sentences to match the context and guide their drawing. For example, '1 more than 5 is 6' or '5 + 1 = 6' or '1 less than 5 is 4' or '5 − 1 = 4'.

DEEPEN Ask children to think of their own sentences to match the number cards given in question ❺. As 0 is included, they could use learning from previous lessons to think of a range of sentences that all have '0' as their answer: for example, '9 less than 9 is 0'.

THINK DIFFERENTLY In question ❹, Ash's question challenges children who think that there can only be one solution. Ensure that children understand that 'the difference' can mean there is one more or one fewer, because we are not told which tower is the taller one. Encourage children to make or draw the original tower and then adapt it either way.

ASSESSMENT CHECKPOINT Assess whether children, if representing any of the questions using physical resources, understand the need to put them in neat rows or a grid. Ensure children do not try to make lines the same length no matter the number of objects. Assess whether children are using the terms 'difference', 'more' and 'fewer' correctly when comparing groups.

ANSWERS Answers for the **Practice** part of the lesson can be found in the *Power Maths* online subscription.

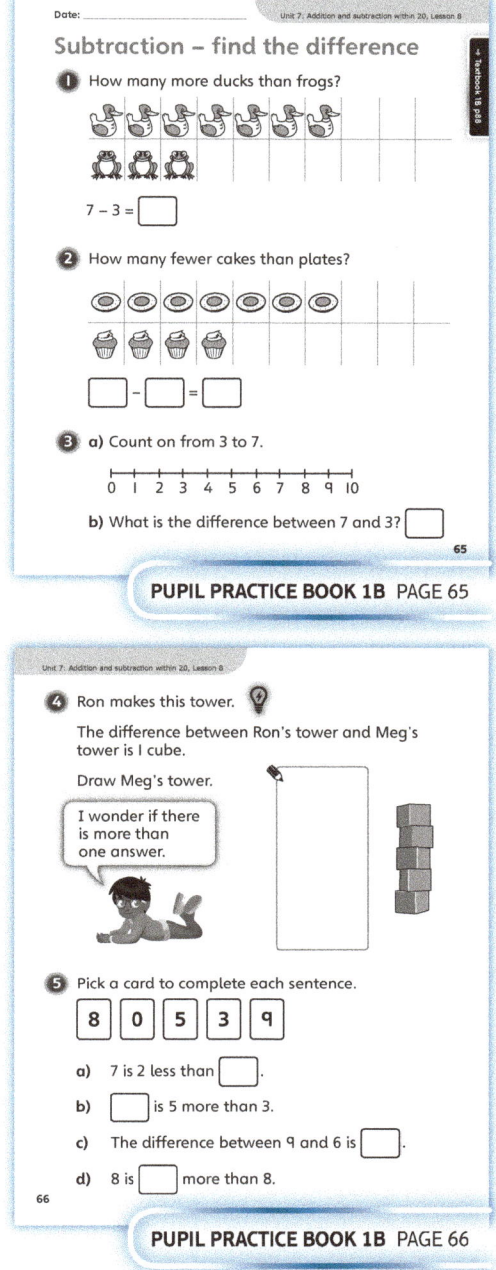

PUPIL PRACTICE BOOK 1B PAGE 65

PUPIL PRACTICE BOOK 1B PAGE 66

Reflect

WAYS OF WORKING Independent thinking

IN FOCUS In this question, children compare their number of counters with a partner's counters. They should use the correct terminology of more and fewer.

ASSESSMENT CHECKPOINT Assess whether children can find the difference between the two sets of counters.

ANSWERS Answers for the **Reflect** part of the lesson can be found in the *Power Maths* online subscription.

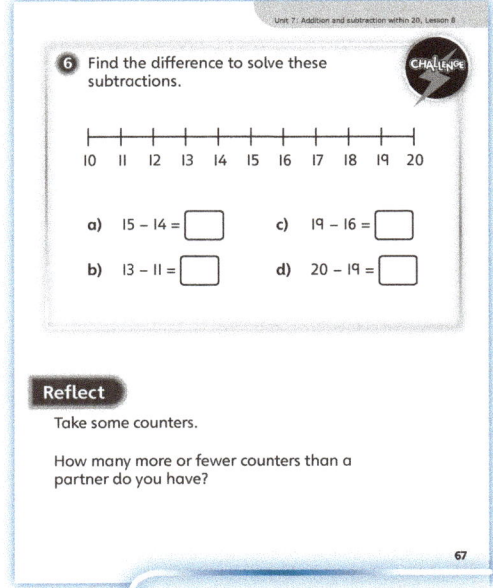

PUPIL PRACTICE BOOK 1B PAGE 67

After the lesson

- Did children understand that the word 'difference' can mean 'more than' or 'fewer than'?
- Did children understand the importance of matching up objects in different groups in order to compare them accurately?
- Were children able to represent difference on a number line and in a subtraction fact?

Unit 7: Addition and subtraction within 20, Lesson 9

Related facts – fact families

Learning focus
In this lesson, children will answer questions worded as 'how many more' and 'how many fewer'. They will compare quantities of objects to find the difference and represent this on a number line.

Before you teach
- How could you make this lesson even more practical?
- Do you need to provide extra time for some addition and subtraction facts practice?

NATIONAL CURRICULUM LINKS
Year 1 Number – addition and subtraction
Represent and use number bonds and related subtraction facts within 20.

ASSESSING MASTERY
Children can quickly recall their addition and subtraction facts to 20 and link facts to those within 10 to support them with higher numbers, for example, 6 + 4 =10 helps with 16 + 4. Children can understand the relationship between addition and subtraction and can write down fact families such as 1 + 2 = 3, 2 + 1 = 3, 3 – 2 = 1, 3 – 1 = 2.

COMMON MISCONCEPTIONS
Children may miscount when trying to make 20. Ask:
- *How could you check your answer?*

Children may not understand the relationship between addition and subtraction. Ask:
- *What do you notice about 1 + 2 = 3 and 3 – 2 = 1?*

STRENGTHENING UNDERSTANDING
Children will need lots of repeated practice to learn their number facts to 20. Quick recall is the end goal. Run some intervention sessions in which children link number bonds to 10 with number bonds to 20; this would make a good matching game.

GOING DEEPER
Challenge children to find addition and subtraction facts that have the same answer and then write them like this: 13 + 6 = 20 – 1. How many can they find?

KEY LANGUAGE
In lesson: fact family, number sentence, ten frames
Other language to be used by the teacher: represent, make, facts, part, whole, bond, add, subtraction

STRUCTURES AND REPRESENTATIONS
Part-whole models, ten frames, number lines, bead strings

RESOURCES
Mandatory: bead strings, counters
Optional: a tin

 In the eTextbook of this lesson, you will find interactive links to a selection of teaching tools.

Quick recap
On the board, write down the numbers 2, 3 and 5 and draw a blank part-whole model. Ask children to put the numbers in the correct place on the part-whole model. Then ask them to write down four facts from the part-whole model.

Unit 7: Addition and subtraction within 20, Lesson 9

Discover

WAYS OF WORKING Pair work

ASK

- Question 1 b): *What number bonds to 10 do you know? How do these number bonds link to the number bonds to 20?*
- After question 1 b): *Why are both answers the same?*

IN FOCUS In question 1 a), observe which children can instantly spot that the child on the right is holding up 10 fingers. Ask them how they knew and share this information with the class.

PRACTICAL TIPS Make these questions more practical by playing a game. Show a certain number of fingers or counters on two incomplete ten frames and ask children to show the number of fingers or counters in the ten frames that will make it total 20.

ANSWERS

Question 1 a): 8 + 10 = 18

Question 1 b): 13 + 7 = 20
7 + 13 = 20

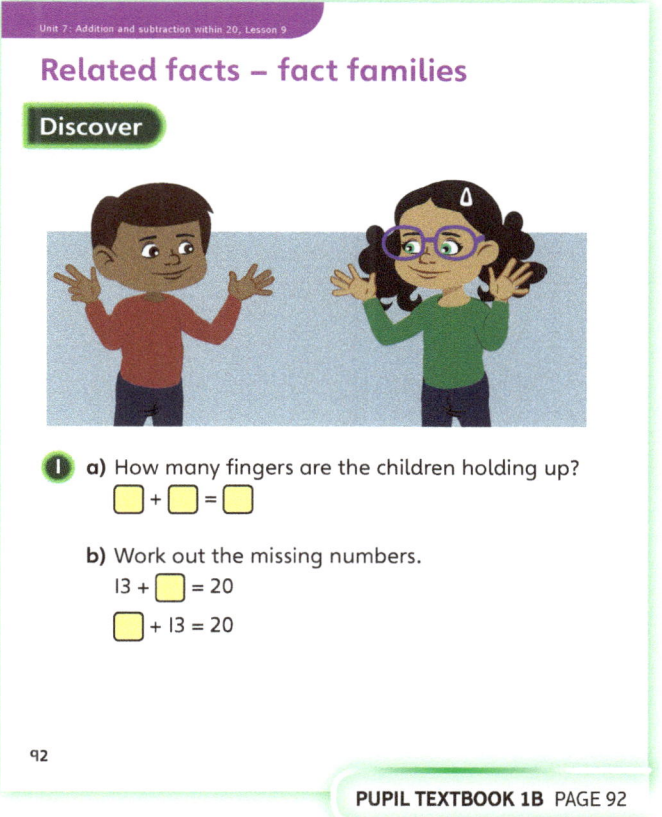

PUPIL TEXTBOOK 1B PAGE 92

Share

WAYS OF WORKING Whole class teacher led

ASK

- Question 1 a): *Why does it not matter if you write 8 + 10 or 10 + 8?*

IN FOCUS Question 1 b) draws on the commutativity of addition. This helps children spot the connections between calculations from the same fact family. Ensure children understand that the order of the addition calculation does not affect the answer.

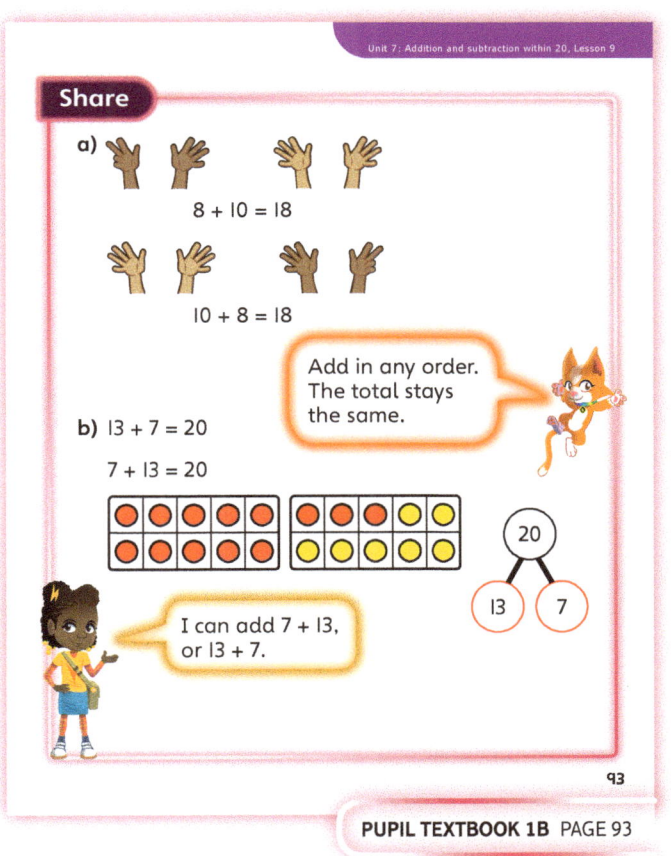

PUPIL TEXTBOOK 1B PAGE 93

Think together

WAYS OF WORKING Whole class teacher led (I do, We do, You do)

ASK
- Question 3 b): *What is the difference between addition and subtraction? How does this help you complete the fact family?*

IN FOCUS Question 3 a) is designed to link a bead string or ten frames with the part-whole model; asking children to find the family of facts.

STRENGTHEN Ask children to represent 20 on ten frames or bead strings in a range of different ways.

DEEPEN Questions 1 and 2 focus on fact families. Challenge children to work out how many families there are up to 20, starting with 0 + 0 = 0 and ending with 20 – 20 = 0. Ask children if there is a method they could use to find out.

ASSESSMENT CHECKPOINT In question 3, assess whether children are able to partition numbers and find the fact family.

ANSWERS

Question 1: 6 + 4 = 10
4 + 6 = 10
10 – 4 = 6
10 – 6 = 4

Question 2: 12 + 3 = 15
3 + 12 = 15
15 – 3 = 12
15 – 12 = 3

Question 3 a): Children should show 1 complete ten frame and 1 ten frame with 4 counters.

Children should show 10 beads and 4 beads on a bead string.

Question 3 b): 8 + 6 = 14
6 + 8 = 14
14 – 6 = 8
14 – 8 = 6

These can be repeated with the answers first.

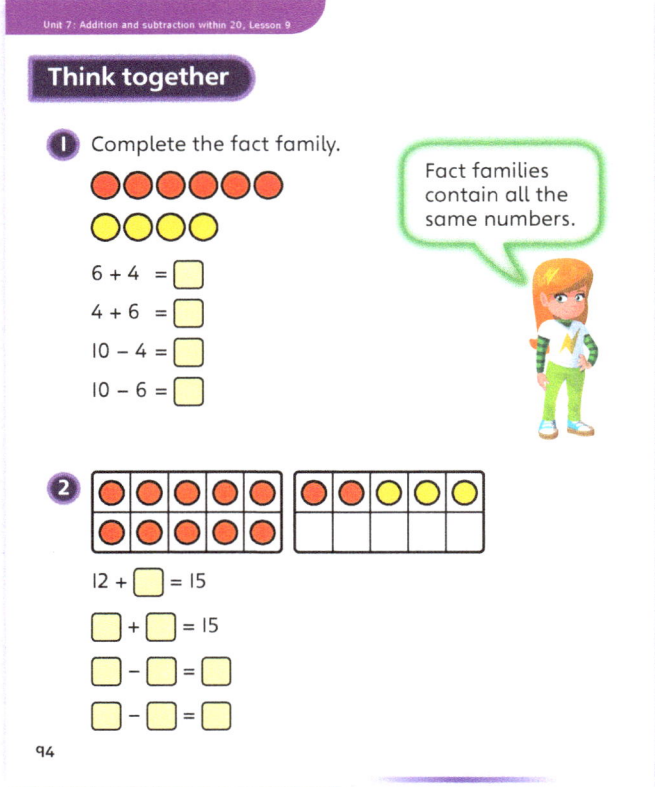

PUPIL TEXTBOOK 1B PAGE 94

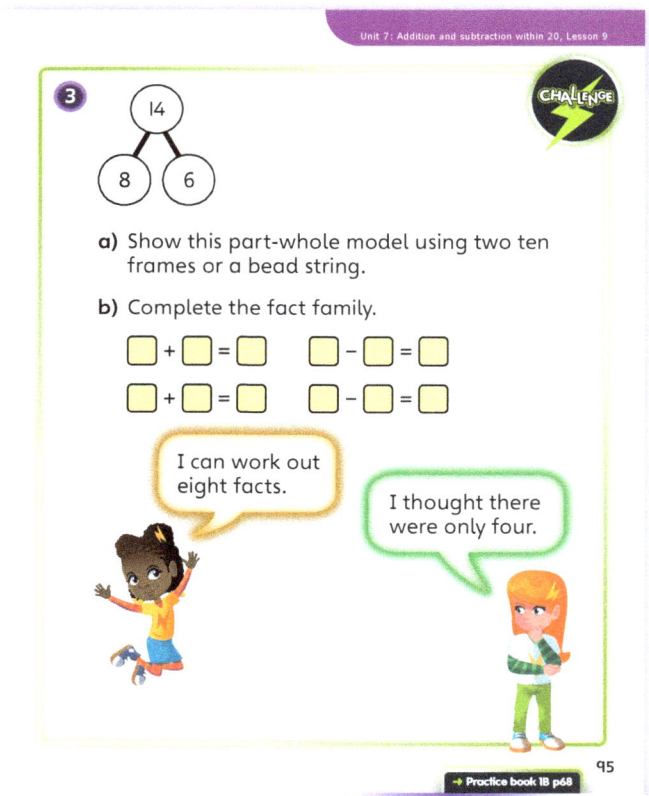

PUPIL TEXTBOOK 1B PAGE 95

Unit 7: Addition and subtraction within 20, Lesson 9

Practice

WAYS OF WORKING Independent thinking

IN FOCUS Question ③ may surprise children because the 'answer' is at the front in each number statement. Children usually see 9 + 6 = 15 rather than 15 = 9 + 6. This is a good opportunity to discuss the meaning of the = sign.

STRENGTHEN Run intervention sessions in which children check subtraction number sentences by counting back. They could use counters to reinforce learning. Ask children to close their eyes and drop the counters into a tin, one by one – the sound will help them with auditory learning.

DEEPEN Can children find three numbers that make 20; for example, 11 + 5 + 4? How many can they find?

THINK DIFFERENTLY Question ④ presents a word problem in which children need to use logic to work out the relevant facts from the fact family to answer it.

ASSESSMENT CHECKPOINT Question ⑤ will allow you to assess whether children can find more than four facts in a fact family. Assess whether they can quickly recall their number facts or whether they need to use a representation for support.

ANSWERS Answers for the **Practice** part of the lesson can be found in the *Power Maths* online subscription.

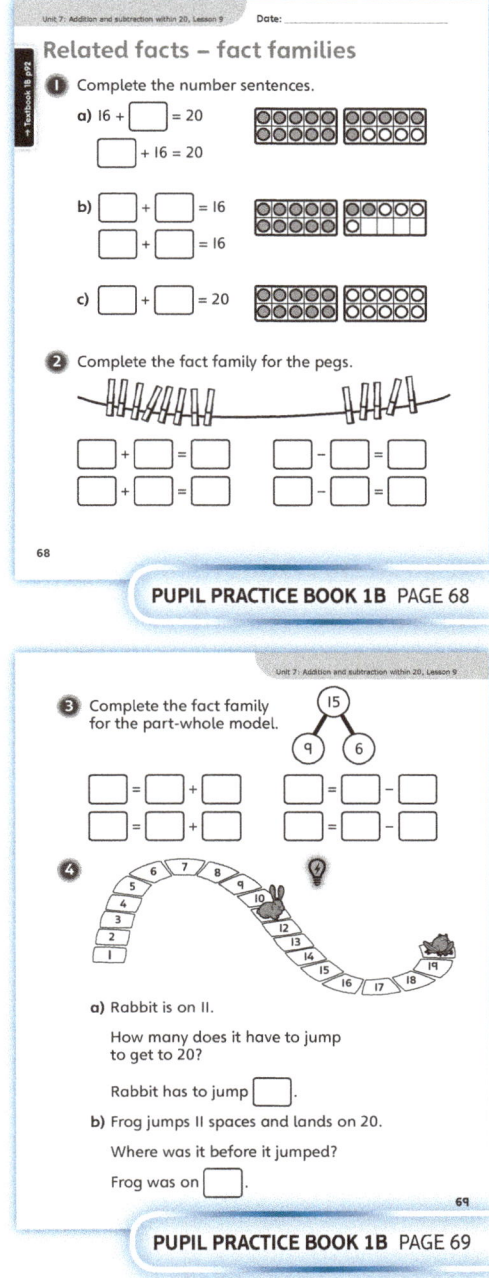

PUPIL PRACTICE BOOK 1B PAGE 68

PUPIL PRACTICE BOOK 1B PAGE 69

Reflect

WAYS OF WORKING Independent thinking

IN FOCUS This is a visual activity that links a visual representation to fact families. Encourage children to explain their answers once they have finished shading in the squares.

ASSESSMENT CHECKPOINT Assess whether children can find the difference between the two sets of counters.

ANSWERS Answers for the **Reflect** part of the lesson can be found in the *Power Maths* online subscription.

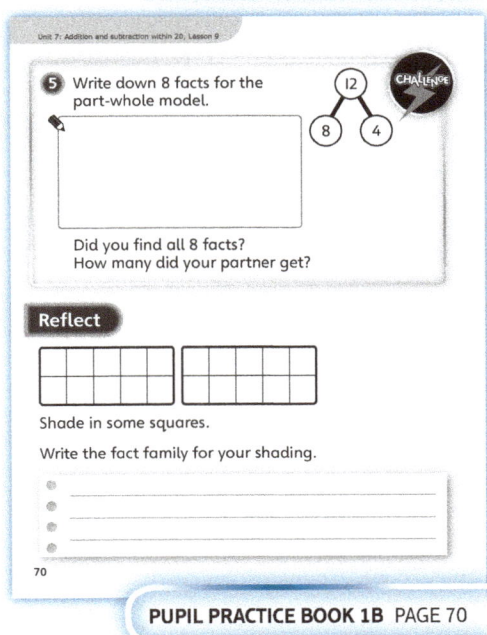

PUPIL PRACTICE BOOK 1B PAGE 70

After the lesson

- Do children need to do some revision of fact families?
- Do children already have instant recall of number facts?
- Where can you fit in some 5 minute revision sessions on number facts?

Unit 7: Addition and subtraction within 20, Lesson 10

Missing number problems

Learning focus

In this lesson, children will solve a variety of missing number problems, such as 5 + ☐ = 9. They will use their knowledge of number bonds and fact families to work out the missing numbers.

Before you teach

- Do children know how to use a number line to count on and count back?
- Do children know their number bonds within 10?
- Can children write down a fact family for a given part-whole model?

NATIONAL CURRICULUM LINKS

Year 1 Number – addition and subtraction

Solve one-step problems that involve addition and subtraction, using concrete objects and pictorial representations, and missing number problems such as 7 = ☐ – 9.

ASSESSING MASTERY

Children can find the missing number in problems such as 5 + ☐ = 9 and 7 – ☐ = 2. They use a number line and/or other visual representations to help them.

COMMON MISCONCEPTIONS

Some children may think they have to use the inverse operation to always find the missing number. For example, 5 + ☐ = 9, children could do 9 – 5 = 4 to find the missing number. However, for problems such as 6 – ☐ = 1, some children will think you have to do 6 + 1 = 7 to find the missing number. Ask:

- *What is the starting number? Do you need to add or subtract? What number do you need to end up on? Can you use a number line to help you?*

STRENGTHENING UNDERSTANDING

Throughout this lesson, use a number line to support children. Take the missing number problem and ask them where they need to start on the number line. Then ask them where they will end up and if they need to count back or count on to work out the missing number. For problems such as ☐ + 3 = 7, ask children to think about where they finish first and then ask them to think about how they can work out where they start. In examples where they are finding the starting number, children will see that they have to do the inverse operation. Children may also hold up fingers or use counters to work out the missing numbers.

GOING DEEPER

Ask children to explain what is the same and what is different about examples such as 4 + ☐ = 9 and ☐ + 4 = 9 and then ask them if the same is the case for 7 – ☐ = 3 and ☐ – 7 = 3.

KEY LANGUAGE

In lesson: add, subtract, count on, count back, missing number

Other language to be used by the teacher: inverse

STRUCTURES AND REPRESENTATIONS

Part-whole models, number lines

RESOURCES

Mandatory: blank number lines
Optional: counters, cubes, blank ten frames

 In the eTextbook of this lesson, you will find interactive links to a selection of teaching tools.

Quick recap

Give children a number line. Ask them to start at 4 and count on 5. What addition have they worked out? Ask them to do other examples, that could cross 10. Once they have completed a few, ask them to start at 11 and count back 6 on the number line. What subtraction have they worked out? Repeat for other examples.

Unit 7: Addition and subtraction within 20, Lesson 10

Discover

WAYS OF WORKING Pair work

ASK

- Question 1 a): *How can you work out the missing number? Do you know a fact that will help you?*
- Question 1 b): *How is this question different to the last one? How can we work out the missing number? How can you use a number line to help you?*

IN FOCUS In questions 1 a) and 1 b), children are given some missing number sentences and are asked to find the missing numbers. They may take different approaches. Some may use their fingers (for example, put 4 fingers up and then work out how many they need to make 6). Others may use cubes or counters or a number line to count on. Children may find question 1 b) a little more difficult as they are not given the starting number.

PRACTICAL TIPS Use counters or cubes to replicate the missing number sentences.

ANSWERS

Question 1 a): 4 + 2 = 6
Question 1 b): 5 + 1 = 6

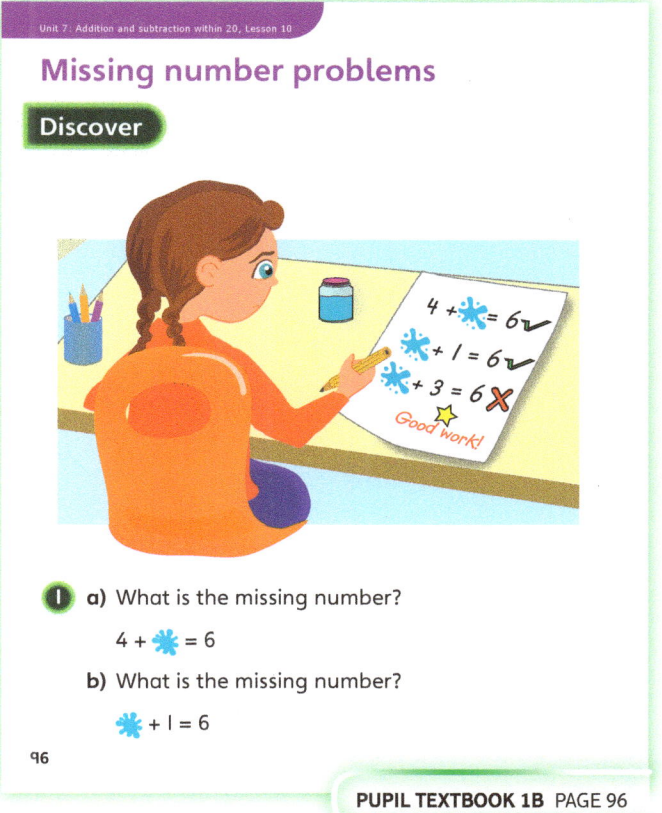

PUPIL TEXTBOOK 1B PAGE 96

Share

WAYS OF WORKING Whole class teacher led

ASK

- Question 1 a): *Can you see how the part-whole model represents the facts? Do you know a number bond that can help you work out the missing number?*
- Question 1 b): *How can you use a number line to help you work out the missing number?*

IN FOCUS In question 1 a), explain to children how the numbers from the missing number problem could be presented in a part-whole model, which they have met many times before. Talk through the method of using fingers, counters or cubes to work out how many more are needed to make 6. Relate this to counting on and counting back on the number line.

In question 1 b), discuss with children how this time they have to work out the starting number if they add 1 and it makes 6. Show how the number line can help them solve this.

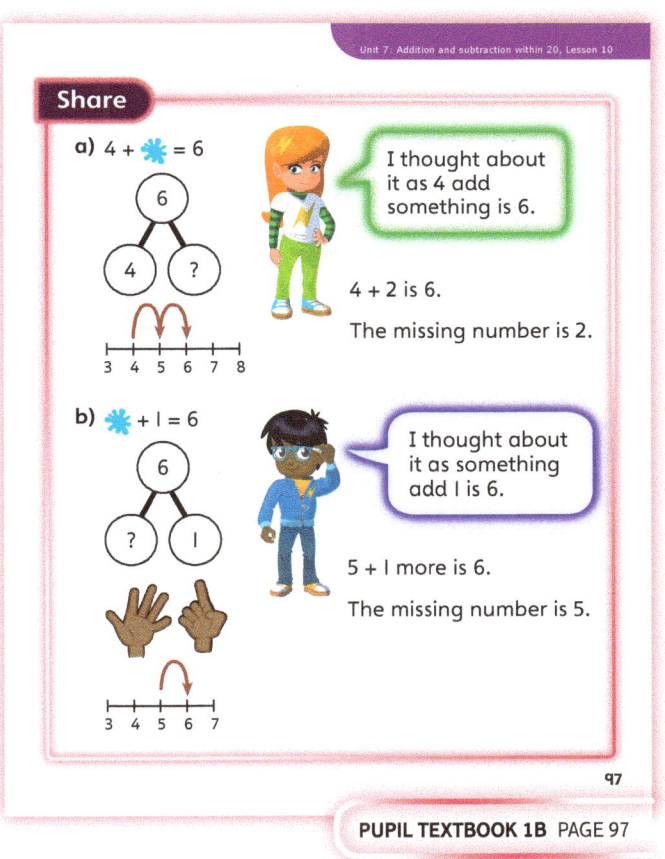

PUPIL TEXTBOOK 1B PAGE 97

135

Unit 7: Addition and subtraction within 20, Lesson 10

Think together

WAYS OF WORKING Whole class teacher led (I do, We do, You do)

ASK
- Question ❶: *What number bond does the part-whole model show? How can you use the number bond to work out the missing numbers?*
- Question ❷: *Where do we need to start each time? Where do we end up? How can you work out how many to count on?*
- Question ❸: *What number is missing from each part-whole model? How can you work that out?*

IN FOCUS In question ❶, discuss the part-whole model with children and that it shows the fact that 11 + 9 = 20. You may want to show this as a representation on two ten frames. Show how you can use the part-whole model to work out the missing numbers. In question ❷, children use a number line to work out the missing numbers. Focus on the start and end numbers and then ask children to count on to help them work out the missing number. When the start number is missing, they may want to think about how many they should have counted on and then count back from the whole by this amount. Question ❸ presents children with a part-whole model and a related fact family. They should use the part-whole model and fact family to work out the missing numbers.

STRENGTHEN Throughout these questions, use a number line to support children. Take the missing number problem and ask children where they need to start on the number line. Ask them where they will end up and if they need to count on or count back to work out the missing number. For problems such as ☐ + 3 = 7, ask children to think about where they finish first and how many they have counted on. They should show these as jumps on the number line to help them work out the starting number.

DEEPEN Children create their own missing number problem for a partner to solve. They create different types using a variety of representations.

ASSESSMENT CHECKPOINT Question ❶ checks that children can use a part-whole model to help them find missing numbers.

Question ❷ checks that children can use a number line to help them find missing numbers. They realise the relationship between the number in the problem and the position on the number line.

ANSWERS
Question ❶: 11 + 9 = 20   9 + 11 = 20   11 + 9 = 20
            20 – 9 = 11   20 – 11 = 9

Question ❷: 6 + 3 = 9   6 + 5 = 11
            12 – 4 = 8   12 – 8 = 4

Question ❸ a): The missing number is 10.
            3 + 7 = 10   7 + 3 = 10
            10 – 3 = 7   10 – 7 = 3

            The missing number is 2.
            9 + 2 = 11   2 + 9 = 11
            11 – 2 = 9   11 – 9 = 2

Question ❸ b): 5 + 4 = 9   13 – 5 = 8
               13 + 4 = 17 19 – 9 = 10

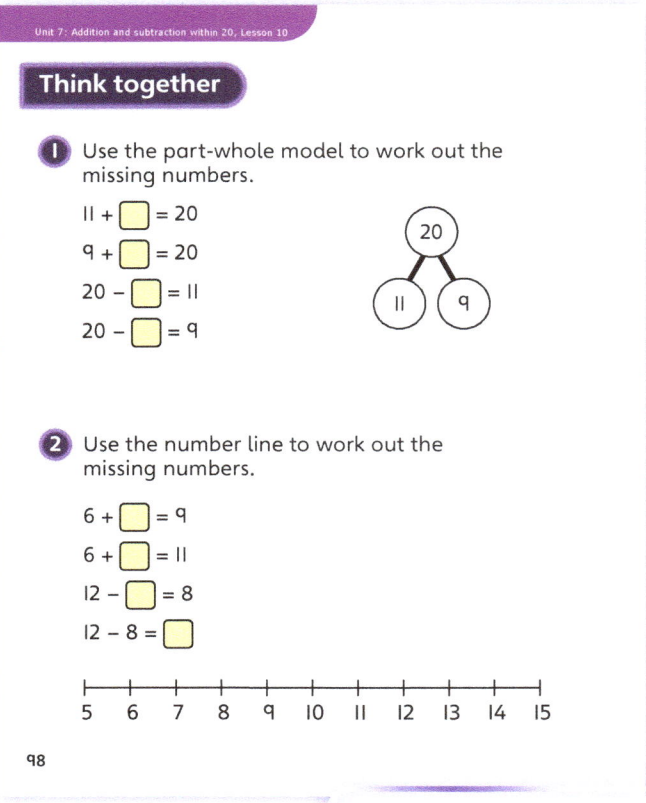

PUPIL TEXTBOOK 1B PAGE 98

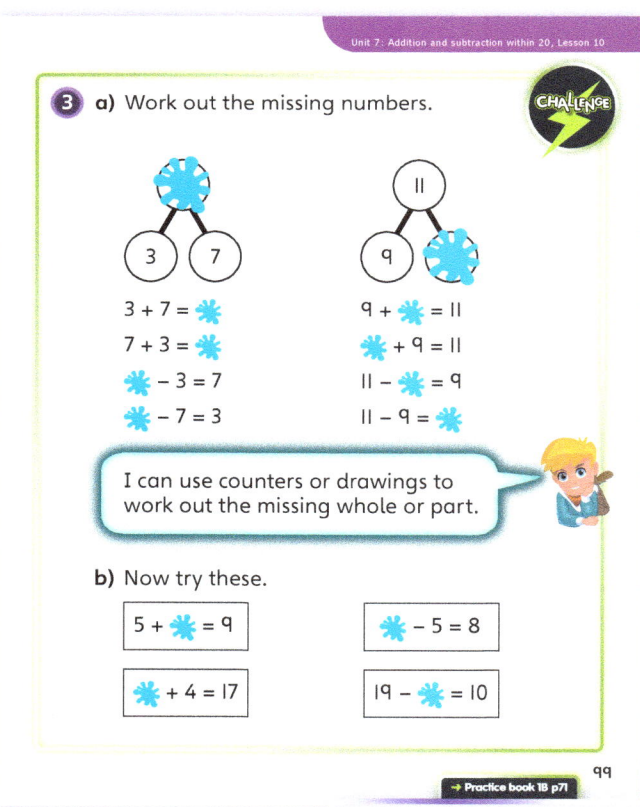

PUPIL TEXTBOOK 1B PAGE 99

136

Practice

WAYS OF WORKING Independent thinking

IN FOCUS Question ❶ uses part-whole models to help children work out missing numbers. They see the connection between the part-whole models and the missing number sentences. Question ❷ focuses on children using a number line to help them work out the missing numbers. They focus on the start number and work out how many they need to add on to work out the end numbers. They may find it helpful to draw jumps on the number lines.

Questions ❸ and ❹ provide examples with part-whole models and number lines, but this time they are finding missing numbers involving subtraction. For question ❹, children may need to count back.

STRENGTHEN Throughout these questions, use a number line to support children. Take the missing number problem and ask children where they need to start on the number line. Ask them where they will end up and if they need to count on or count back to work out the missing number. For problems such as ☐ + 3 = 7, ask children to think about where they finish first and how many they have counted on. They should show these as jumps on the number line to help them work out the starting number.

DEEPEN In question ❺, children are asked to find two missing numbers. Ask them what they know about the two numbers. They should say that they know the difference between the two numbers is 2. Ask them how they can use a number line to help them. How many different answers can they find using numbers within 20?

ASSESSMENT CHECKPOINT Questions ❶ and ❸ will help you see if children can use part-whole models and their knowledge of fact families to work out missing numbers. Questions ❸ and ❹ will help you see if children can count on and count back on a number line to work out missing numbers.

ANSWERS Answers for the **Practice** part of the lesson can be found in the *Power Maths* online subscription.

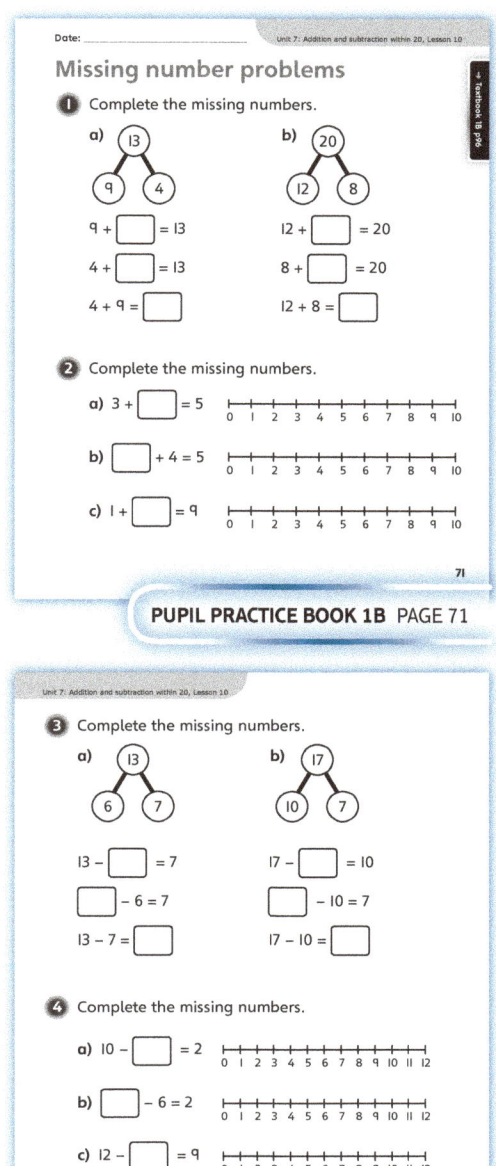

PUPIL PRACTICE BOOK 1B PAGE 71

PUPIL PRACTICE BOOK 1B PAGE 72

Reflect

WAYS OF WORKING Independent thinking

IN FOCUS Children may use part-whole models or a number line to work out the missing numbers. Some children may know what the answers are straight away, because of their knowledge of number bonds. Look for the methods that children use. Provide them with a number line and equipment to help them.

ASSESSMENT CHECKPOINT This question will help you determine how confident children are at working out missing numbers. The problems increase in difficulty and include examples involving addition and subtraction. Observe which ones they struggle with most and the methods that they use to find the answers.

ANSWERS Answers for the **Reflect** part of the lesson can be found in the *Power Maths* online subscription.

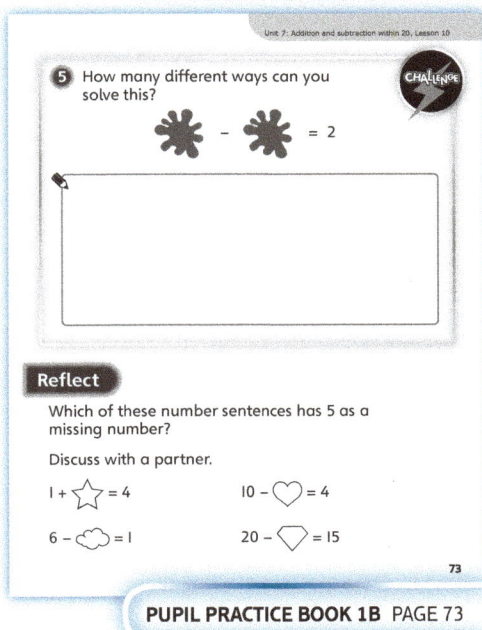

PUPIL PRACTICE BOOK 1B PAGE 73

After the lesson

- Can children find the missing number in a variety of different problems?

137

Unit 7: Addition and subtraction within 20, Lesson 11

Solve word and picture problems – addition and subtraction

Learning focus
In this lesson, children will use a range of strategies to find the solutions to word and picture problems involving addition and subtraction within 20.

Before you teach
- Have you prepared any additional multi-step word problems?
- Do children need mathematical language support?

NATIONAL CURRICULUM LINKS

Year 1 Number – addition and subtraction

Solve one-step problems that involve addition and subtraction, using concrete objects and pictorial representations, and missing number problems such as 7 = ☐ – 9.

ASSESSING MASTERY

Children can select an efficient strategy to solve word and picture problems involving addition and subtraction within 20. Children can explain their answers clearly and concisely using mathematical reasoning, and some may go on to solving multi-step problems.

COMMON MISCONCEPTIONS

Children may not use the most efficient method and may instead count back in 1s. Ask:
- *Can you think of a quicker way of working that out?*

Children may not know whether to add or subtract. Ask:
- *What information or words might help you understand whether you need to add or subtract?*

STRENGTHENING UNDERSTANDING

The key to this lesson is for children to recognise whether they have to do an addition or a subtraction. The characters encourage children to look for words that will help them work out which calculation they need to do. Support children by giving them word problems and asking them to highlight the key words. They can then sort the problems into additions and subtractions.

GOING DEEPER

Challenge children to solve multi-step problems to deepen learning in this lesson.

KEY LANGUAGE

In lesson: subtraction, total, more, fewer, most, difference, altogether, how many?

Other language to be used by the teacher: addition, less, greater, least, calculation

STRUCTURES AND REPRESENTATIONS

Part-whole model, number line, ten frame

RESOURCES

Mandatory: counters, additional word problems

Optional: cardboard tubes

In the eTextbook of this lesson, you will find interactive links to a selection of teaching tools.

Quick recap

Practise the addition and subtraction methods used within this lesson. Ask children questions such as 15 + 2, 15 – 2, 15 – 8, 8 + 6 and 6 + 7. What methods did they use to work out the answers?

Unit 7: Addition and subtraction within 20, Lesson 11

Discover

WAYS OF WORKING Pair work

ASK

- Question 1 b): *What does each number in the calculation represent? How can you represent this? What calculation do you need to do?*

IN FOCUS For question 1 b), encourage children to read the question aloud and discuss what it is asking. Some children may want to draw the 19 and 6 next to each other, so they can visualise it more clearly. When children realise they must subtract, encourage them to put 19 in their head and count back 6 using their fingers or counters. Alternatively, they might draw the 19 and cross out 6.

PRACTICAL TIPS Recreate the scenario using counters. You could use cardboard tubes to represent the logs.

ANSWERS

Question 1 a): 8 + 4 = 12

There are 12 ants in total.

Question 1 b): 19 − 6 = 13

There are 13 more snails inside the log than on the log.

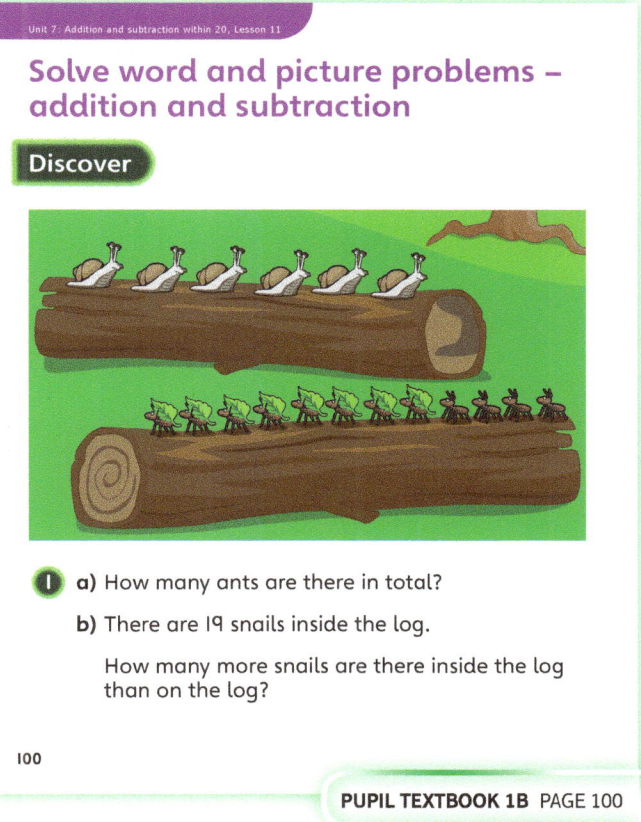

PUPIL TEXTBOOK 1B PAGE 100

Share

WAYS OF WORKING Whole class teacher led

ASK

- Question 1 b): *Can you explain how you got your answer? What method did you use?*

IN FOCUS For both questions, discuss which methods children used. Ask children to model these on the board to the whole class. Some children may need a representation, such as a number line, to demonstrate their method. Promote the use of efficient strategies such as jumping in 2s for question 1 a).

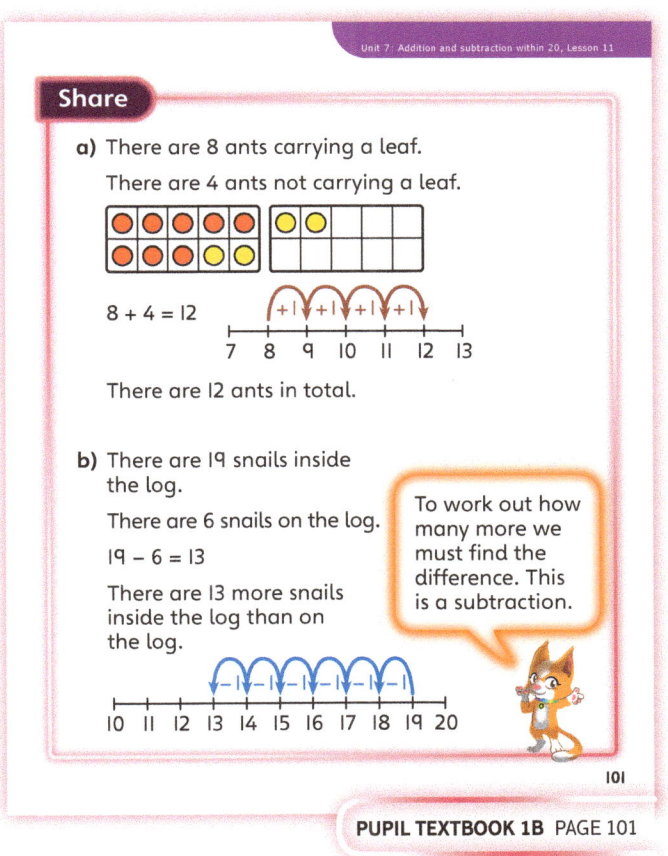

PUPIL TEXTBOOK 1B PAGE 101

139

Unit 7: Addition and subtraction within 20, Lesson 11

Think together

WAYS OF WORKING Whole class teacher led (I do, We do, You do)

ASK
- Question ③: *Do you remember how to cross the 10? Would a part-whole model help? Would a number line help?*

IN FOCUS In question ①, children may need a quick recap of how to cross the 10. They may need to draw a part-whole model first, to break the 6 into 4 and 2.

STRENGTHEN Work with children who need support with question ③. Remind them about fact families. Can they write down the fact family for this question? (5 + 12 = 17, 12 + 5 = 17, 17 – 12 = 5, 17 – 5 = 12). They will need to select the correct fact for each part of the question.

DEEPEN Put children in pairs and ask them to write a word problem for each other. Give them some key words to use, such as 'more than' and 'difference'.

ASSESSMENT CHECKPOINT Question ③ will allow you to assess whether children can solve a range of calculations, by using different strategies.

Question ② checks that children can use a number line to help them find missing numbers. They realise the relationship between the number in the problem and the position on the number line.

ANSWERS

Question ①: 14 – 6 = 8

There are 8 bees left in the hive.

Question ②: 6 + 8 = 14

There are 14 spots altogether.

Question ③ a): 17 – 12 = 5

There are 5 stars left in the box.

Question ③ b): 17 – 12 = 5

There are 5 silver stars.

Question ③ c): 17 – 12 = 5

There are 5 more stars than coins.

Question ③ d): All of the questions are solved by the subtraction 17 – 12 = 5.

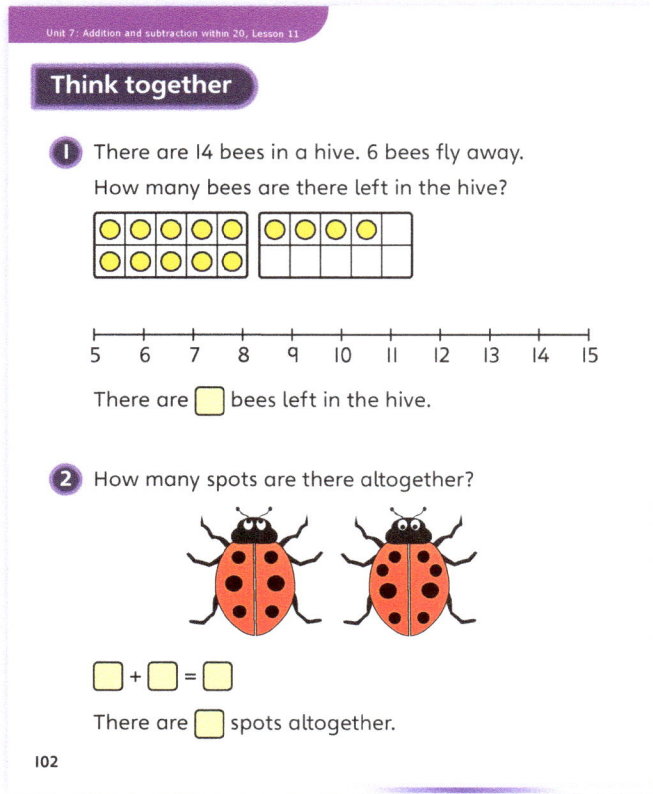

PUPIL TEXTBOOK 1B PAGE 102

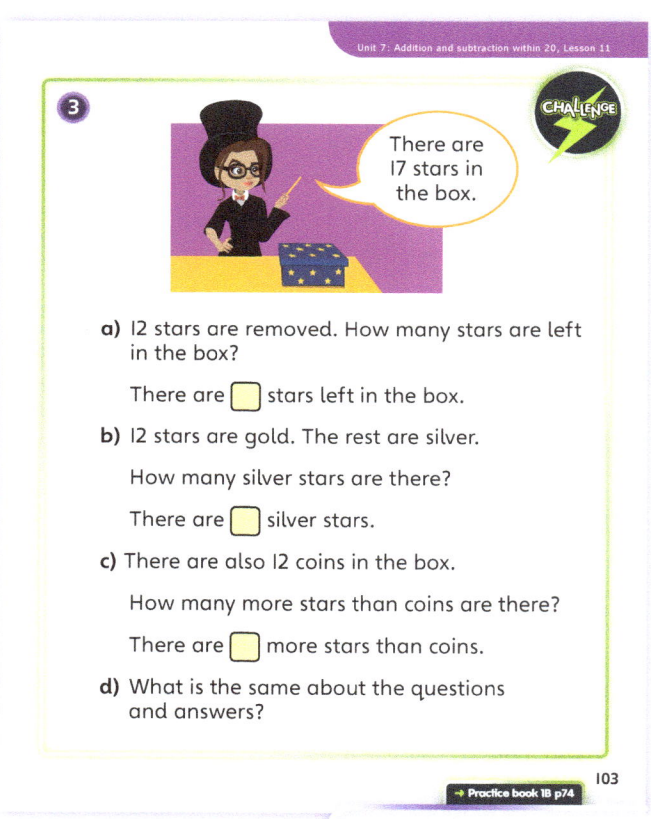

PUPIL TEXTBOOK 1B PAGE 103

Unit 7: Addition and subtraction within 20, Lesson 11

Practice

WAYS OF WORKING Independent thinking

IN FOCUS In these questions, children need to identify whether the picture/word problems require an addition or subtraction and then write the numbers in the correct places in the number sentences to solve the problem.

STRENGTHEN For questions ❶, ❷, ❸ and ❹, help children identify the key words that indicate addition (total, more) or subtraction (fewer). Allow them to model the questions using apparatus or drawings.

DEEPEN Give children some multi-step word problems in which they must add and then subtract or vice versa.

ASSESSMENT CHECKPOINT Question ❺ will give you a good indication of who has mastered the lesson. It is a two-step problem that will require children to interpret and then find the solution by using a range of strategies.

ANSWERS Answers for the **Practice** part of the lesson can be found in the *Power Maths* online subscription.

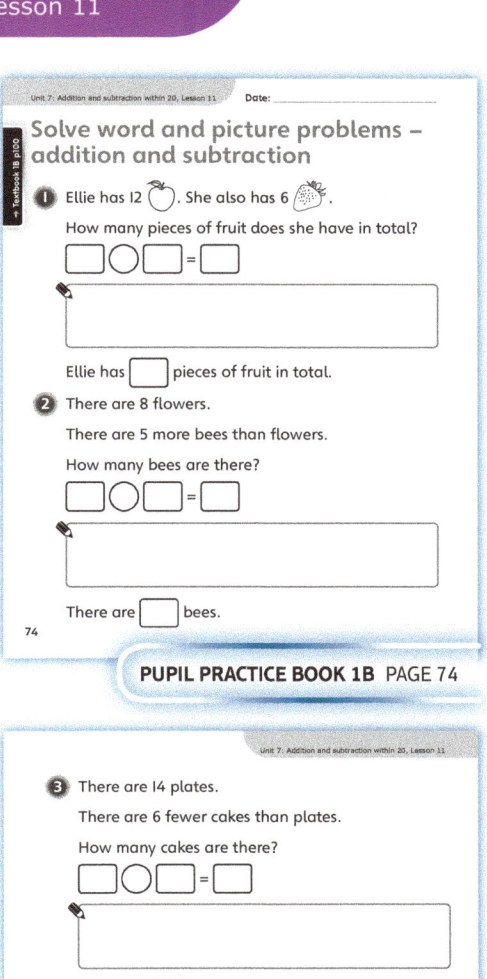

PUPIL PRACTICE BOOK 1B PAGE 74

PUPIL PRACTICE BOOK 1B PAGE 75

Reflect

WAYS OF WORKING Independent thinking

IN FOCUS Ask children to write their question independently and then come together as a group and swap their sentences. Children must solve a partner's question and verbally evaluate it. Have some prompts on the board, such as, 'Did the question make sense?' 'Did it make you think deeply?'

ASSESSMENT CHECKPOINT This question focuses on children's understanding of number sentences and will allow you to assess their ability to formulate effective questions to match these statements.

ANSWERS Answers for the **Reflect** part of the lesson can be found in the *Power Maths* online subscription.

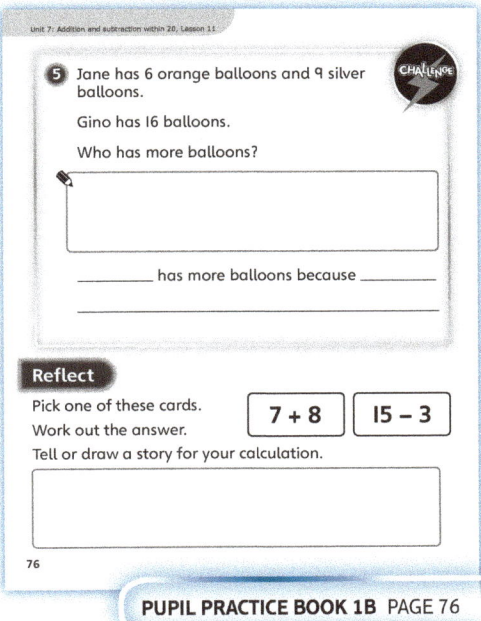

PUPIL PRACTICE BOOK 1B PAGE 76

After the lesson

- Can children find the missing number in a variety of different problems?

End of unit check

Don't forget the unit assessment grid in your *Power Maths* online subscription.

WAYS OF WORKING Group work adult led

IN FOCUS The questions are designed to prompt children to use different addition and subtraction methods.
- Question ① could use a count on or add by making 10 method.
- Question ② suits a count back method.
- Question ③ requires finding the difference between two numbers, crossing the 10.
- Question ④ prompts an understanding of place value fact families and asks children to find the fact that is not correct.
- Question ⑤ requires children to demonstrate an understanding of the relationships between numbers by solving missing number problems.

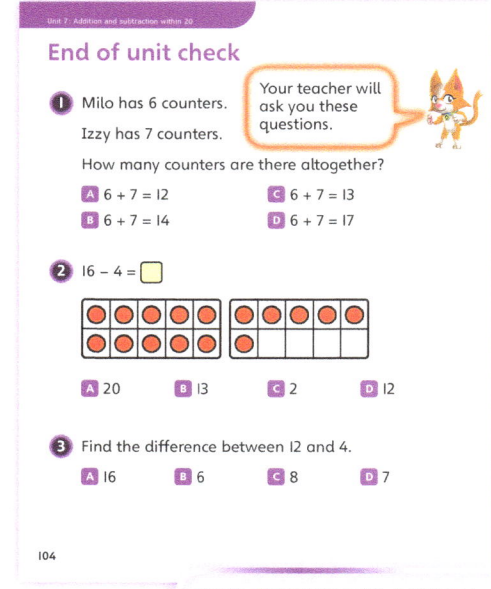

PUPIL TEXTBOOK 1B PAGE 104

Think!

WAYS OF WORKING Pair work or small groups

IN FOCUS In this section, different shapes stand for a hidden number. Children must use different ways of thinking about addition and subtraction within 20 and reason logically. They need to decide which number they can know for sure and then use that to find the remaining values.

Encourage children to think through the different kinds of addition and subtraction needed, including number bonds to 10 and 20. Discuss which number they can work out first, before they write their answer in **My journal**. Some children may think that they can choose any pairs of numbers that add to 20, but they need to recognise that the value of ☆ can only be one number and so is the starting point.

ANSWERS AND COMMENTARY

☆ = 3   △ = 17   ☐ = 18

Encourage children to notice which shape they can know for certain. Working this out will allow them to follow a chain of reasoning to complete the puzzle.

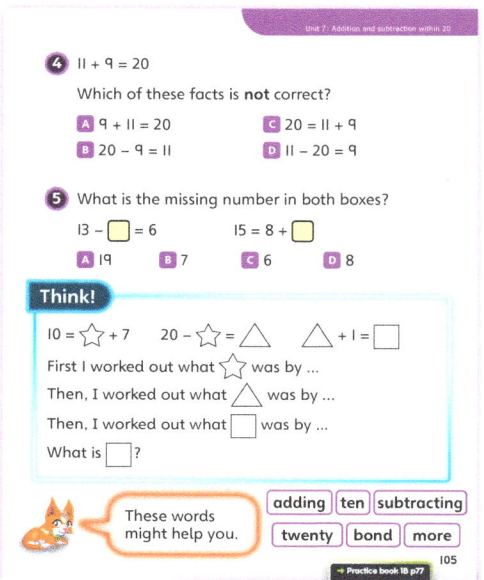

PUPIL TEXTBOOK 1B PAGE 105

Q	A	WRONG ANSWERS AND MISCONCEPTIONS	STRENGTHENING UNDERSTANDING
1	C	A suggests that children have counted on but included the starting number.	Encourage children to predict if the addition or subtraction will cross the 10.
2	D	A indicates that children have used addition instead of subtraction. B suggests they have included the starting number in their count back. C suggests they have not understood place value.	Use ten frames and part-whole models to support using known number bonds rather than a counting strategy.
3	C	A suggests children do not understand the phrase 'find the difference' and have added the two numbers together instead of subtracting them.	Remind children that the start number is not included in the count. Use a number line and cubes to illustrate this.
4	D	A might tell you that children have not read the question correctly. They read A and chose it because it is correct.	
5	B	A suggests that children have added the numbers together in the first calculation. C and D suggest that children have started at the wrong number.	

Unit 7: Addition and subtraction within 20

My journal

WAYS OF WORKING Independent thinking

ANSWERS AND COMMENTARY Children will need to use their journal to explain the different methods they used, including using number bonds or adding and subtracting 1s. If children can explain this verbally but struggle to put it into a written sentence, encourage them to use the word bank to choose a key word to show their thinking.

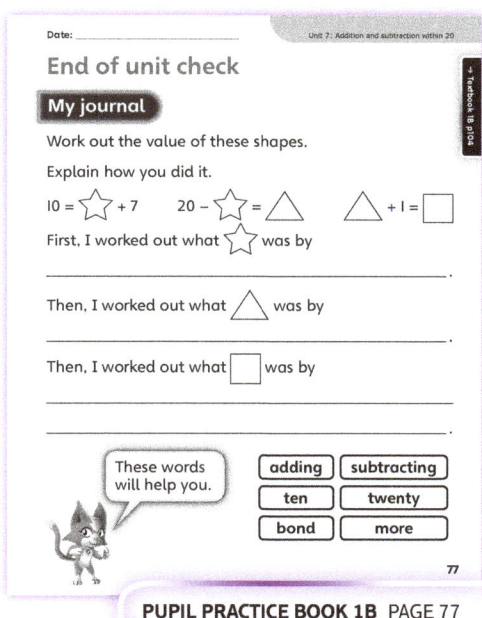

PUPIL PRACTICE BOOK 1B PAGE 77

Power check

WAYS OF WORKING Independent thinking

ASK
- Which method of adding do you find easiest? Which method of subtracting do you find easiest?
- Is there a method for adding/subtracting that you find difficult?
- Which method did you use? Why did you choose that method?

Power puzzle

WAYS OF WORKING Pair work

IN FOCUS Use this **Power puzzle** to see if children can find missing numbers in additions and subtractions using their knowledge of number bonds. Children may need to use a trial and improvement method – which is where they think they might have an idea for a solution, but when they check the calculation, find they need to change the answer slightly. This is a good opportunity to reinforce that a wrong answer is a step on the way to better understanding.

ANSWERS AND COMMENTARY If children can complete the **Power puzzle**, then this demonstrates that they can use knowledge of number bonds to solve additions and subtractions. If they struggle to know if their missing numbers are correct, then they need some help to strengthen their confidence in using addition and subtraction methods to check their solutions.

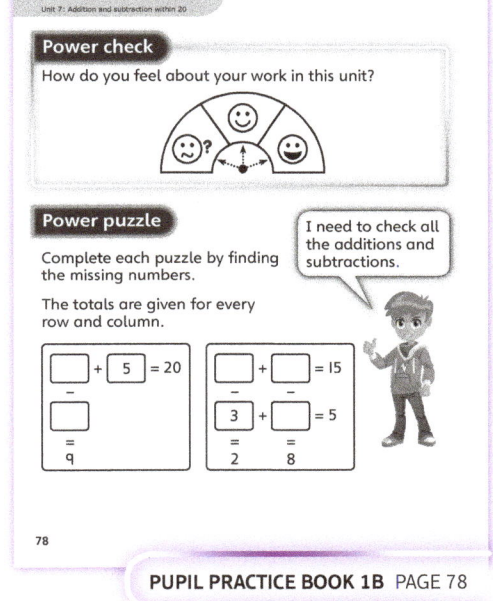

PUPIL PRACTICE BOOK 1B PAGE 78

After the unit

- Are children confident recognising calculations in the same fact family?
- What opportunities can you identify across the curriculum to reinforce and apply this unit's learning?

Strengthen and **Deepen** activities for this unit can be found in the *Power Maths* online subscription.

Unit 8
Numbers to 50

Mastery Expert tip! 'When I taught this unit, I ensured children were secure with the order of numbers to 50 and partitioning them into 10s and 1s. This really helped children with the subsequent lessons in this unit.'

Don't forget to watch the Unit 8 video!

WHY THIS UNIT IS IMPORTANT

In this unit, children will count beyond 20 to 50, counting objects as they go. They will learn that it is easier to count in 10s and 1s, rather than just in 1s. They will see common representations to help them count easily, such as ten frames, counters, cubes and egg boxes, etc. Children will notice the patterns in the number count beyond 20, for example, 20, 21, 22 … 30, 31, 32 … 40, 41, 42 … They will look at what digits are the same and what digits are different in these patterns of numbers.

Using their knowledge of counting in 10s and 1s will help children begin to partition numbers beyond 20 into 10s and 1s, representing them on a part-whole model. They will become increasingly confident in partitioning numbers as they work through the unit. Finally, they will use their knowledge of the counting sequence to find one more and one less than a number less than 50.

WHERE THIS UNIT FITS

→ Unit 7: Addition and subtraction within 20
→ **Unit 8: Numbers to 50**
→ Unit 9: Introducing length and height

This unit builds on children's knowledge of numbers to 10 and then 20 developed so far in Year 1, as well as their knowledge of the part-whole model. Going forward, children will apply their knowledge of numbers to 50 in units involving measure and numbers to 100.

Before they start this unit, it is expected that children:
- know how to count to 20, using correct number names and numerals
- can partition numbers below 20 into 10s and 1s confidently, using different mathematical equipment.

ASSESSING MASTERY

Children who master this unit can count confidently and fluently from 1 to 50. They can count objects in 1s using one-to-one correspondence. They count more efficiently by counting in 10s and 1s and realise, by grouping in 10s, it makes it easier to count numbers to 50. They are able to use this knowledge to partition a number into 10s and 1s and represent this partition using different mathematical equipment and also using ten frames. Finally, they use their knowledge of counting numbers to work out one more and one less, without the aid of equipment.

COMMON MISCONCEPTIONS	STRENGTHENING UNDERSTANDING	GOING DEEPER
Children may count across 10s boundaries without exchanging, for example, twenty-nine/29, twenty-ten/2010.	Use a range of representations, including counters and ten frames, to help children understand the nature of our number system and the need to exchange 10 ones for 1 ten.	Ask children to compare numbers, for example, looking at the number of 10s and 1s.
Children may count in 1s to find the total, even though they have made 10s.	Explain to children the reason they made 10s was to make it easier to count and they can now count in 10s instead of 1s.	Give children objects or images in 10s and 1s and ask them to make a number from 20 to 50, for example, 27, 45.

Unit 8: Numbers to 50

UNIT STARTER PAGES

Use the unit starter pages to remind children of counting from 0 to 20 and the different representations and language they used earlier in the year.

STRUCTURES AND REPRESENTATIONS

Ten frame: This model helps children to develop a sense of 10. It helps children know when they move to the next 10. It also makes counting in 10s and 1s much easier.

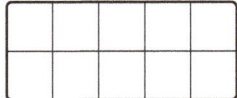

Number grid: This helps children count to 50 and partition numbers into blocks of 10.

1	2	3	4	5	6	7	8	9	10
11	12	13	14	15	16	17	18	19	20
21	22	23	24	25	26	27	28	29	30
31	32	33	34	35	36	37	38	39	40
41	42	43	44	45	46	47	48	49	50

Number track: Like a number line, a number track helps children represent the order of numbers. A number track can support children in counting and finding one more and one less.

1	2	3	4	5	6	7	8	9	10

Bead string: The bead string offers children the opportunity to manipulate different numbers.

Part-whole model: This model helps children understand that two or more things combine to make a whole. In this unit, it is especially useful for showing how a number is made up of 10s and 1s.

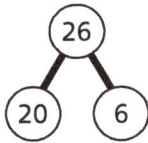

Rekenrek: A rekenrek is a tool that helps children count in 10s and 1s.

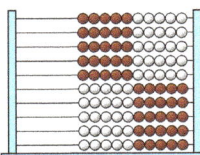

KEY LANGUAGE

There is some key language that children will need to know as part of the learning in this unit:

- tens, ones
- one more, one less
- partition
- count
- compare, order
- less than (<), greater than (>)

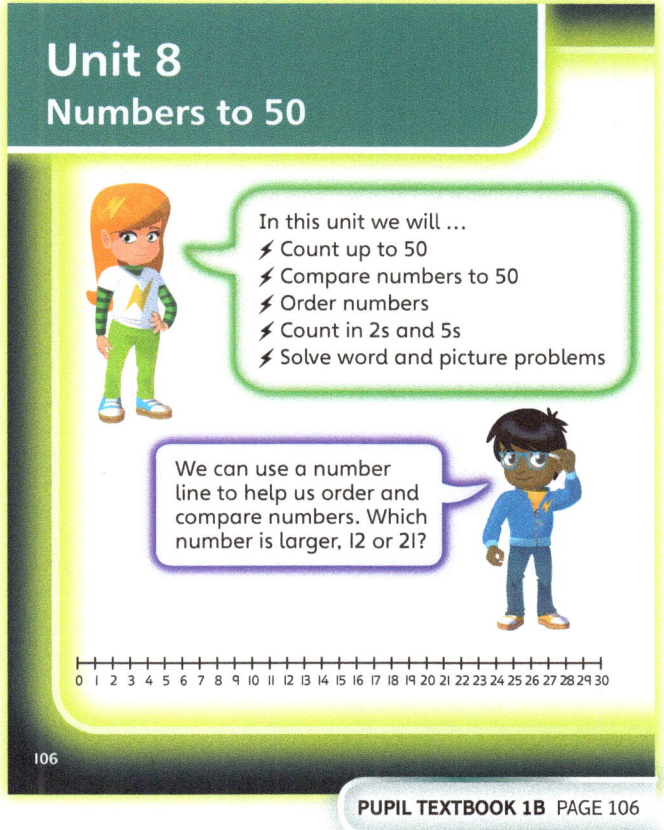

PUPIL TEXTBOOK 1B PAGE 106

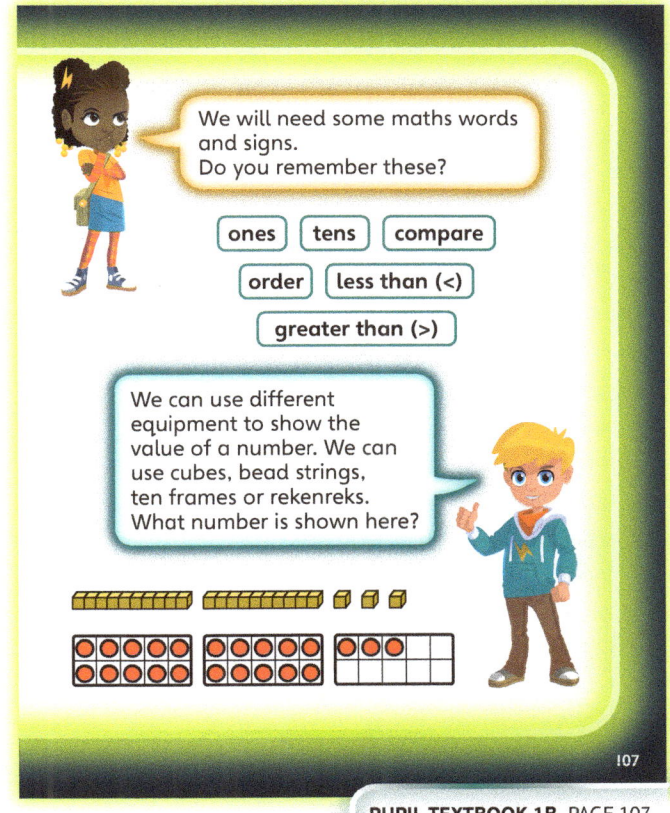

PUPIL TEXTBOOK 1B PAGE 107

Unit 8: Numbers to 50, Lesson 1

Count to 50

Learning focus
In this lesson, children will recognise numbers to 50 and count objects to 50. They will develop their use of the number line beyond 20 and up to 50, and be able to count on or back from any number up to 50.

Before you teach
- Are children secure with number names up to 20, including 'teen' numbers?
- Are there any additional misconceptions that need to be assessed?

NATIONAL CURRICULUM LINKS

Year 1 Number – number and place value

Count to and across 100, forwards and backwards, beginning with 0 or 1, or from any given number.

Count, read and write numbers to 100 in numerals; count in multiples of twos, fives and tens.

ASSESSING MASTERY

Children can accurately count to 50 in 1s and can count up to 50 objects. Children can find one more or one less than any number up to 50, using number tracks and base 10 equipment to help as needed.

COMMON MISCONCEPTIONS

Children may not use the concept of exchange when adding 1 to a number with 9 ones. For example, when adding 1 to 29 they may say 'twenty-ten', and write this either as 2010 or 210. Ask:
- *What number comes after 29 when you are counting in 1s?*

STRENGTHENING UNDERSTANDING

Strengthen understanding by using a range of representations. Base 10 equipment can be used to revisit the concept of exchange: 10 ones = 1 ten. 100 squares (10 × 10 grids with squares labelled 1 to 100) and number lines that are marked in ones enable children to notice patterns in our number system and the numerals used, and can also be used to support counting on and back in 1s.

GOING DEEPER

Deepen understanding by asking children to count back to 0 from a number above 20.

KEY LANGUAGE

In lesson: count, number

Other language to be used by the teacher: numeral, place value, all number names from 0–50, count back

STRUCTURES AND REPRESENTATIONS

Number tracks, number lines

RESOURCES

Mandatory: base 10 equipment

Optional: number lines marked in ones, 100 squares, digit cards, counters, cubes, blank number tracks, marbles or counters in a tub

 In the eTextbook of this lesson, you will find interactive links to a selection of teaching tools.

Quick recap

Spend time looking at the numbers from 11 to 16. Show children representations of these numbers where the 10 is clearly visible. Ask: *How many altogether?* Ask children to explain how they know, ensuring they do not count on from 1 each time.

Unit 8: Numbers to 50, Lesson 1

Discover

WAYS OF WORKING Pair work

ASK

- Question 1 a): *How can you count the jars? How can you make sure you do not miss any?*
- Question 1 b): *Let's keep counting. What comes after 30? What comes after 39?*

IN FOCUS The purpose of questions 1 a) and 1 b) is to ask children to count beyond 20. Encourage children to point to the jars as they count. Look for them to use a strategy where they do not miss any. Pay particular attention to what children say after numbers like 29, 39 and 49, as these are the ones that they often find difficult. Children may find it useful to use number tracks or a 100 square to help them.

PRACTICAL TIPS Replicate the scenario in small groups by using counters or cubes to represent the jars.

ANSWERS

Question 1 a): Children should count the jars together from 1 to 30.

Question 1 b): Children should count together from 30 to 50.

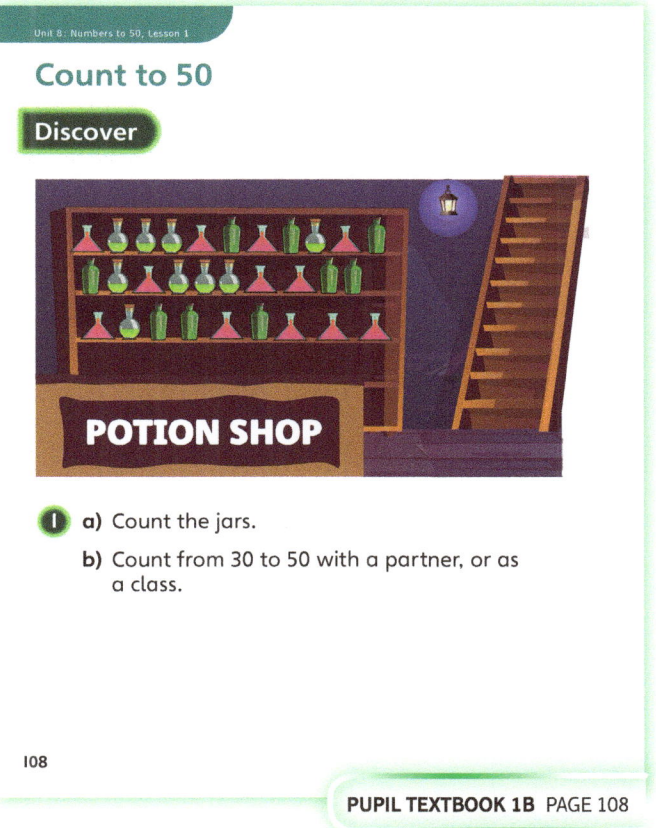

PUPIL TEXTBOOK 1B PAGE 108

Share

WAYS OF WORKING Whole class teacher led

ASK

- Question 1 a): *How will you count the jars? Where will you start?*
- Question 1 b): *How does the number track help you count? Can you see a pattern? Which numbers are the same? Which are different?*

IN FOCUS In question 1 a), count the jars with children. Show a strategy to help them with their counting, focusing on counting the top shelf first from left to right and then moving to the second shelf, and so on. Ask children to point to the jars in the textbook and count with you. In question 1 b), use the number tracks that they are familiar with to help children count. Slow down when you read 39 and 49 to make sure that children move to the next 10.

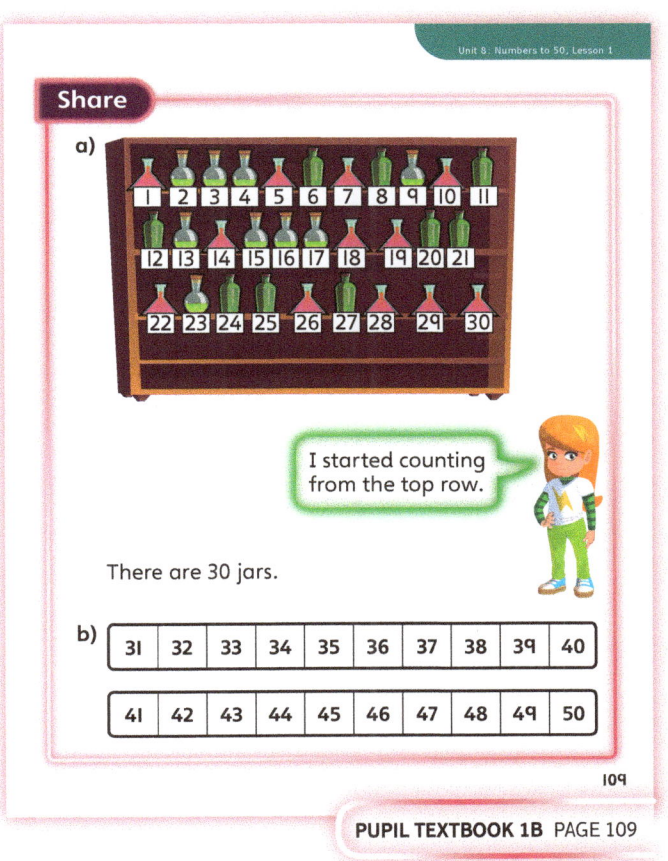

PUPIL TEXTBOOK 1B PAGE 109

Think together

WAYS OF WORKING Whole class teacher led (I do, We do, You do)

ASK

- Question 1: *How can you count the ladybirds? Where will you start? How will you know when you have counted them all?*
- Question 2: *How can you work out what numbers are missing? Do all the counts go up?*
- Question 3: *How many sweets are in the jar at the start? How do you know? What happens when you take 1 out? How many are there now?*

IN FOCUS The focus of question 1 is for children to have further practice of counting more than 20 objects. Children should point to each ladybird as they count them, to make sure they do not count too many or too few. As a class, focus on counting efficiently from left to right so that children do not miss any. If using the whiteboard, you may want to cross out each ladybird as you go. Question 2 asks children to work out missing numbers. To help them work out the missing numbers in the counting pattern, encourage them to say the numbers out loud. In question 3, we cannot see the 50 objects so children cannot count the sweets one by one. Children are asked to count back from 50. Model the situation with marbles or counters in a tub. Start slowly as children learn to count back from 50.

STRENGTHEN Use concrete objects in question 1 instead of images. Children can then move the objects from one side of their table to the other as they count each one. For questions 2 and 3, provide children with number tracks that they can use to help them count on and back. The number tracks may be numbered or may be blank, in which case children will need to fill in the numbers before counting back from 50.

DEEPEN Ask children to count out a number of objects from a larger group of objects.

ASSESSMENT CHECKPOINT It is important that children can work out the missing numbers in question 2. This shows that children are secure in the counting pattern.

ANSWERS

Question 1: Toshi had £160. Amal had £400.

Question 2 a): 23, 24, 25, 26, 27, 28

Question 2 b): 40, 41, 42, 43, 44

Question 2 c): 28, 29, 30, 31, 32, 33

Question 3: When 1 sweet is taken out there are 49 left in the jar. When another sweet is taken out there are 48 left in the jar.

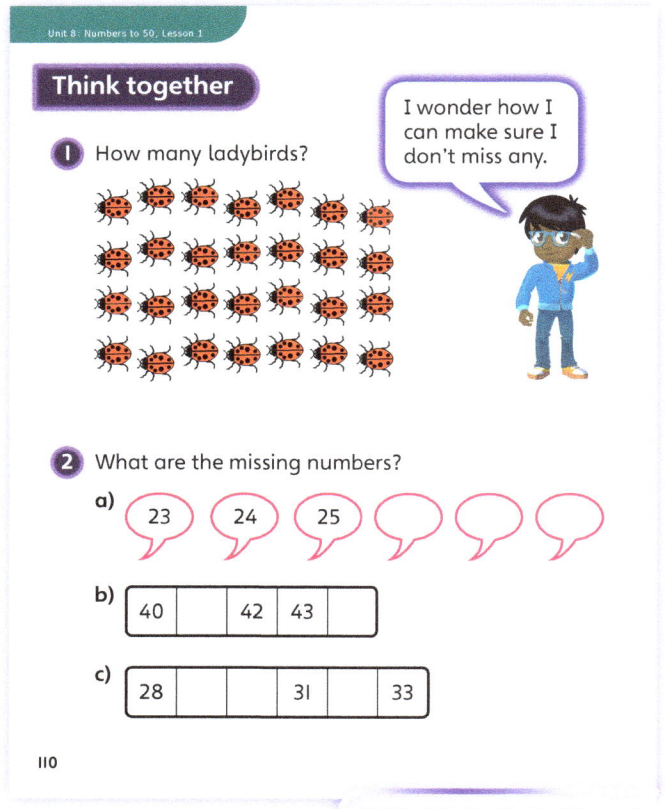

PUPIL TEXTBOOK 1B PAGE 110

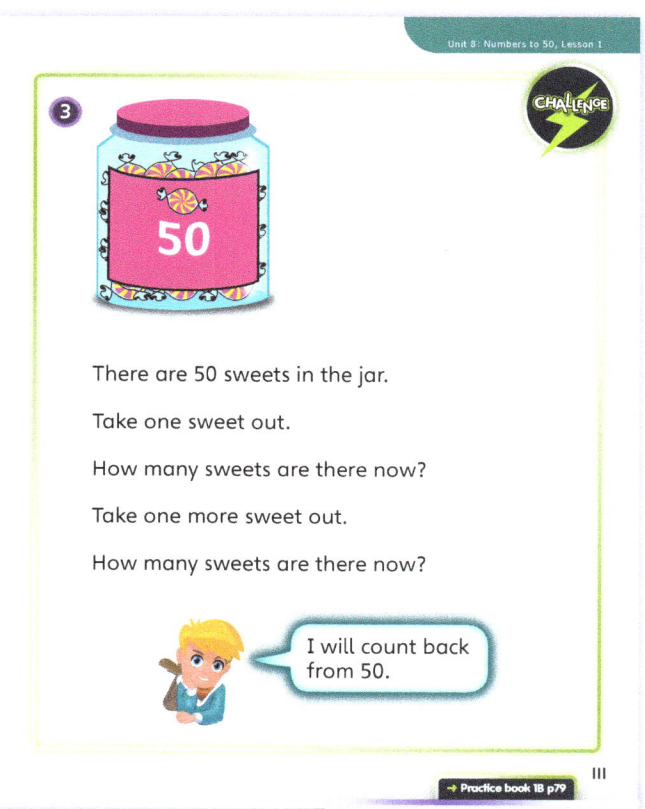

PUPIL TEXTBOOK 1B PAGE 111

Practice

WAYS OF WORKING Independent thinking

IN FOCUS Question 4 asks children to fill in missing numbers in number tracks, including one number track counting back from 50. This encourages children to apply their conceptual understanding.

In question 5, children are asked to spot the mistake and discuss the common misconception with a partner. They should be able to identify that Anna has not crossed the 10. In question 6, children should spot a pattern in the way that Jack has drawn his circles (he has drawn one row of 10 circles and three rows of 8 circles). Children should see that they can complete three rows of 10 by drawing two additional circles on two rows, and then draw one additional circle on the fourth row to make 39.

STRENGTHEN In questions 3 and 4, children are required to fill in adjacent missing numbers. Children should be able to use their answer for the first gap to help them fill the second gap. Number lines or 100 squares can be used to reinforce the connection between adding 1 and the corresponding numerals. Use base 10 equipment to help children partition numbers and to identify how many 1s are in each number (for example, to see that 39 is made up of 3 tens (30) and 9 ones (9)).

DEEPEN Children should be encouraged to explore different ways in which numbers to 50 could be represented – either physically or using drawings. For example, children could identify that in question 6, Jack has drawn circles in groups of 8 and 10.

THINK DIFFERENTLY In question 4, it is not always the first numbers in each track that are given. Children will need to work from the numbers they are given and either count on or back to fill in the gaps.

ASSESSMENT CHECKPOINT Assess whether children are secure in their use of the number names and numerals up to 50, and can confidently count on and back in 1s. Use children's explanations of Anna's mistake in question 5 to assess whether they have grasped the concept of exchange when one more is added to 9 ones.

ANSWERS Answers for the **Practice** part of the lesson can be found in the *Power Maths* online subscription.

Reflect

WAYS OF WORKING Pair work

IN FOCUS The **Reflect** part of the lesson allows children to practise their knowledge of the number system up to 50 and the numerals and number names used. Children also need to apply this knowledge in order to check their partner's counting.

ASSESSMENT CHECKPOINT Listen to children as they count. Are they using the correct number names? Pay particular attention when they are crossing a 10s boundary (such as 19, 20, 21). Can children also count back from their number to 0, as suggested by Sparks?

ANSWERS Answers for the **Reflect** part of the lesson can be found in the *Power Maths* online subscription.

After the lesson

- What opportunities can you find to apply the knowledge of numbers to 50 to contexts within the school day?
- Are children secure with number names and numerals when crossing the 10s boundary?

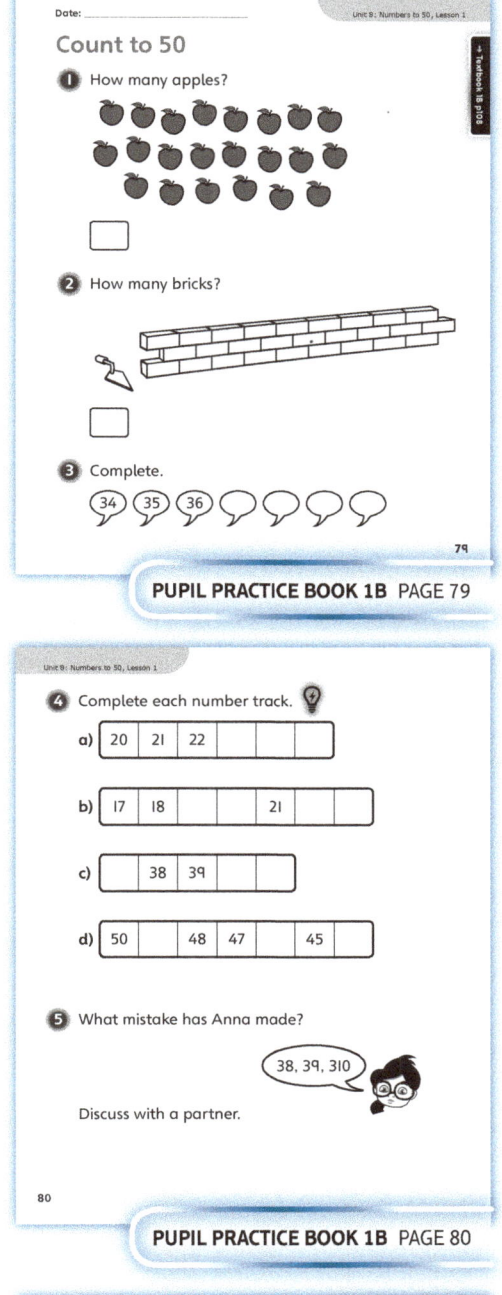

PUPIL PRACTICE BOOK 1B PAGE 79

PUPIL PRACTICE BOOK 1B PAGE 80

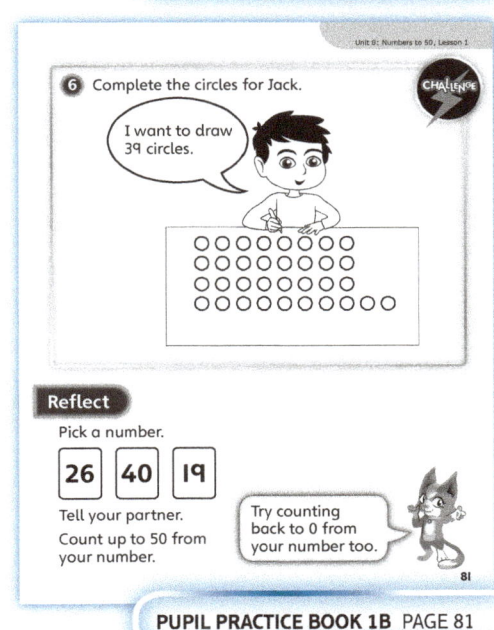

PUPIL PRACTICE BOOK 1B PAGE 81

Unit 8: Numbers to 50, Lesson 2

Numbers to 50

Learning focus

In this lesson, children will develop their knowledge of numbers between 20 and 50. They will be able to count on and back up to 50 and identify missing numbers in a sequence.

Before you teach

- Are children secure with number names up to 50, including 'teen' numbers?
- Are children secure with counting on and back in steps of 1?
- Are there any additional misconceptions that need to be addressed in this lesson?

NATIONAL CURRICULUM LINKS

Year 1 Number – number and place value

Count to and across 100, forwards and backwards, beginning with 0 or 1, or from any given number.

Count, read and write numbers to 100 in numerals; count in multiples of twos, fives and tens.

ASSESSING MASTERY

Children can accurately count on and back to 50 in 1s. Children can identify missing numbers in a sequence by counting on and back in 1s from the numbers in the sequence that are known.

COMMON MISCONCEPTIONS

This lesson expands children's knowledge of counting to 50 from Lesson 1. Therefore, the same misconceptions may still be present and children may not use the concept of exchange when adding 1 to a number with 9 ones. For example, when adding 1 to 29 they may say 'twenty-ten', and write this as either 2010 or 210. Ask:
- *What number comes after 29 when you are counting in 1s?*

STRENGTHENING UNDERSTANDING

Use number tracks (marked in ones) and 100 squares to help children identify missing numbers up to 50: encourage them to look at the numbers either side of the missing number and count on or back by 1. Number tracks can also be used to help children notice patterns in our number system and the numerals used.

As in the previous lesson, use base 10 equipment to revisit the concept of exchange: 10 ones = 1 ten. This will help support children who are not secure with the number names and numerals used, particularly when crossing a 10s boundary (for example, 39, 40, 41).

GOING DEEPER

Can children apply their knowledge of the number system and number names to 50 to a range of unfamiliar and non-standard applications, such as game boards and circular scales?

KEY LANGUAGE

In lesson: count, number, count back, in order

Other language to be used by the teacher: sequence, numeral, place value, all number names from 0–50

STRUCTURES AND REPRESENTATIONS

Number track, number grid

RESOURCES

Mandatory: number tracks, 100 squares

Optional: base 10 equipment

In the eTextbook of this lesson, you will find interactive links to a selection of teaching tools.

Quick recap

Ask children to count how many tables there are in the classroom, how many chairs, and how many children.

Unit 8: Numbers to 50, Lesson 2

Discover

WAYS OF WORKING Pair work

ASK

- Question 1 a): *Where will you start your count? When will you stop?*
- Question 1 b): *How can you work out the numbers the frogs are sat on? Do you always need to count from 1 each time? Do you know any of them without counting?*

IN FOCUS In this lesson, we are building on the previous lesson's learning, of children counting to 50. This time, the focus is on writing the numbers. In question 1 a), practise as a class counting to 50. Show the number and say it. In question 1 b), children use their knowledge of counting to help them work out the missing numbers. Look for different methods. Some children may know the missing numbers immediately, others may need to count from 1.

PRACTICAL TIPS Use 50 concrete objects as children count to 50.

ANSWERS

Question 1 a): Children should count from 1 to 50.

Question 1 b): The frogs are on numbers 8, 17, 26, 30 and 32.

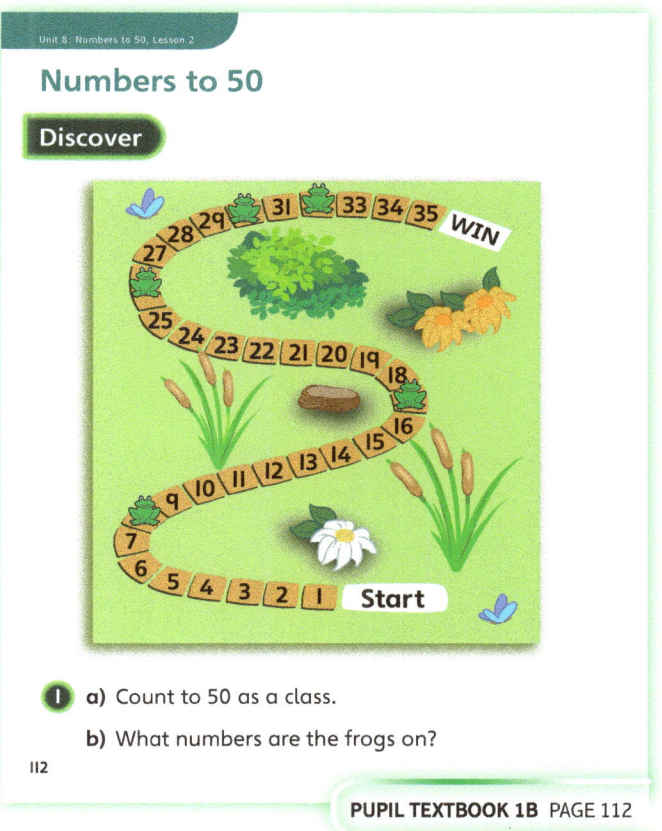

PUPIL TEXTBOOK 1B PAGE 112

Share

WAYS OF WORKING Whole class teacher led

ASK

- Question 1 a): *How can you use the number grid to help you count? What happens when you reach the end of each row?*
- Question 1 b): *How can you find the numbers that the frogs are sat on? Can you work out any without counting? Can you write down the missing numbers?*

IN FOCUS In question 1 a), use the number grid to 50 to help children count from 1 to 50. Encourage children to point to the correct number as they count. When they reach the end of a row, explain they move to the start of the next row. Some children may move down, like they would move on a snakes and ladders board, so it is important to count with them. In question 1 b), encourage different methods. Some children may know the answers straight away. Counting on from 1 will help them check whether they are correct. Once they have found a number, encourage them to count from that number to get to the next frog.

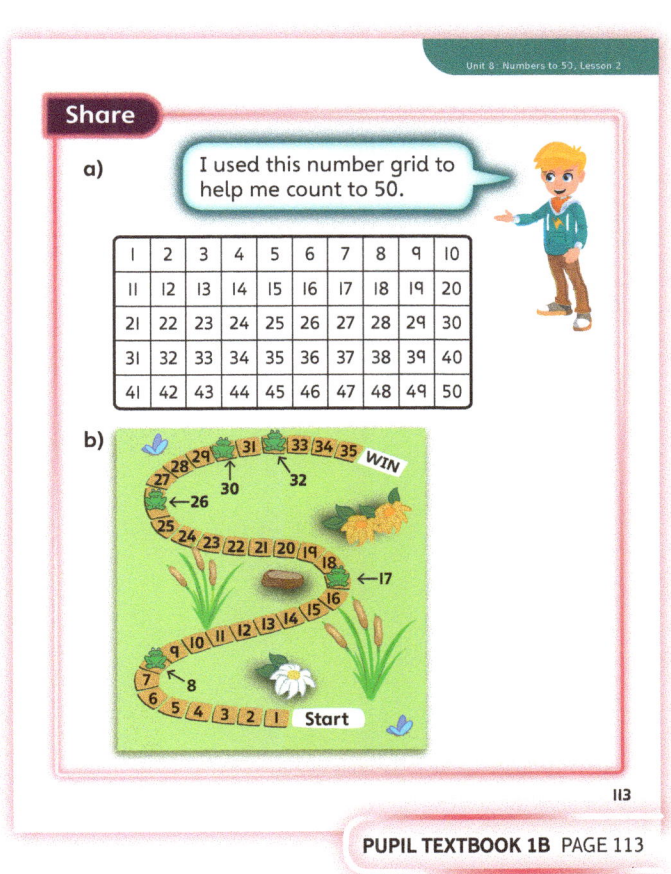

PUPIL TEXTBOOK 1B PAGE 113

Unit 8: Numbers to 50, Lesson 2

Think together

WAYS OF WORKING Whole class teacher led (I do, We do, You do)

ASK

- Question ❶: *Which numbers can you use to help you work out the missing numbers? What is one more than [for example] 15?*
- Question ❶: *Are there any patterns on the number grid that you could use to help you check your answers?*
- Question ❸: *What number comes after 37? How do you know this is the correct route?*

IN FOCUS Questions ❶ and ❷ require children to identify missing numbers between 1 and 50, allowing them to apply their conceptual understanding of the number system, including number names and numerals. Question ❶ a) is presented as a section from a 100 square and question ❷ as sections from number tracks.

Question ❸ encourages children to apply their knowledge of the number system to 50 to find different routes and possibilities, checking each against the correct order of numbers between 34 and 45.

STRENGTHEN In questions ❶ and ❷, use 100 squares and number tracks to help children identify the missing numbers by counting on and back. You can further support children in writing the correct numerals by representing the missing numbers with base 10 equipment.

DEEPEN Encourage children to reason using their knowledge of numbers up to 50. For example, say: *Convince me that 37 comes after 34.* Children may count up from 34 to 37 in order to do this.

ASSESSMENT CHECKPOINT Check whether children can identify the missing numbers in questions ❶ and ❷. If they can do this with confidence, including when presented with a number track as in question ❷, this suggests that they have a secure knowledge of the order of numbers, and the number names and numerals, up to 50.

ANSWERS

Question ❶: The missing numbers are 16, 22, 40 and 43.

Question ❷ a): 17

Question ❷ b): 23

Question ❷ c): 38, 41

Question ❸:

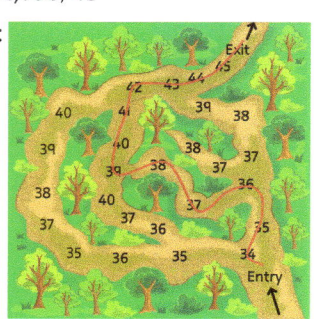

Think together

❶ What are the missing numbers?

1	2	3	4	5	6	7	8	9	10
11	12	13	14	15		17	18	19	20
21		23	24	25	26	27	28	29	30
31	32	33	34	35	36	37	38	39	
41	42		44	45	46	47	48	49	50

❷ What are the missing numbers?

a) | 14 | 15 | 16 | | 18 |

b) | 21 | 22 | | 24 | 25 |

c) | 37 | | 39 | 40 | | 42 | 43 | 44 | 45 |

PUPIL TEXTBOOK 1B PAGE 114

❸ Ola is lost in the forest.

She must find her way from 34 to 45, in order.

Can you help her to find her way out?

How did you help her find the way?

PUPIL TEXTBOOK 1B PAGE 115

Unit 8: Numbers to 50, Lesson 2

Practice

WAYS OF WORKING Independent thinking

IN FOCUS Questions ❶, ❷ and ❸ require children to fill in sequences of missing numbers. This encourages children to apply their knowledge of the order of numbers, and how these are written in numerals.

Question ❹ asks children to count back from 48 to 37 in order. Children may notice that they could find the route between 37 and 48 counting on instead of counting back. Question ❺ introduces children to a new representation of the numbers to 50: a game board on which the numbers 'snake' upwards.

Question ❻ encourages children to apply their knowledge of the number system to 50 to find two different ways to complete number tracks that both start with 35.

STRENGTHEN In questions ❸ and ❹, encourage children to use a 100 square to support their counting and identification of the missing numbers and sequences. Ensure that they do not simply copy from the 100 square by asking them to count aloud, using their knowledge of the number names between 0 and 50.

DEEPEN Encourage children to find different ways of solving the same problem – for example, with questions ❹ and ❻, children could be asked if they could solve the problems in more than one way (such as counting on and back).

ASSESSMENT CHECKPOINT Assess whether children can apply their knowledge of the number system to the different representation shown in question ❺. If they cannot, it suggests that their conceptual understanding of the numbers between 0 and 50 needs to be strengthened.

ANSWERS Answers for the **Practice** part of the lesson can be found in the *Power Maths* online subscription.

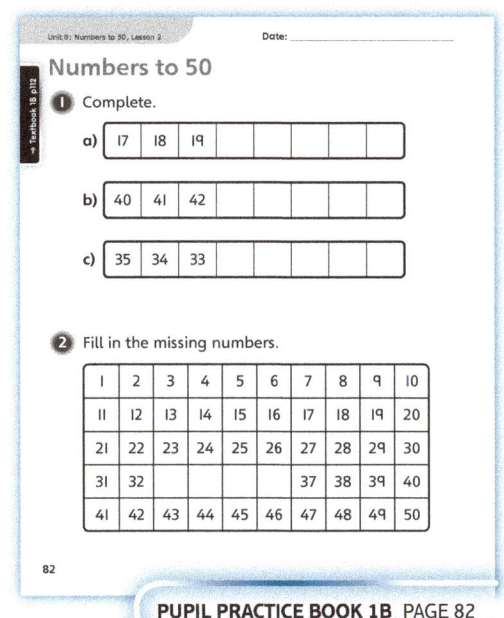

PUPIL PRACTICE BOOK 1B PAGE 82

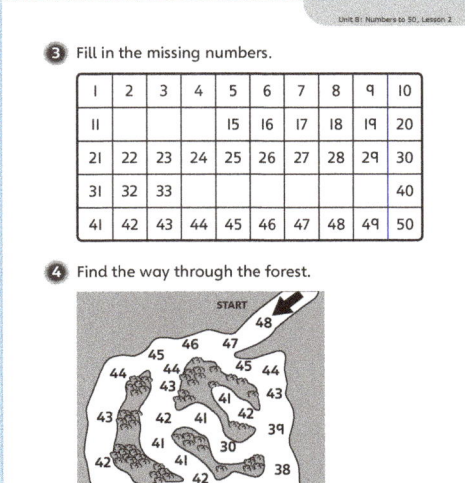

PUPIL PRACTICE BOOK 1B PAGE 83

Reflect

WAYS OF WORKING Pair work

IN FOCUS The **Reflect** questions allow children to practise their conceptual understanding of the number system up to 50 by counting on and back. Both questions involve crossing a 10s boundary, requiring children to apply their knowledge of exchange.

ASSESSMENT CHECKPOINT Listen to children as they count to a partner. Are they using the correct number names in the correct order? Are they able to count both on and back?

ANSWERS Answers for the **Reflect** part of the lesson can be found in the *Power Maths* online subscription.

After the lesson

- What opportunities can you find to apply the knowledge of numbers to 50, including counting on and back, to contexts within the school day?
- Are all children secure with both counting on and back with numbers to 50? If any children are not, focus on these children as part of your same-day intervention.

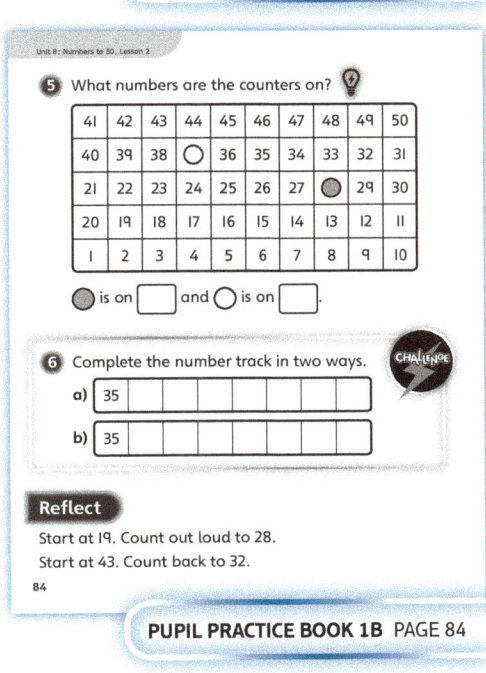

PUPIL PRACTICE BOOK 1B PAGE 84

Unit 8: Numbers to 50, Lesson 3

20, 30, 40 and 50

Learning focus
In this lesson, children count objects by making groups of 10. They focus on just making 10s and know the count 10, 20, 30, 40 and 50.

Before you teach
- Can children can count from 1 to 50?
- Do children know how to make a group of 10?
- Do children know that a full ten frame means there are 10 counters?

NATIONAL CURRICULUM LINKS

Year 1 Number – number and place value

Identify and represent numbers using objects and pictorial representations including the number line, and use the language of: equal to, more than, less than (fewer), most, least.

Year 2 Number – number and place value

Recognise the place value of each digit in a two-digit number (tens, ones).

ASSESSING MASTERY

Children can make 10s to help them count and they should be able to count in 10s to 50. They use common manipulatives to help them make 10 such as ten frames and towers of cubes.

COMMON MISCONCEPTIONS

When making 10s, the 10s are not separated and are unclear. It is important that children can see the 10s easily. Ask:
- How can the 10s be shown clearly? Can you use a ten frame to help you?

Children may mix up the 1s count with the 10s count, for example, 10, 20, 21. Ask:
- How can you make sure you stay counting in 10s? Will counting out loud help?

Children may count in 1s to find the total, even though they have made 10s. Explain to children the reason they made 10s was to make it easier to count and they can now count in 10s. Ask:
- Is it easier to count in 1s or count in 10s? Why?

STRENGTHENING UNDERSTANDING

Use ten frames to help children see the 10. Count out loud with children in 10s once they have made the 10s. This will help make counting in 10s to 50 familiar.

GOING DEEPER

Children count back in 10s from 50 to 10.

KEY LANGUAGE

In lesson: count, 10s

STRUCTURES AND REPRESENTATIONS

Ten frames, bundle of 10 straws, tower of 10 cubes

RESOURCES

Mandatory: multilink cubes, counters, straws, ten frames

 In the eTextbook of this lesson, you will find interactive links to a selection of teaching tools.

Quick recap
Ask children to count out 20 counters (or other objects) by putting the counters onto 2 ten frames.

Unit 8: Numbers to 50, Lesson 3

Discover

WAYS OF WORKING Pair work

ASK

- Question 1 a): *How many counters can go on a ten frame? How many ten frames can you fill? How many counters are there? How did you count them?*
- Question 1 b): *Can you make a tower of 10 cubes? How do you know there are 10 cubes in the tower? Make another tower of 10 cubes – did you have to count them? How can you check if all your towers of cubes have 10 in them easily? How many cubes are there in total?*

IN FOCUS In question 1 a), children make 10 using counters on a ten frame. Children should see they do not need to count 10 when filling a ten frame as it has 10 in it when full. In question 1 b), they make towers of 10 cubes. Unlike with the ten frames, when making towers children need to count the cubes to check there are 10 in each tower. They can check if other towers have 10 cubes in by making them the same height. Children can use the 10s to help them count easily. Make sure that they do not count in 1s.

PRACTICAL TIPS Ensure children have the exact number of counters and cubes.

ANSWERS

Question 1 a): Danny can fill 3 ten frames with 30 counters.

Question 1 b): Meg can make 4 towers of 10 cubes with 40 cubes.

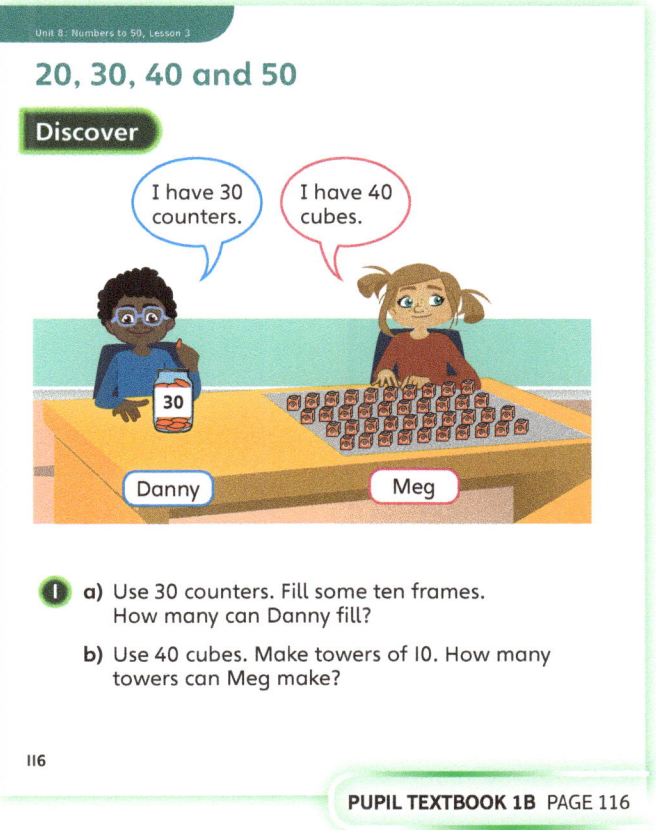

PUPIL TEXTBOOK 1B PAGE 116

Share

WAYS OF WORKING Whole class teacher led

ASK

- Question 1 a): *How many counters are on each ten frame? How do you know? How can you find how many counters there are in total?*
- Question 1 b): *How many cubes are in each tower? How many cubes are there in total? How do you know?*

IN FOCUS In question 1 a), children see that there are 10 counters on each ten frame, because the ten frames are full, and that they can find the total number by counting in 10s. Count with children in 10s to ensure they understand this. You may want to show how much easier it is to count in 10s than in 1s. In question 1 b), point and count with children to check there are 10s cubes in each tower. Talk to children about why they do not need to check the other towers as they are all the same height. Count in 10s to 50 again as a class.

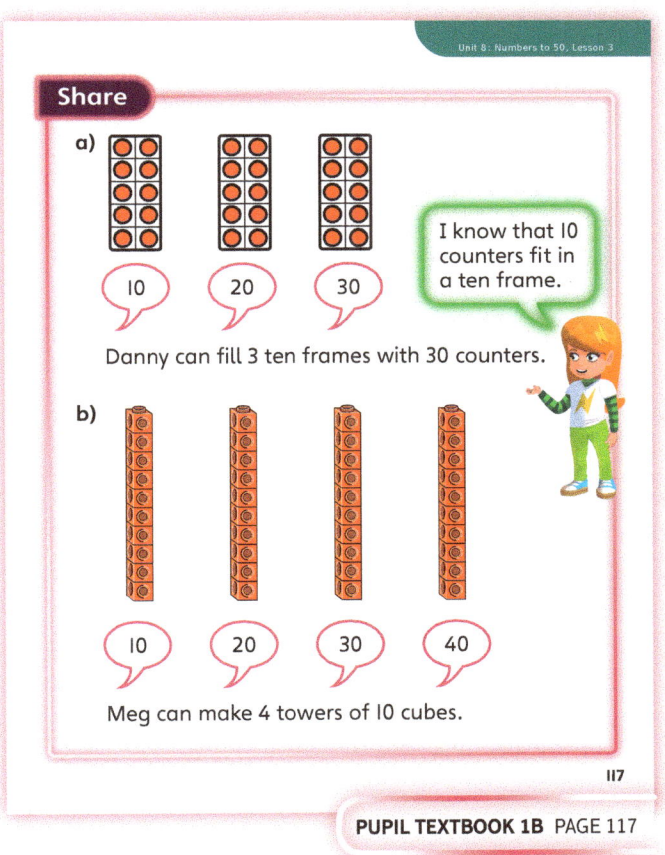

PUPIL TEXTBOOK 1B PAGE 117

155

Unit 8: Numbers to 50, Lesson 3

Think together

WAYS OF WORKING Whole class teacher led (I do, We do, You do)

ASK
- Question ❶: *What can you see? How many straws are in each bundle? How can we find the total number of straws?*
- Question ❷: *How many counters are there in total? How many cubes are there in total? How did you count them?*
- Question ❸: *How many of each coloured counter are there? How do you know? What patterns do you notice?*

IN FOCUS Throughout these questions, children are counting common manipulative objects in 10s. They strengthen their understanding of counting in 10s to 50, avoiding counting in 1s. In question ❶, they are introduced to bundles of 10 straws that they may not have seen before. In question ❷ a) children should be able to recognise that a full ten frame holds 10 counters, and so immediately count in 10s. In question ❷ b), children will need to count the first tower to see that it has 10 cubes. They should see that each tower is the same height, and so each one has 10 cubes. Children can further practise counting in 10s using similar objects to those they used in the **Discover** section. In question ❸, children see the 10s from 10 to 50 in ten frames. Encourage them to look for patterns. The images should let them see that each time there is 10 more.

STRENGTHEN Give children 30 straws and ask them to put them in bundles of 10s. Count the straws in 10s with children. Count out loud with them, pointing to or taking the straws as you count.

DEEPEN Give children five bundles of 10 straws. Ask: *How many straws do you have?* Ask them to give you a bundle of straws. Then ask: *How many straws do you have now?* Keep going until there are no bundles left. This encourages children to count back in 10s.

ASSESSMENT CHECKPOINT Throughout these questions, check that children can confidently count in 10s and do not need to resort to counting in 1s.

ANSWERS

Question ❶: Kat has 50 straws.

Question ❷ a): 40

Question ❷ b): 50

Question ❸: Children should count 10, 20, 30, 40, 50. Children should notice that the number of ten frames is the same as the tens digit.

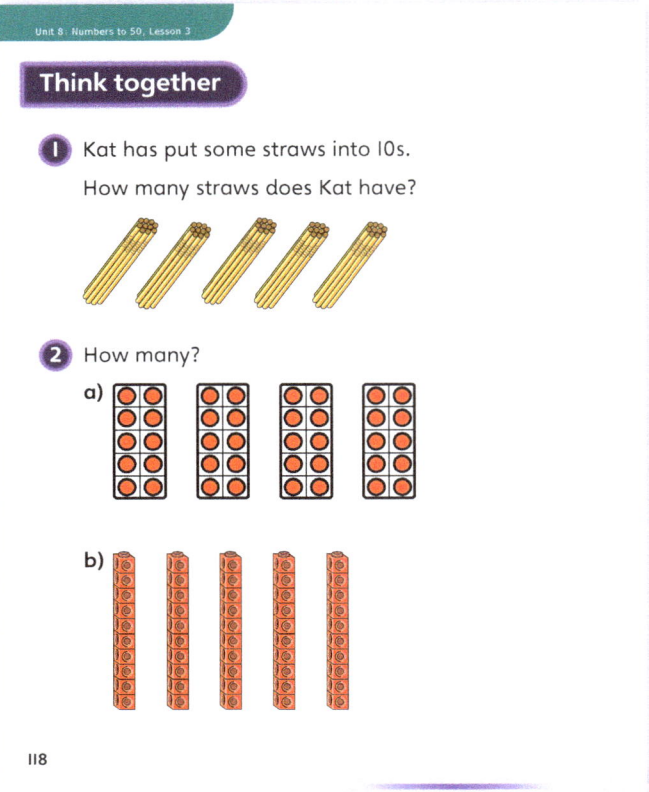

PUPIL TEXTBOOK 1B PAGE 118

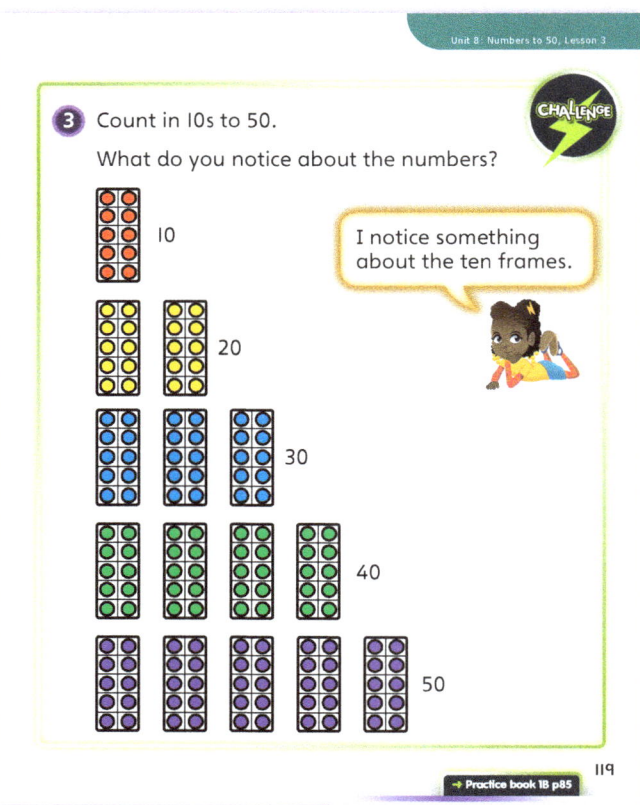

PUPIL TEXTBOOK 1B PAGE 119

Unit 8: Numbers to 50, Lesson 3

Practice

WAYS OF WORKING Independent thinking

IN FOCUS In questions ❶ and ❷, children count boxes of 10 eggs or trays of 10 biscuits. Children should recognise straight away that there are 10 eggs in each box, because they look like ten frames. In question ❸, children use common mathematical objects to count in 10s. In question ❹, children have to circle the correct number of beads. Children may circle the beads in 10s or circle 40 beads in one go as they realise they need 4 lots of 10 beads. In question ❺, children complete the count abstractly and write down the numbers.

STRENGTHEN Ask children to make 10s using bundles of straws or counters on ten frames. Count out loud with them, pointing to each bundle of straws/each ten frame as you count in 10s.

DEEPEN Ask children to make multiple representations of different 10s to 50. What do they notice about the next 10 compared to the previous 10?

ASSESSMENT CHECKPOINT All these questions assess whether children can count confidently in 10s from 10 to 50.

ANSWERS Answers for the **Practice** part of the lesson can be found in the *Power Maths* online subscription.

PUPIL PRACTICE BOOK 1B PAGE 85

PUPIL PRACTICE BOOK 1B PAGE 86

Reflect

WAYS OF WORKING Pair work

IN FOCUS Children are asked to make or draw 20. Share the different representations that children have made or drawn. This will help children see that there are many different ways to show 20.

ASSESSMENT CHECKPOINT Check that children can represent 10s using common mathematical equipment.

ANSWERS Answers for the **Reflect** part of the lesson can be found in the *Power Maths* online subscription.

After the lesson
- Check children can count in 10s from 10 to 50.

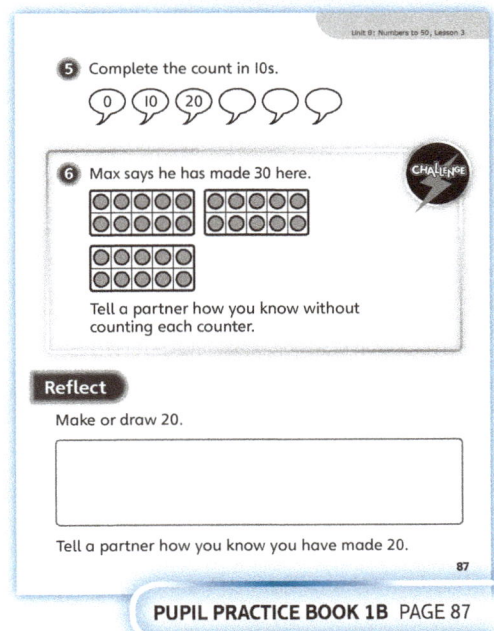

PUPIL PRACTICE BOOK 1B PAGE 87

Unit 8: Numbers to 50, Lesson 4

Count by making groups of 10s

Learning focus
In this lesson, children will learn that numbers up to 50 are made up of some 10s and some 1s through simple partitioning.

Before you teach
- Are children secure with the number names and numerals between 1 and 50?
- Do you need to address any misconceptions identified during children's work on partitioning numbers to 20 in Unit 6?

NATIONAL CURRICULUM LINKS

Year 1 Number – number and place value

Identify and represent numbers using objects and pictorial representations including the number line, and use the language of: equal to, more than, less than (fewer), most, least.

ASSESSING MASTERY

Children can express a number up to 50 as 10s and 1s (for example, they know that 43 is made up of 4 tens and 3 ones). Children can demonstrate this using a range of different representations.

COMMON MISCONCEPTIONS

Children may not realise the value of each place in a number. For example, they may think of the number 34 as 3 and 4 rather than 3 tens (30) and 4 ones (4). Ask:
- *If you have three counters and four counters, how many do you have? Is this 34?*

STRENGTHENING UNDERSTANDING

Encourage children to represent the numbers using ten frames and counters. It is important that children are encouraged to write each number in numerals so that they see the link between the number of 10s and 1s in the representation and the two digits in the written number.

GOING DEEPER

Encourage children to transfer their understanding between multiple representations. For example, ask children to represent the number 46 using groups of straws, base 10 equipment, ten frames and bead strings.

KEY LANGUAGE

In lesson: tens, ones, in total

Other language to be used by the teacher: numerals, digits, place value

STRUCTURES AND REPRESENTATIONS

Ten frames

RESOURCES

Mandatory: ten frames, counters

Optional: straws, base 10 equipment, bead strings

In the eTextbook of this lesson, you will find interactive links to a selection of teaching tools.

Quick recap

Practise as a class counting on and back in 10s from 0. Do it as a full class or go from child to child.

Discover

WAYS OF WORKING Pair work

ASK
- Question 1 a): *What could you use to represent the eggs so you can put them into groups of 10?*
- Question 1 b): *How many eggs can you see?*
- Question 1 b): *How could you count the eggs?*

IN FOCUS Question 1 a) introduces children to the idea of grouping objects into groups of 10 (which represent the 10s in a number), with some left over (which represent the 1s).

ANSWERS

Question 1 a): You can fill 3 egg boxes.

Question 1 b): There are 35 eggs in total.

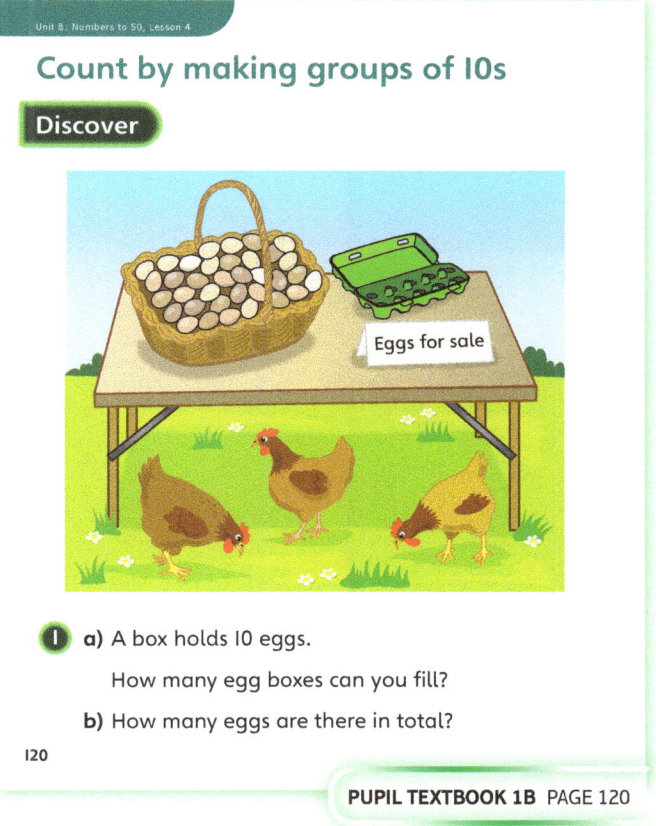

PUPIL TEXTBOOK 1B PAGE 120

Share

WAYS OF WORKING Whole class teacher led

ASK
- Question 1 a): *How could you represent the eggs using a ten frame?*
- Question 1 b): *How many eggs did you count?*
- Question 1 b): *How many 10s are there? How many 1s are there?*

IN FOCUS In question 1 a), children are encouraged to transfer their counting onto ten frames, using one counter to represent each egg. In question 1 b), children are able to see that 35 is made up of 3 tens and 5 ones.

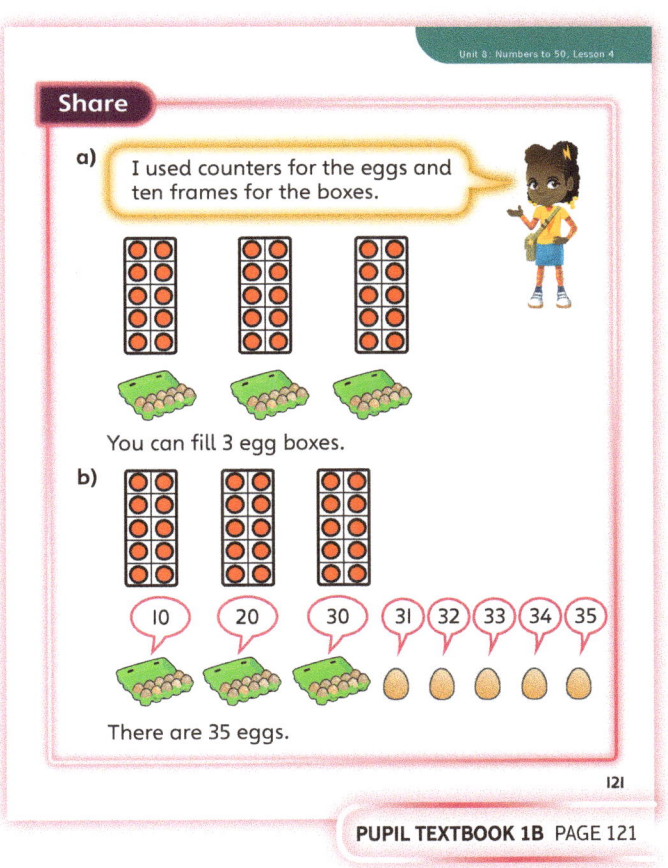

PUPIL TEXTBOOK 1B PAGE 121

159

Think together

WAYS OF WORKING Whole class teacher led (I do, We do, You do)

ASK
- Question ①: *How can you represent the eggs on the ten frames? How many eggs are there in total?*
- Question ②: *What should you count first? What should you count next? How many apples are on each ten frame? How do you know? How many apples are there in total?*
- Question ③: *How many balls are there in total? Did you have to count the 10s? Did you have to count the 1s?*

IN FOCUS Children continue consolidating their knowledge of first counting in 10s and then counting the 1s to work out how many objects there are. Question ① asks children to use counters to represent the eggs and put them on to ten frames. They make 10s and have some 1s left over. In question ②, the apples have already been grouped into 10s. Children should realise that they should first count the 10s and then the 1s. As children move onto question ③, they may be able to work out how many balls there are by subitising rather than by counting.

STRENGTHEN Help children make 10s using counters, cubes or straws. Count the 10s and then the 1s aloud with children.

DEEPEN Ask children to make 24 without counting. Look at how children approach this. Ask children to count in 1s and then count in 10s. Ask: *Which is easier? Why do you think we count the 10s first?*

ASSESSMENT CHECKPOINT By the end of this section, children should be able to count a group of objects first by counting the 10s and then counting the 1s.

ANSWERS

Question ①: There are 26 eggs in total.

Question ②: There are 37 apples.

Question ③: There are 46 balls.

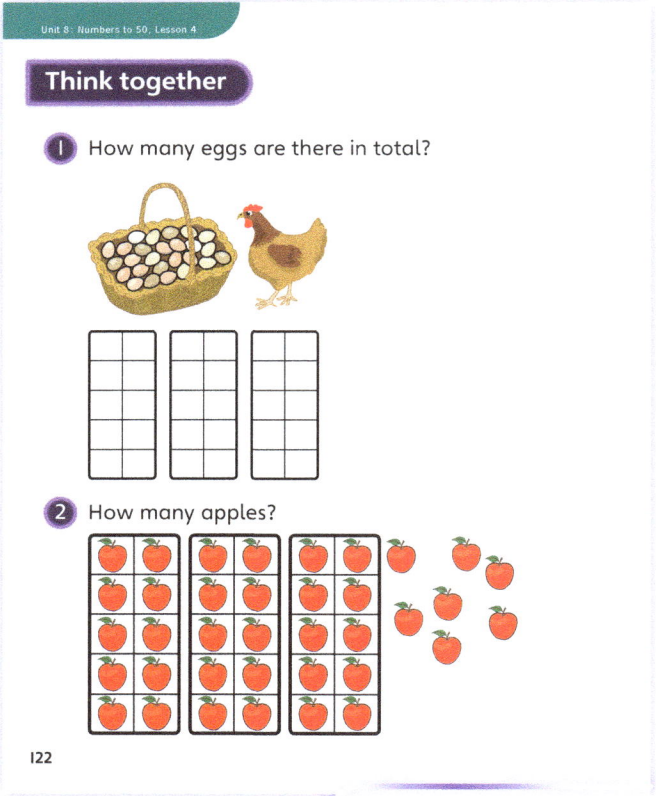

PUPIL TEXTBOOK 1B PAGE 122

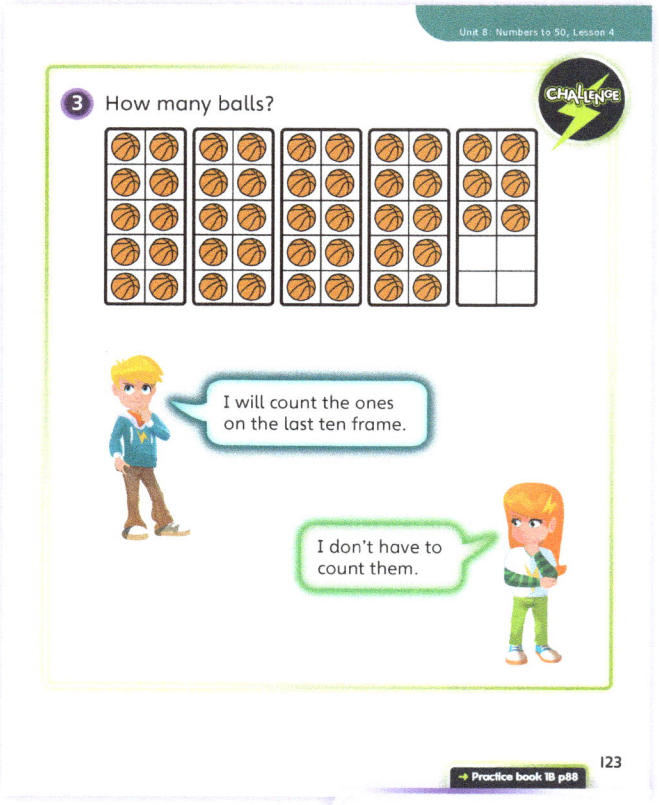

PUPIL TEXTBOOK 1B PAGE 123

160

Unit 8: Numbers to 50, Lesson 4

Practice

WAYS OF WORKING Independent thinking

IN FOCUS The main purpose of these questions is for children to count groups of objects by first counting the 10s and then counting the 1s. In question ❶, children work out the total number of eggs. They should recognise that a full egg box is like a full ten frame, and so should not need to count each egg individually when it is part of a full box. In question ❷, children count the biscuits using the same method. They should recognise that the biscuits are arranged like a ten frame. In question ❸, the steps involved in the count are now removed and children have to keep track of their count. Some may need to record their answers as they go, others will do this in their head. Encourage children to do what works for them. In question ❺, children start to see that numbers can be represented using counters on ten frames.

STRENGTHEN Work with children to count the 10s and 1s aloud. In question ❸, encourage children who struggle to keep track in their heads, to write down all the steps in the count. Point and count slowly until children have grasped the concept.

DEEPEN Ask children to make numbers using ten frames and counters. See if children can make the numbers without counting, by recognising the number of 10s needed and then the number of 1s, which they may subitise.

THINK DIFFERENTLY In question ❹, children are asked to circle a given number of footballs. This reverses the process that children have been doing in the previous questions.

ASSESSMENT CHECKPOINT Check that children can confidently count in 10s and 1s. Questions ❷ and ❸ assess the key understanding that children need to have secured by the end of the lesson.

ANSWERS Answers for the **Practice** part of the lesson can be found in the *Power Maths* online subscription.

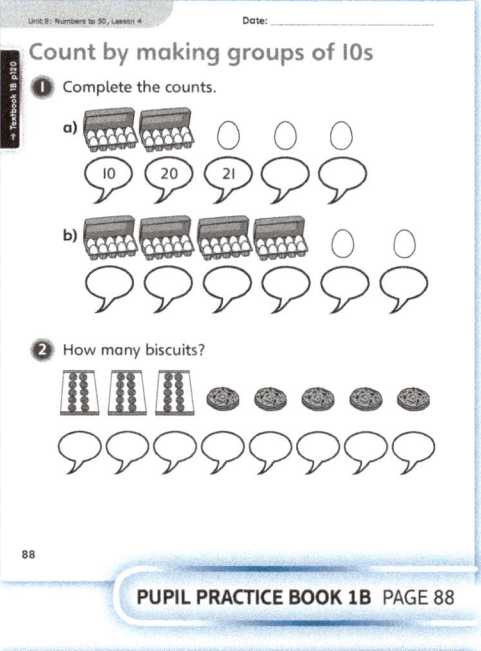

PUPIL PRACTICE BOOK 1B PAGE 88

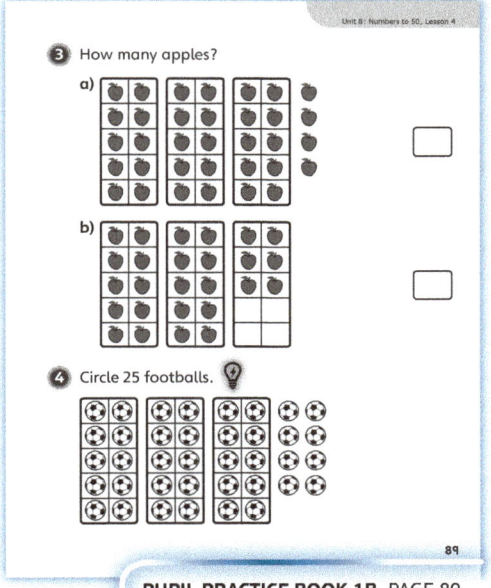

PUPIL PRACTICE BOOK 1B PAGE 89

Reflect

WAYS OF WORKING Pair work

IN FOCUS Provide children with counters, ten frames and/or cubes and ask them to make the number 37. Children start to realise they need 3 tens and 7 ones. They may not make that explicit at this stage, but this is the approach they take.

ASSESSMENT CHECKPOINT Check children can make the number 37.

ANSWERS Answers for the **Reflect** part of the lesson can be found in the *Power Maths* online subscription.

After the lesson

- Children know how to count objects by first counting the 10s and then the 1s.
- Children can make any number between 1 and 50 using concrete objects.

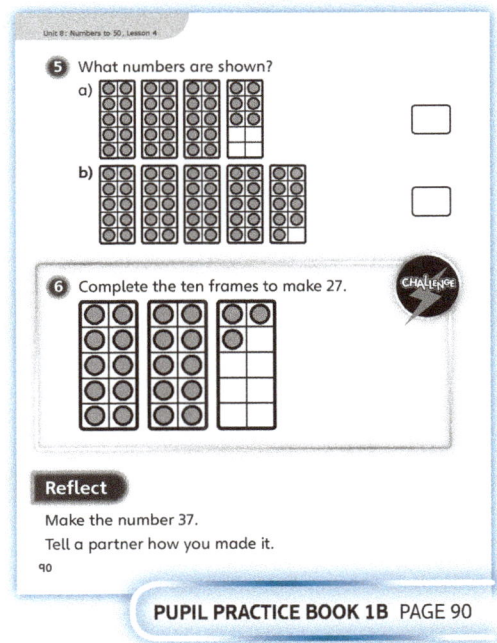

PUPIL PRACTICE BOOK 1B PAGE 90

Unit 8: Numbers to 50, Lesson 5

Groups of 10s and 1s

Learning focus
In this lesson, children will learn how to count groups of objects in 10s and 1s where the 10s cannot always be seen.

Before you teach
- Can children count confidently to 50?
- Can children work out how many objects there are by counting in 10s and 1s?
- Can children represent numbers to 50 on ten frames?

NATIONAL CURRICULUM LINKS

Year 1 Number – number and place value

Identify and represent numbers using objects and pictorial representations including the number line, and use the language of: equal to, more than, less than (fewer), most, least.

ASSESSING MASTERY

Children can count objects to 50 in 10s and 1s where the 10s cannot be seen. They can also represent numbers to 50 using different objects and manipulatives.

COMMON MISCONCEPTIONS

Children may mistake a group of 10 for 1 rather than 10. Ask:
- *How many 1s are there? Can you put the 1s into a ten frame?*

When it is not possible to see the 10 and it is written as a word, children may just think this is 1. To support children's understanding, count aloud with them slowly pointing as you count. Ask:
- *How many 1s are in a group of 10?*

STRENGTHENING UNDERSTANDING

Ask children to represent 25 counters on ten frames and explain how this shows 25. Put each group of 10 counters into a cup. Ask children how many counters there are now. To support children's understanding, count aloud with them slowly pointing as you count. Explain to children that there are still 25 counters even though you might not be able to see the 2 lots of 10 counters on the ten frame.

GOING DEEPER

Children may use a rekenrek to represent numbers to 50. They learn how to use the rekenrek and make numbers for each other to work out.

KEY LANGUAGE

In lesson: tens, ones, ten frames, count

Other language to be used by the teacher: rekenrek, groups

STRUCTURES AND REPRESENTATIONS

Ten frames, rekenreks

RESOURCES

Mandatory: counters, ten frames

Optional: straws, cubes, rekenreks, cups to hide counters in

 In the eTextbook of this lesson, you will find interactive links to a selection of teaching tools.

Quick recap

In pairs, ask children to make the number 32 using counters and ten frames. Count aloud as a class to show where the 32 has come from. Repeat with other 2-digit numbers less than 50.

Unit 8: Numbers to 50, Lesson 5

Discover

WAYS OF WORKING Pair work

ASK

- Question 1 a): *How many boxes of 10 cakes are there? How many cakes are not in boxes? How can you count the total number of cakes?*
- Question 1 b): *How many cakes are in each box? How do you know? How many boxes of 10 cakes are there? How can you count the total number of cakes? What is the same and what is different about the two questions?*

IN FOCUS The main purpose of these questions is for children to understand that the number of cakes does not change even if you count that there are 10 cakes in a box. In question 1 a), children will notice that the cakes are all countable. In question 1 b), the cakes in the boxes cannot be seen, but we know there are 10 cakes in each box. Children may use counters and ten frames to represent the number of cakes. They could place the counters in cups for question 1 b) so they cannot count them. They should realise the answer stays the same.

PRACTICAL TIPS Replicate the activity using counters and ten frames. Use cups or sheets of paper to hide the counters for question 1 b).

ANSWERS

Question 1 a): There are 25 cakes.

Question 1 b): There are still 25 cakes: 20 inside the closed boxes and 5 on the shelf.

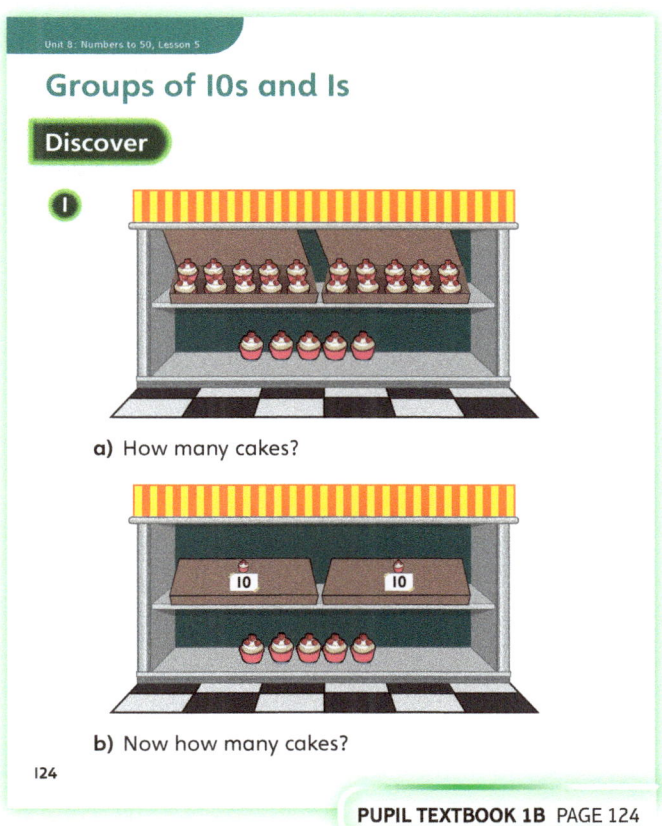

PUPIL TEXTBOOK 1B PAGE 124

Share

WAYS OF WORKING Whole class teacher led

ASK

- Question 1 a): *Count the number of cakes with me. Can you see there are 25 cakes without counting? How do you know?*
- Question 1 b): *Count the number of cakes with me. Is it easy to see that there are 25 cakes?*

IN FOCUS Count the number of cakes in questions 1 a) and 1 b) aloud with children. In question 1 a), ask them to point to the cakes in their books as they are counted. In question 1 b), you will not be able to point to each cake in the boxes, but you can still count from 1 to 10 in the first box and from 11 to 20 in the second box. Compare the two counts and ask children what was the same and what was different. Explain that the answer is the same, we just cannot see the 10 cakes in each box in question 1 b).

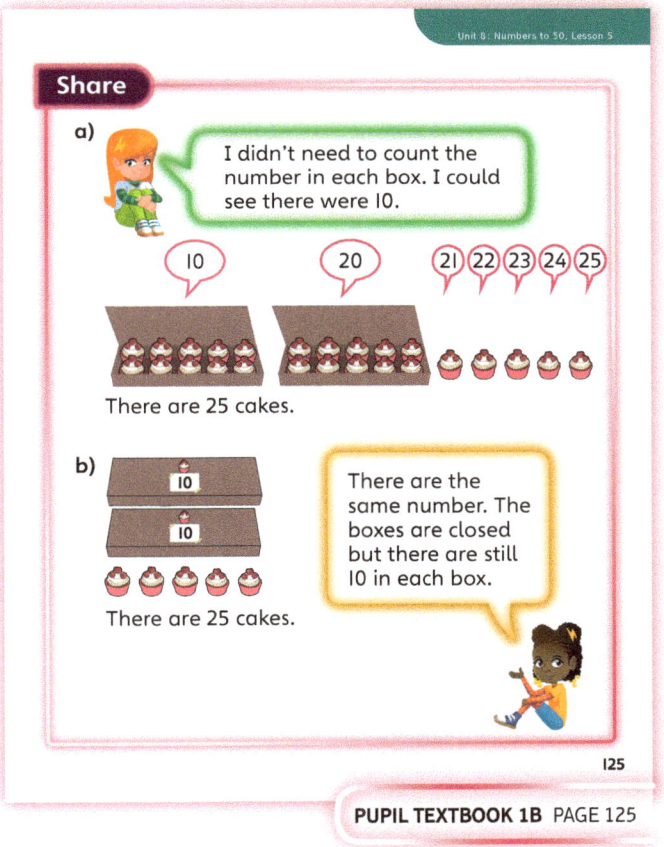

PUPIL TEXTBOOK 1B PAGE 125

Unit 8: Numbers to 50, Lesson 5

Think together

WAYS OF WORKING Whole class teacher led (I do, We do, You do)

ASK

- Question ❶: *How will you count the eggs? What will you count first? How many eggs are in each box? How do you know?*
- Question ❷: *How many marbles and pens are there? How many marbles are in each jar/pens in each box?*
- Question ❸: *Have you seen one of these before? Can you see a 10? Can you see another 10? What number do you think is being shown?*

IN FOCUS Children practise counting objects first in 10s and then 1s. This time, children cannot see the 10 individual ones. They count the 10s first and then the 1s. Some children may no longer need to count, but will instead use their knowledge of 10s and 1s (for example, they see 3 tens and 5 ones and know this is 35). Count aloud with children and point as you count to show the total in each question. Question ❸ introduces the 100 rekenrek, which some children may have seen before. Talk through where the 10s are and how to show a number on a rekenrek.

STRENGTHEN If children are still struggling with not being able to see the 10 objects, use counters to represents the eggs or pens in questions ❶ and ❷ and put them on a ten frame. Cover the ten frame and explain there are still 10, even though you cannot see them.

DEEPEN Children may use a rekenrek to represent numbers to 50. They learn how to use the rekenrek and make numbers for each other to work out.

ASSESSMENT CHECKPOINT Questions ❶ and ❷ assess that children can work out the total by counting in 10s and 1s.

ANSWERS

Question ❶: There are 42 eggs.

Question ❷ a): There are 31 marbles.

Question ❷ b): There are 28 pens.

Question ❸: Tim has made the number 34.

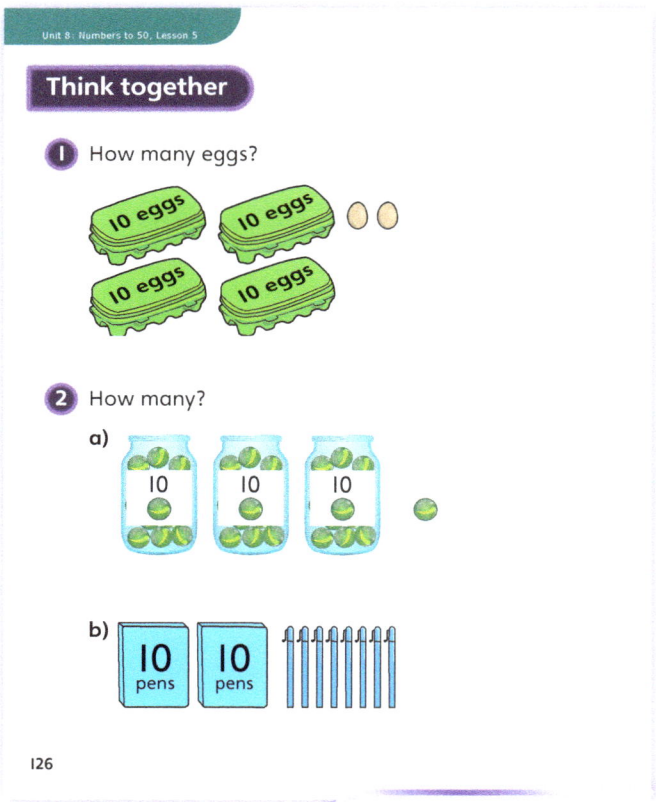

PUPIL TEXTBOOK 1B PAGE 126

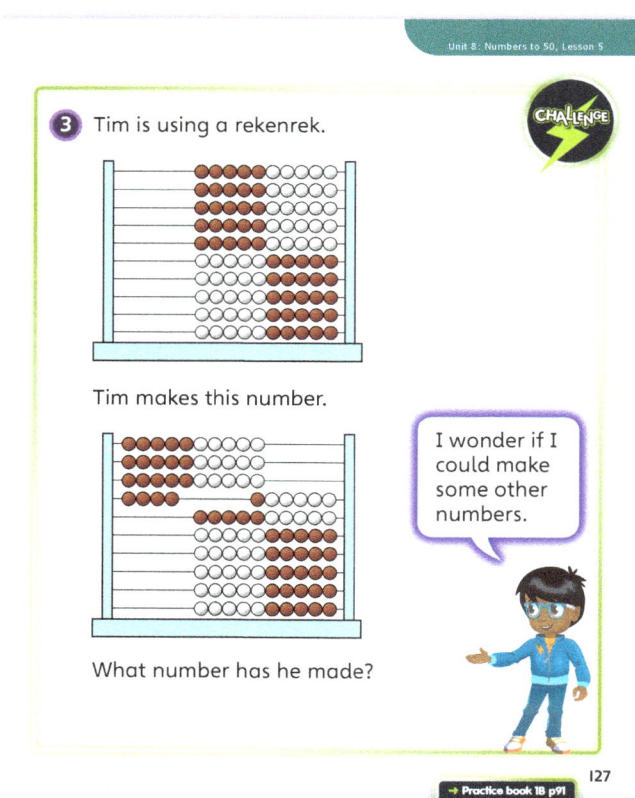

PUPIL TEXTBOOK 1B PAGE 127

164

Unit 8: Numbers to 50, Lesson 5

Practice

WAYS OF WORKING Independent thinking

IN FOCUS Questions ① and ② provide practice for children to work out how many there are of each object in total, counting in 10s and then 1s. They cannot see the 10 individual ones. Questions ③ and ④ ask children to circle the correct number of each object. They should first circle the correct number of 10s and then the correct number of 1s. Some children may be able to circle all the required 10s at once rather than individually. In question ⑤, children recognise numbers on a rekenrek. Question ⑦ provides the first example where children realise they may need to exchange 1 ten into 10 ones. They should explain to a partner that they only need some apples from the last bag.

STRENGTHEN If children are still struggling with not being able to see 10 objects, use counters to represents them and put them on a ten frame. Cover the ten frame and explain there are still 10 counters, even though you cannot see them.

DEEPEN Children may use a rekenrek to represent numbers to 50. They learn how to use the rekenrek and make numbers for each other to work out.

ASSESSMENT CHECKPOINT Children should display a level of confidence with questions ① to ④ as this will show understanding of counting objects using 10s and 1s.

ANSWERS Answers for the **Practice** part of the lesson can be found in the *Power Maths* online subscription.

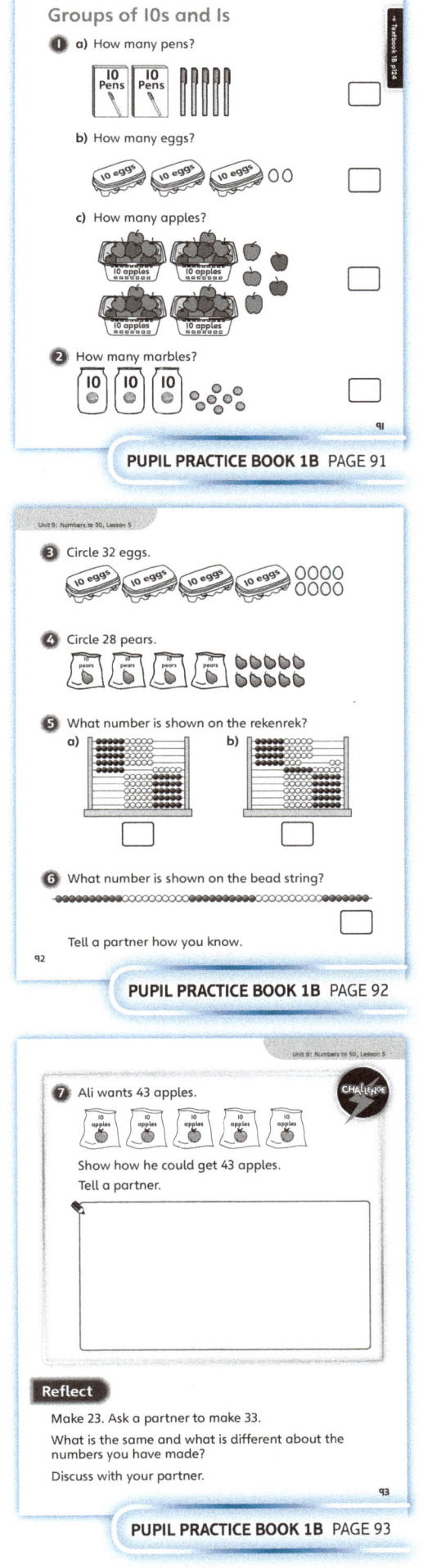

PUPIL PRACTICE BOOK 1B PAGE 91

PUPIL PRACTICE BOOK 1B PAGE 92

PUPIL PRACTICE BOOK 1B PAGE 93

Reflect

WAYS OF WORKING Pair work

IN FOCUS In pairs, children represent the numbers 23 and 33. Some children may use counters and ten frames or their fingers. Others may draw bags to represent 10, similar to what they have been doing in this lesson. Discuss with children what they notice about the two numbers. They should see that 33 has 1 more 10.

ASSESSMENT CHECKPOINT Check that children can represent 23 and 33 by drawing or making the numbers.

ANSWERS Answers for the **Reflect** part of the lesson can be found in the *Power Maths* online subscription.

After the lesson

- Children can count a group of objects in 10s without being able to see every object.

Unit 8: Numbers to 50, Lesson 6

Partition into 10s and 1s

Learning focus
In this lesson, children will explore different ways to represent numbers to 50, using objects such as counters and cubes, and mathematical models such as the part-whole model. They will see that a number is made up some 10s and 1s.

Before you teach
- Can children count a group of objects made up of 10s and 1s?
- Can children select the correct number of 10s and 1s to make a number of objects, such as for 32 apples, they select three bags of 10 apples and 2 apples?

NATIONAL CURRICULUM LINKS

Year 1 Number – number and place value

Identify and represent numbers using objects and pictorial representations including the number line, and use the language of: equal to, more than, less than (fewer), most, least.

ASSESSING MASTERY

Children can represent numbers up to 50 in different ways, including using the part-whole model. Children are beginning to understand how the position of a digit in a number impacts its value. They will see that a number is made up of some 10s and 1s.

COMMON MISCONCEPTIONS

Children may not realise the value of each place in a number. For example, they may think of the number 34 as 3 and 4 rather than 3 tens (30) and 4 ones (4). Ask:
- *If you have three counters and four counters, how many do you have? Is this 34?*

If they use base 10 equipment, children may use 10 ten rods to represent the number 10, rather than 1 ten rod. To address this, draw children's attention to the number of 1s cubes that make up the 10s rod. Ask:
- *How many 10s are in this number? How many 1s are in each 10 rod? How many 10 rods do you need?*

STRENGTHENING UNDERSTANDING

Use a range of different representations to represent 2-digit numbers up to 50, including bead strings, base 10 equipment and ten frames. Draw attention to how these representations show groups of 10 and groups of 1.

GOING DEEPER

Encourage children to explain why, for example, 47 is made up of 4 tens (40) and 7 ones (7) rather than a 4 and a 7. Ask children to show different representations of the number and ask children what is the same and what is different.

KEY LANGUAGE

In lesson: tens, ones, partition

Other language to be used by the teacher: part-whole model

STRUCTURES AND REPRESENTATIONS

Ten frames, part-whole models

RESOURCES

Mandatory: counters, blank ten frames

Optional: base 10 equipment, rekenreks, cubes

 In the eTextbook of this lesson, you will find interactive links to a selection of teaching tools.

Quick recap

Ask children to make a number from 11 to 19. Ask children how the number is made up, for example, 17 = 10 + 7 or 1 ten and 7 ones. Repeat for other numbers.

Discover

WAYS OF WORKING Pair work

ASK
- Question 1 a): *How can you show 34? Can you show it with counters and ten frames/cubes/base 10 equipment? How many 10s can you see? How many 1s can you see?*
- Question 1 b): *What is the whole? What do you think the parts could be?*

IN FOCUS In question 1 a), children use mathematical equipment to make the number 34. They should be able to see the 10s and the 1s, building on their work from previous lessons. Children should be able to see and point to where the 3 tens and 4 ones are. They should consider how they might then complete the part-whole model in question 1 b). Watch out for children who put 3 and 4 as opposed to 30 and 4 as the parts.

PRACTICAL TIPS Have plenty of mathematical equipment available, such as counters, ten frames, base 10 equipment and cubes that children can use to support their learning.

ANSWERS

Question 1 a): Children should show 34 on ten frames or using bead strings or by making 3 towers of 10 cubes and 4 single cubes.

Question 1 b):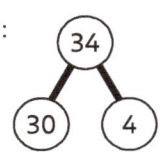

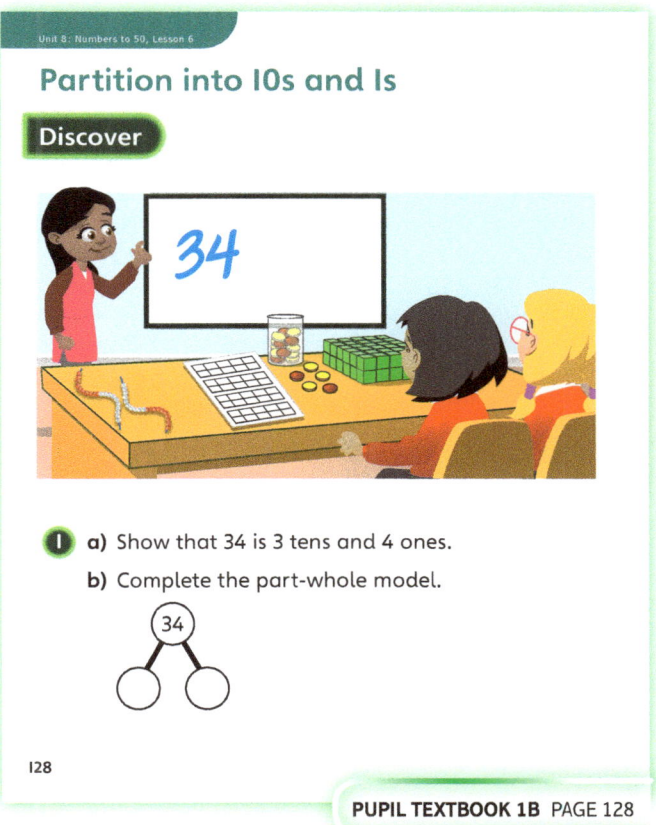

PUPIL TEXTBOOK 1B PAGE 128

Share

WAYS OF WORKING Whole class teacher led

ASK
- Question 1 a): *Can you see 30? How many 10s is this? Can you see 4 ones? Can you see why 34 is the same as 3 tens and 4 ones?*
- Question 1 b): *What is the whole? Where is the whole? What do you think the 30 and 4 represent?*

IN FOCUS In question 1 a), show children why 34 is made up of 3 tens and 4 ones. They should be able to see the ten frames showing 3 tens and 4 ones. Ask children if they can see the same on the bead string. Share with the class any other representations that children may have made. Remind children that this is called partitioning a number. Earlier in the year, the numbers that children experienced just had 1 ten and now these numbers have more than 1 ten. In question 1 b), children see that they can use a part-whole model to show how a number is partitioned. One way to do this is to partition into 10s and 1s.

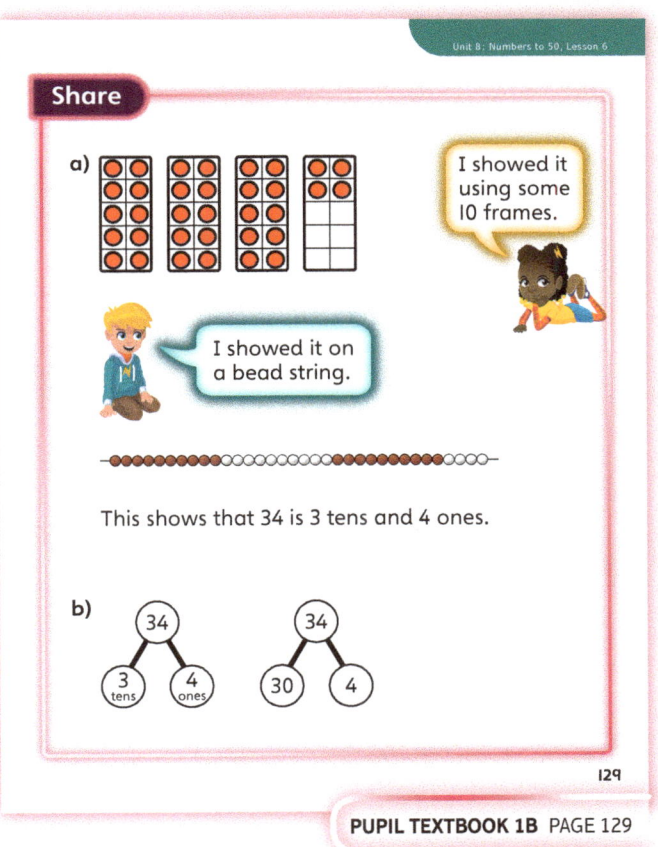

PUPIL TEXTBOOK 1B PAGE 129

Unit 8: Numbers to 50, Lesson 6

Think together

WAYS OF WORKING Whole class teacher led (I do, We do, You do)

ASK

- Question ❶: *How many 10s are there in each number? How many 1s are there in each number? What numbers are represented? Can you draw a part-whole model for each number?*
- Question ❷: *What numbers are missing? Can you make the numbers using mathematical equipment?*

IN FOCUS In question ❶, children write out the partition of 2-digit numbers. The ten frames help them to see the number of 10s and 1s that make up a number. In question ❷, children work on just the abstract number. Some children may need to make the number using 10s and 1s to be able to complete the part-whole models. Question ❸ brings together the last few lessons of work and asks children to write down all they know about a 2-digit number. They represent the number, make the number, partition it and write it in different ways.

STRENGTHEN For questions ❷ and ❸, encourage children to represent the numbers using counters and ten frames. This will help them see how many 10s and 1s make up each number.

DEEPEN Ask children to make different numbers using base 10 equipment.

ASSESSMENT CHECKPOINT Children should be able to draw a part-whole model for a 2-digit number between 20 and 50. Question ❷ helps you to check this. Give more examples if necessary to check children have fully grasped this concept.

ANSWERS

Question ❶ a): 2 tens and 3 ones is 23.

Question ❶ b): 3 tens and 7 ones is 37.

Question ❷ a): The parts are 20 and 6.

Question ❷ b): The whole is 45.

Question ❸ a): 24

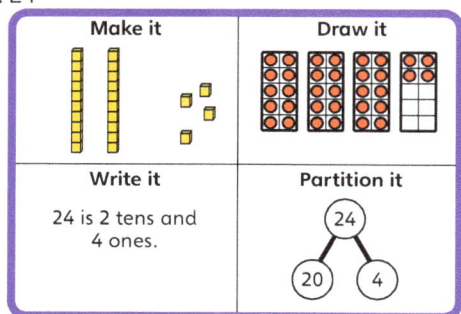

Question ❸ b): 42

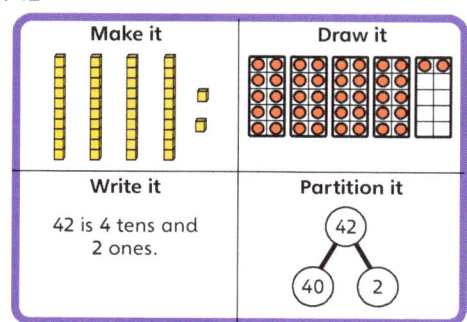

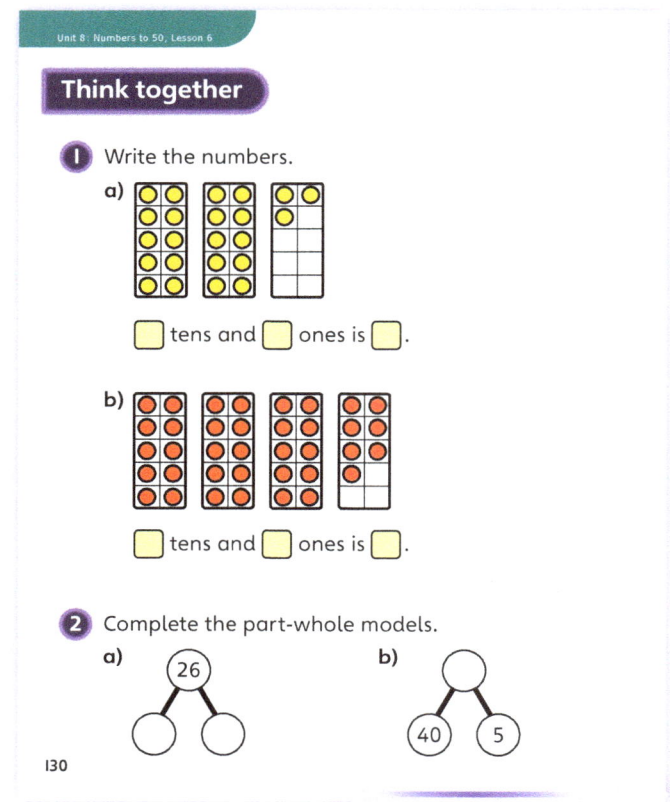

PUPIL TEXTBOOK 1B PAGE 130

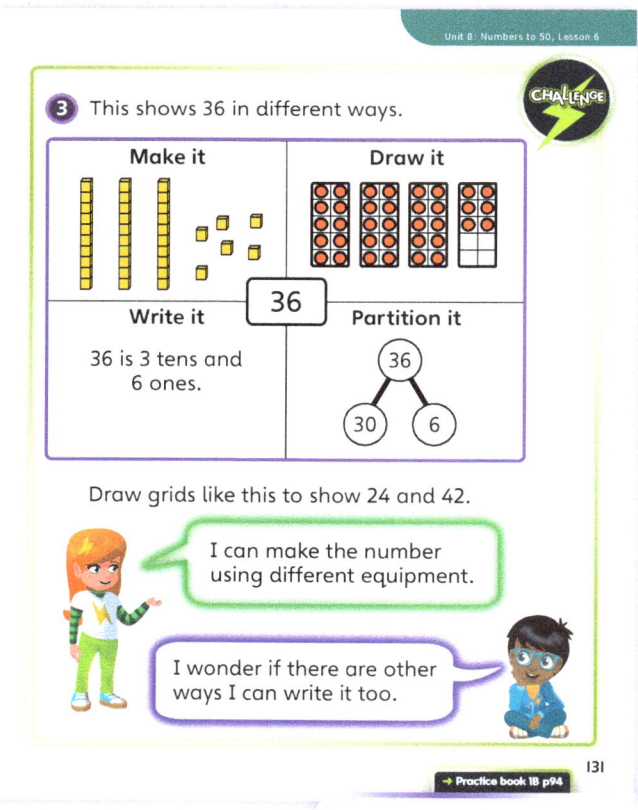

PUPIL TEXTBOOK 1B PAGE 131

Unit 8: Numbers to 50, Lesson 6

Practice

WAYS OF WORKING Independent thinking

IN FOCUS These questions provide opportunities for children to make connections between different concrete representations and the part-whole model, showing the partitioning of numbers up to 50. This helps to ensure that children have a secure conceptual understanding of a 2-digit number as being made up of 10s and 1s. Questions ❶ and ❷ ask children to write the number of 10s and 1s that make up a number. Question ❸ asks children to complete part-whole models for 2-digit numbers. Question ❹ provides an opportunity for children to show a more abstract understanding of the concept. This will demonstrate whether children understand how 10s and 1s make up a 2-digit number.

STRENGTHEN For questions ❸ and ❹, encourage children to represent the numbers using counters and ten frames. This will help them see how many 10s and 1s make up each number.

DEEPEN To deepen understanding, ask children to explain their reasoning, thinking about their representations. For example, ask: *Why does this part-whole model show 49? Why have you chosen to represent 24 using base 10 equipment?*

ASSESSMENT CHECKPOINT Questions ❶ and ❷ check that children can partition a 2-digit number into 10s and 1s. Question ❸ checks that children can represent this partition in a part-whole model and, finally, question ❹ demonstrates children's abstract understanding.

ANSWERS Answers for the **Practice** part of the lesson can be found in the *Power Maths* online subscription.

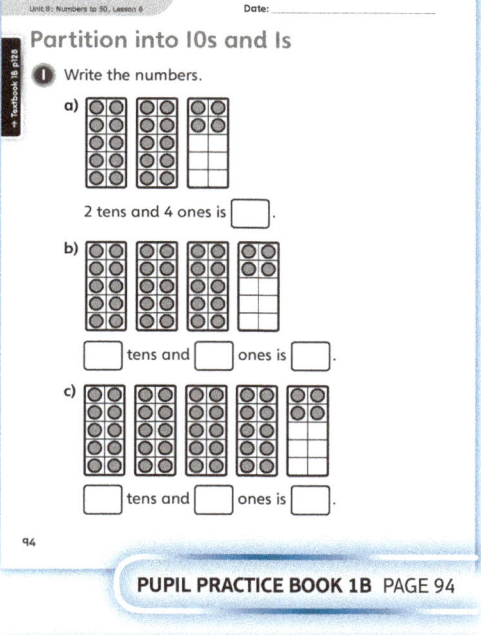

PUPIL PRACTICE BOOK 1B PAGE 94

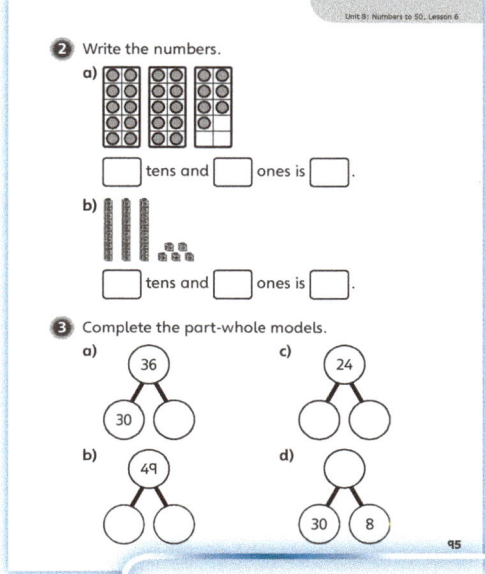

PUPIL PRACTICE BOOK 1B PAGE 95

Reflect

WAYS OF WORKING Independent thinking

IN FOCUS This question mirrors question ❸ in the **Think together** section of the main lesson. Children consolidate their knowledge of the last few lessons, representing a 2-digit number in different ways and showing how they can partition it.

ASSESSMENT CHECKPOINT Check that children can represent and make the given number. Check they can also partition it and draw a part-whole model.

ANSWERS Answers for the **Reflect** part of the lesson can be found in the *Power Maths* online subscription.

After the lesson

- Children can partition a 2-digit number less than 50 into 10s and 1s.
- Children can draw a part-whole model for a number less than 50.

PUPIL PRACTICE BOOK 1B PAGE 96

169

Unit 8: Numbers to 50, Lesson 7

One more, one less

Learning focus

In this lesson, children will find one more and one less of any number to 50.

Before you teach

- Can children find one more and one less than a number to 20?
- Can children confidently count from 1 to 50?

NATIONAL CURRICULUM LINKS

Year 1 Number – number and place value

Given a number, identify one more and one less.

ASSESSING MASTERY

Children can find one more and one less of numbers less than 50 with and without supporting equipment.

COMMON MISCONCEPTIONS

Children may struggle when they have to complete sentences such as '34 is one more than ☐'. Children see the words 'one more' and a number and think they have to find one more than that number. Ensure children are confident completing sentences such as 'X is one more/one less than Y'. Ask:
- *Can you use equipment to make 34? Remove 1 to make 33. Have you found one less or one more than 34? What is 34 one more than?*

STRENGTHENING UNDERSTANDING

To help children work out one more and one less abstractly, ask them to make the numbers using counters and ten frames. Ask them to count the counters in 10s and 1s. Then ask them to either add a counter or remove a counter, depending on whether they are working out one more or one less. Ask them to count the resulting number of counters.

GOING DEEPER

Ask children to answer questions such as '34 is one more than ☐' or 'what is one more than 27?' This will demonstrate a deeper understanding.

KEY LANGUAGE

In lesson: one more, one less, less than

Other language to be used by the teacher: count

STRUCTURES AND REPRESENTATIONS

Counters, ten frames

RESOURCES

Mandatory: counters, ten frames

Optional: cubes, rekenreks

 In the eTextbook of this lesson, you will find interactive links to a selection of teaching tools.

Quick recap

Play a counting game with children where they select a number between 1 and 50 randomly and, as a class, count up to the selected number and back to 0 in 1s.

Unit 8: Numbers to 50, Lesson 7

Discover

WAYS OF WORKING Pair work

ASK

- Question 1 a): *How many awards are on the wall? How can you count them quickly?*
- Question 1 b): *When Amy puts the one in her hand on the wall, how many awards are there now on the wall?*

IN FOCUS In question 1 a), children should use their knowledge of counting in 10s and 1s to find the number of awards already on the wall, as opposed to counting in 1s. In question 1 b), children are shown concrete objects to work out one more than a given number. Encourage children to use the language of 'one more' by asking them what they think Amy has done (i.e. she has won one more award). Children may represent the objects using counters and ten frames or other equipment.

PRACTICAL TIPS Instead of using the context, you might use other equipment in the classroom that is to hand or class points (for example, table X wins one more point – how many do they have now?)

ANSWERS

Question 1 a): There are 34 awards on the wall.

Question 1 b): One more than 34 is 35.

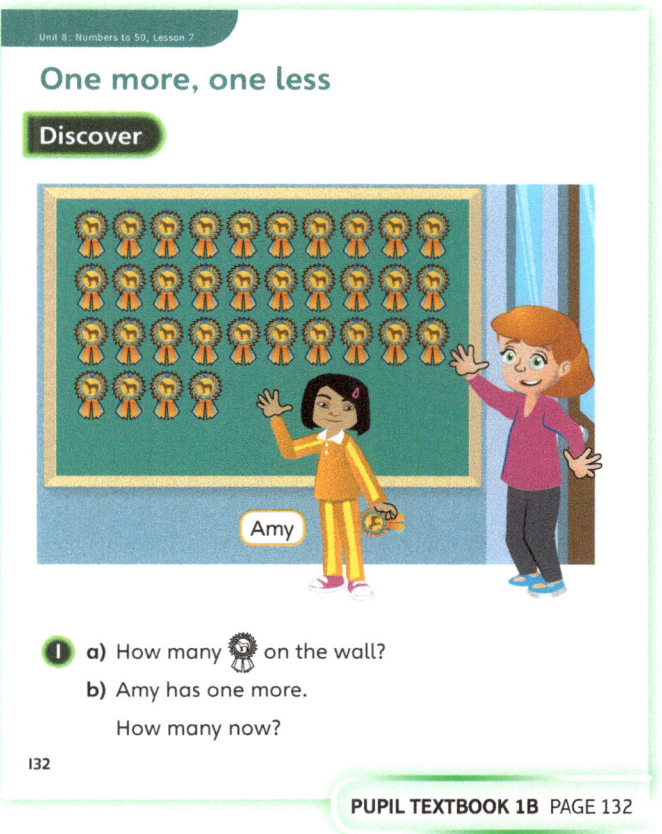

PUPIL TEXTBOOK 1B PAGE 132

Share

WAYS OF WORKING Whole class teacher led

ASK

- Question 1 a): *Can you see the 30? Can you see the 4? How did you count them quickly?*
- Question 1 b): *How many awards are there now? How do you know? Did you need to count them again or could you work it out another way?*

IN FOCUS For question 1 a), remind children of how they can count the number of awards in 10s and 1s. As children count, get them to point to the 10s and 1s so it is clear what they are counting. In question 1 b), use the language 'one more'. Discuss with children whether they need to count the awards again or whether they know from their counting that one more than 34 is 35. Ask children answer in out full sentences such as 'One more than 34 is 35.'

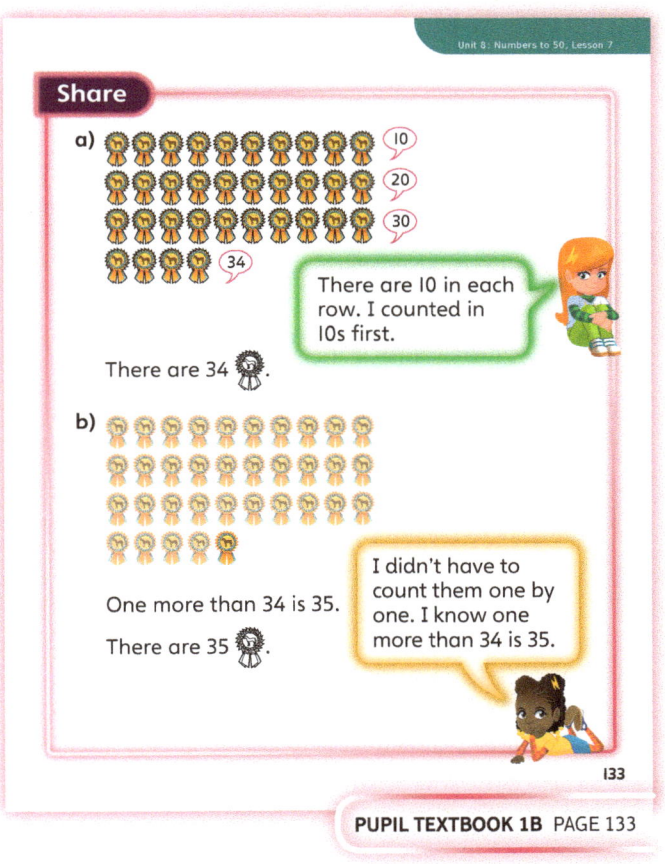

PUPIL TEXTBOOK 1B PAGE 133

171

Think together

WAYS OF WORKING Whole class teacher led (I do, We do, You do)

ASK
- Question ❶: *How many coconuts are there in total? What happens if you knock one coconut down? Why is this one less?*
- Question ❷: *How many full ten frames are there? How many counters are in the last ten frame? How does this show 43? How can you show one more than 43? How can you show one less than 43?*
- Question ❸: *What do the images show? What do you think you have to work out?*

IN FOCUS Question ❶ allows children to practise their skills from the **Discover** questions. This time, they have to work out one less. Question ❷ provides children with a strategy of finding one more and one less using a ten frame and counters. It is important that children realise that they can add 1 counter to find one more and remove 1 counter to find one less. In question ❸, children are presented with abstract diagrams to find one more and one less. They should realise that they can use their knowledge of the count from 1 to 50 or a number line to find one more and one less. Discuss with children how they do not need to use counters and ten frames all the time.

STRENGTHEN In question ❸, some children may need to use ten frames and counters to help them work out one more and one less. Explain clearly what the diagrams are asking them to find. Strengthen understanding further by asking children to work through plenty of examples involving one more and one less, helping them to develop their recall of the number count.

DEEPEN Ask children to answer questions such as '34 is one more than ☐' or 'what is one more than 27?' This will demonstrate a deeper understanding.

ASSESSMENT CHECKPOINT Children need to be confident in finding one more and one less using their knowledge of the number count to help them, rather than always relying on mathematical equipment. Question ❸ helps assess whether children have secured this understanding.

ANSWERS

Question ❶ a): There are 28 coconuts.

Question ❶ b): When 1 coconut is knocked over, there are 27 coconuts left.

Question ❷ a): One more than 43 is 44.

Question ❷ b): One less than 43 is 42.

Question ❸: One more than 37 is 38.
One more than 49 is 50.
One less than 25 is 24.

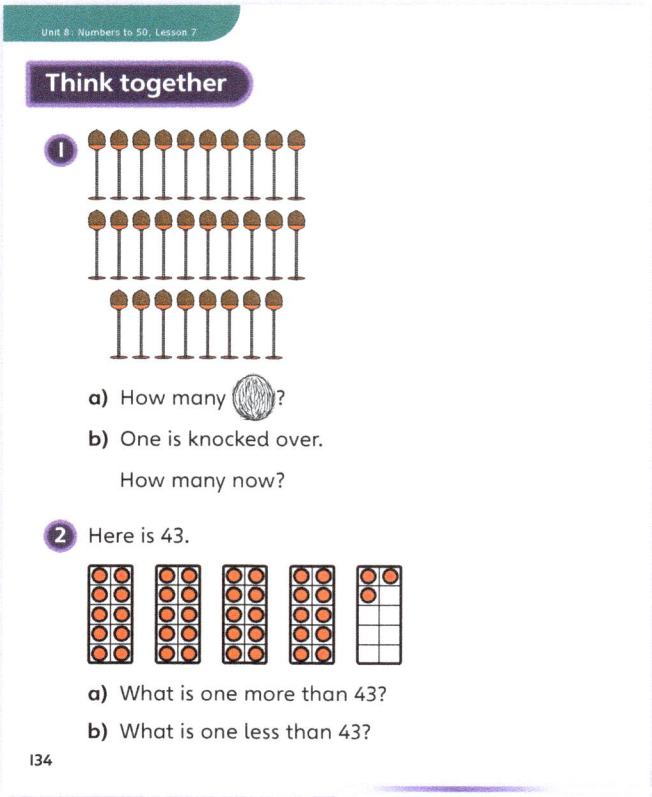

PUPIL TEXTBOOK 1B PAGE 134

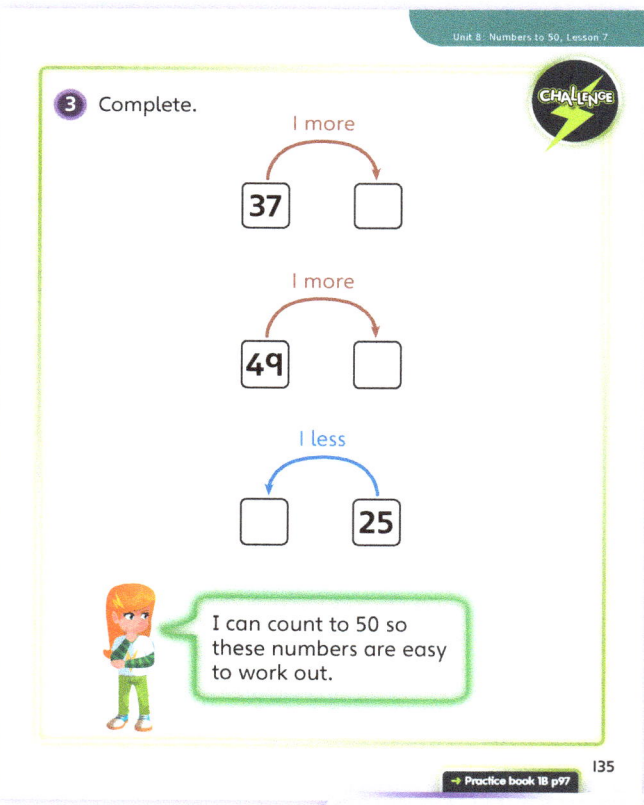

PUPIL TEXTBOOK 1B PAGE 135

Unit 8: Numbers to 50, Lesson 7

Practice

WAYS OF WORKING Independent thinking

IN FOCUS In questions ① and ②, children draw or cross out counters to work out one more or one less. In questions ③ and ④, children use their knowledge of counting numbers to work out one more and one less. Encourage them to start the count from the closest 10. Question ⑤ shows the diagrams that children met in question ③ of the **Think together** section. Question ⑥ provides an example where children have to find one more and one less within the same diagram. Children may not have met a diagram like this before. Ask them to explain what they think it means first.

STRENGTHEN In questions ③ and ④, some children may need to use ten frames and counters to help them work out one more and one less. Explain clearly what the diagrams are asking them to find. Strengthen understanding by asking children to work through plenty of examples of one more and one less, which will help them develop their recall of the number count.

DEEPEN Question ⑦ provides abstract sentences for children to complete, which initially they may find a bit trickier. Encourage them to write their own number sentences involving one more and one less.

ASSESSMENT CHECKPOINT Questions ③ to ⑤ are key questions that children need to answer correctly and confidently. They assess their understanding of using the counting sequence to find one more and one less. This indicates whether children are fluent in this concept.

ANSWERS Answers for the **Practice** part of the lesson can be found in the *Power Maths* online subscription.

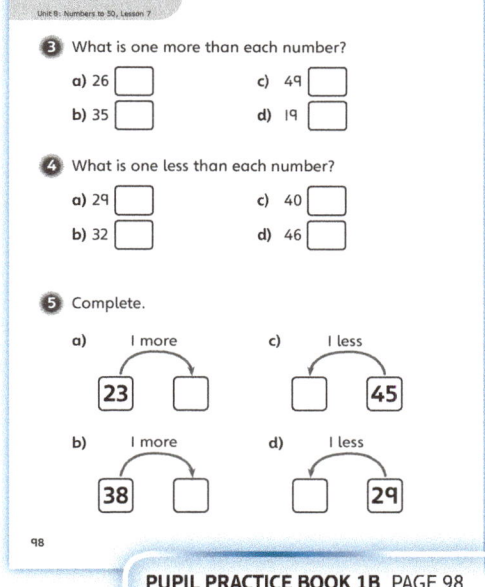

PUPIL PRACTICE BOOK 1B PAGE 97

PUPIL PRACTICE BOOK 1B PAGE 98

Reflect

WAYS OF WORKING Pair work

IN FOCUS Children pick their own number and put it in the box in the middle. They work out one more and one less than their chosen number. You may need to work through an example as a class first. Ask children to complete their own for a number of their choice and then ask a partner to check.

ASSESSMENT CHECKPOINT Check that children can find one more and one less than their number.

ANSWERS Answers for the **Reflect** part of the lesson can be found in the *Power Maths* online subscription.

After the lesson

- Children can find one more than a given number between 0 and 49.
- Children can find one less than a given number between 1 and 50.

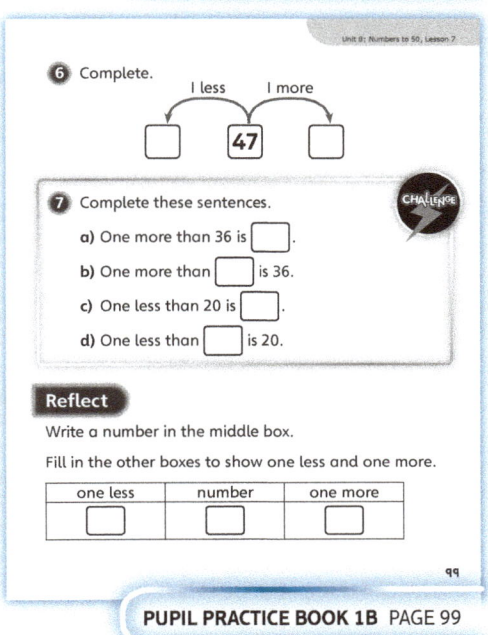

PUPIL PRACTICE BOOK 1B PAGE 99

Unit 8: Numbers to 50

End of unit check

Don't forget the unit assessment grid in your *Power Maths* online subscription.

WAYS OF WORKING Group work adult led

IN FOCUS Question 1 checks that children know the correct count to 50. Question 2 checks that children can confidently recognise groups of 10 and then count in 10s. Question 3 provides a common representation of a 2-digit number using ten frames and counters. Question 4 asks children to apply their knowledge they learnt on finding one more and one less.

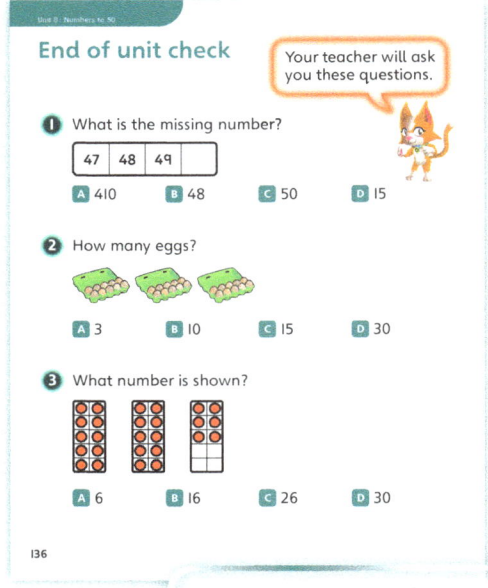

PUPIL TEXTBOOK 1B PAGE 136

Think!

WAYS OF WORKING Pair work

IN FOCUS This question allows children to explore many of the key concepts from this unit, including counting, representing numbers and partitioning 2-digit numbers into 10s and 1s. They could base their drawing on their concrete representation of 32. If children struggle with this, support them by helping them to arrange their representation, or use ten frames and counters, on a blank part-whole model. Children will also have the opportunity to revisit key vocabulary and representations, including 'tens' and 'ones', and the part-whole model. Encourage children to think through or discuss the number and what they know about it before writing their answer in **My journal**.

ANSWERS AND COMMENTARY Children who master this unit are ones that can count confidently and fluently from 1 to 50. They can count objects in 1s using one-to-one correspondence. They count more efficiently by counting in 10s and 1s and realise, by grouping in 10s, it makes it easier to count. They are able to use this knowledge to partition a number into 10s and 1s and represent this partition using different mathematical equipment and also in a ten frame. Finally, they use their knowledge of counting numbers to work out one more and one less without the aid of equipment.

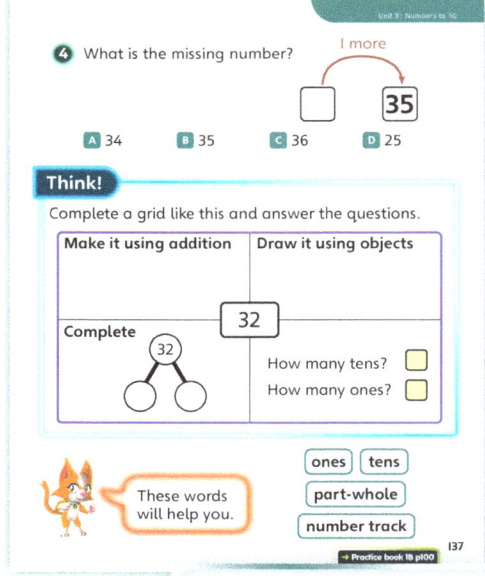

PUPIL TEXTBOOK 1B PAGE 137

Q	A	WRONG ANSWERS AND MISCONCEPTIONS	STRENGTHENING UNDERSTANDING
1	C	A shows a common misconception when children do not move to the next 10.	When counting on ten frames and other objects, count aloud with children, asking them to point as they count. Use a range of representations, including ten frames and counters to help children represent numbers and identify the number of 10s and 1s that make up a number.
2	D	A indicates that children have counted the number of egg boxes rather than the number of eggs. B suggests that children have just counted the number of eggs in one egg box.	
3	C	Some children may think that D is the correct answer as they can see three ten frames. Explain that the third ten frame is not full.	
4	A	C suggests that children have misinterpreted the question as what is one more than 35. Explain that the arrow is coming from the blank box. To go back, they need to work out one less than 35.	

Unit 8: Numbers to 50

My journal

WAYS OF WORKING Independent thinking

ANSWERS AND COMMENTARY Children should choose a clear, concrete method (for example, base 10 equipment, place value counters or ten frames) to represent 32, showing that there are 3 tens and 2 ones.

They should draw a clear representation of 32, ideally using the part-whole model.

They should identify that there are 3 tens and 2 ones in 32.

Children who struggle with this task need further support before comparing measurements in Unit 9 and Unit 10, and need further work on comparing numbers in Unit 14. Encourage children to use a number line to help them order and compare numbers.

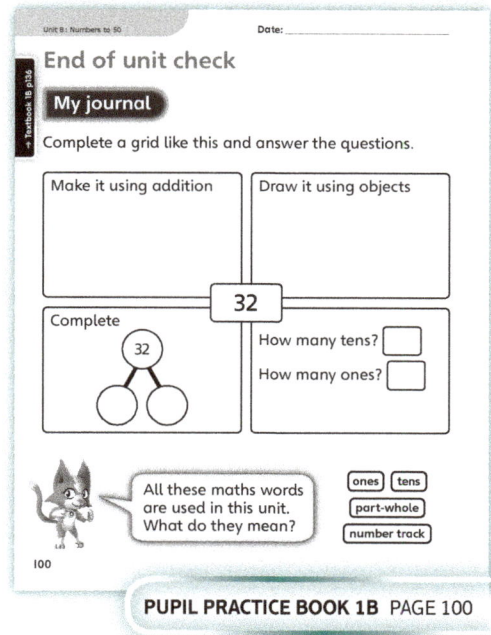

PUPIL PRACTICE BOOK 1B PAGE 100

Power check

WAYS OF WORKING Independent thinking

ASK
- What do you know now that you did not know at the start of this unit?
- How confident do you feel?

Power play

WAYS OF WORKING Pair work or small groups

IN FOCUS Use this **Power play** to see if children can use a range of representations to make and identify numbers between 20 and 50.

ANSWERS AND COMMENTARY Children should be able to accurately make numbers between 20 and 50 using a range of representations, some of which (for example, base 10 equipment, ten frames, bead strings) will show the number of 10s and 1s that make up the number. If children can complete this **Power play** using a range of representations, it indicates that they are secure with the place value and relative size of numbers between 20 and 50. If they are not able to complete this **Power play**, they are likely to need further support.

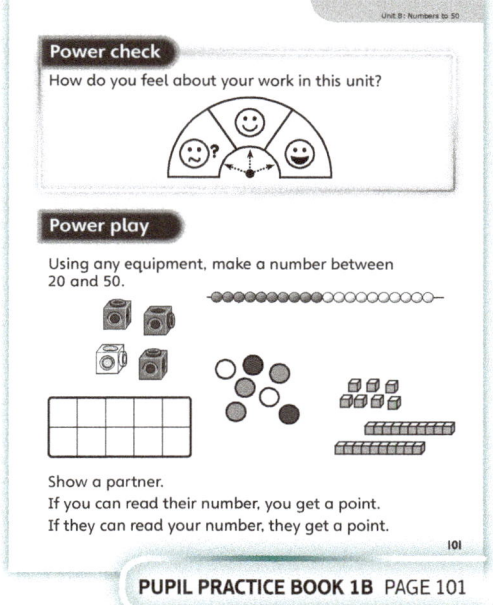

PUPIL PRACTICE BOOK 1B PAGE 101

After the unit

- How did the unit assessment go?
- Do children recognise the place value in numbers to 50?
- What would you do differently next time?

Strengthen and **Deepen** activities for this unit can be found in the *Power Maths* online subscription.

Unit 9
Introducing length and height

> **Mastery Expert tip!** 'Wherever possible, I provided practical opportunities for children to measure using a range of non-standard and standard units of length. This really helped children to understand how to align measuring tools and objects to be measured as well as how to use scales to determine length.'

Don't forget to watch the Unit 9 video!

WHY THIS UNIT IS IMPORTANT

This unit focuses on comparing and measuring the height and length of objects using non-standard and standard units of measure. Children will learn how to accurately compare and measure and will understand the importance of aligning starting points. Children will draw on their knowledge of number, particularly ordering and comparing numbers. Children will also learn the relationship between number lines and scales on a ruler and use this understanding to calculate differences in length. Children will use key language such as longer, longest, shorter, shortest, taller and tallest when comparing length and height.

WHERE THIS UNIT FITS

→ Unit 8: Numbers to 50
→ **Unit 9: Introducing length and height**
→ Unit 10: Introducing mass and capacity

This unit builds on children's previous work on number, and children will apply their understanding of number within practical contexts relating to height and length. Children's previous use of number lines will help them understand how to use scales to calculate the difference between two or more lengths.

Before they start this unit, it is expected that children:
- know how to compare and order numbers to 50
- understand that subtraction means finding the difference between two quantities
- know how to add and subtract using a number line.

ASSESSING MASTERY

Children who have mastered this unit will be able to align objects to make direct comparisons of length and height. Children will be able to suggest suitable non-standard units to measure a range of distances and will use these units accurately. Children will use a ruler with precision to measure length in centimetres and they will be able to solve 1- and 2-step problems asking them to find the total or the difference in the context of length and height.

COMMON MISCONCEPTIONS	STRENGTHENING UNDERSTANDING	GOING DEEPER
Children may fail to align the starting points of objects to make direct comparisons or fail to align the starting point of the measuring device with the object being measured.	When using non-standard units or directly comparing the length of two objects, encourage children to use a flat surface such as a wall, floor or book to help them align the starting points.	Ask children to create their own problems involving length and height. Can children develop a problem that involves addition and one that involves subtraction? How about a problem that requires both addition and subtraction? Children can then test their problems on a partner and discuss how to make their problems easier or more challenging.
Children may struggle to identify the calculations needed in order to solve a contextualised problem.	Where possible, allow children to carry out the problems practically and encourage them to talk about their thinking and what they are doing. This will help them to see the required mathematics.	

Unit 9: Introducing length and height

UNIT STARTER PAGES

Use Astrid's comment to help introduce this unit. Discuss with children how the multilink cubes and the ruler are used to measure and compare length. Look at the vocabulary introduced by Flo and discuss with children what they understand these terms to mean.

STRUCTURES AND REPRESENTATIONS

Ruler:
In this unit, children are introduced to measuring lengths in cm. It is best if they use rulers where the only scale is in cm to avoid any confusion with other units (such as inches).

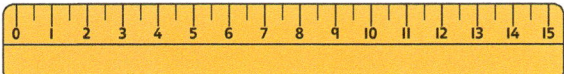

Number line:
Children can recognise how a ruler is similar to a number line. This can help them to apply skills such as counting on and back, or adding and subtracting, to length and measurement.

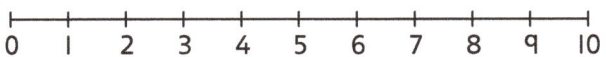

KEY LANGUAGE

There is some key language that children will need to know as part of the learning in this unit:
- long, longer, longest
- short, shorter, shortest
- tall, taller, tallest
- wide, wider, widest
- thin, thinner, thinnest
- length, height
- compare, comparison
- measure
- distance
- unit, non-standard units
- ruler
- centimetre (cm)
- total
- difference

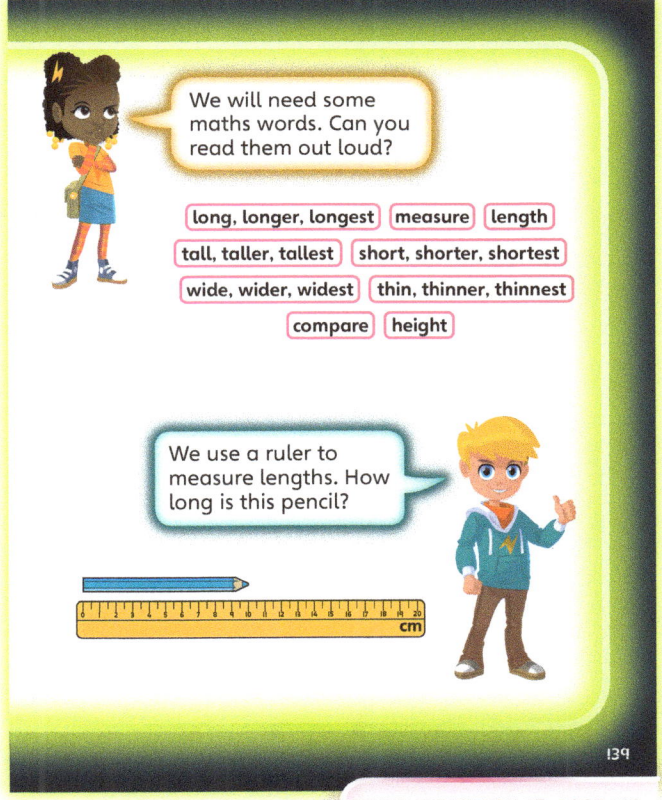

PUPIL TEXTBOOK 1B PAGE 138

PUPIL TEXTBOOK 1B PAGE 139

Unit 9: Introducing length and height, Lesson 1

Compare lengths and heights

Learning focus
In this lesson, children will compare lengths and heights of objects and make accurate comparisons.

Before you teach
- What experience do children have comparing lengths and heights?
- Do you have physical objects to compare to help support children's understanding?
- How will you introduce, explain and reinforce the key vocabulary?

NATIONAL CURRICULUM LINKS

Year 1 Measurement

Compare, describe and solve practical problems for: lengths and heights (for example, long/short, longer/shorter, tall/short, double/half).

ASSESSING MASTERY

Children can compare and order two or more objects based on their height or length. Children can compare length accurately by ensuring that both objects start at the same point.

COMMON MISCONCEPTIONS

When comparing objects, children may believe that one object is taller when it is raised up. Ask:
- *How do you know which one is taller? Why does this one look taller? Why does this one look shorter?*

STRENGTHENING UNDERSTANDING

Provide children with physical objects to compare height and length. Encourage children to place the items on a table or the floor when comparing height and against a wall or book when comparing length.

GOING DEEPER

Ask children to compare more than two objects and put them in order from shortest to longest and vice versa. Can they find an object which is twice as long or half as long as another?

KEY LANGUAGE

In lesson: length, **heights**, **longer**, **longest**, **shorter**, shortest, **taller**, **tallest**, **wider**, **thinner**, **compare**

Other language used by the teacher: comparison, accurate

RESOURCES

Mandatory: a variety of classroom objects to compare heights and lengths

Optional: three skipping ropes of different lengths

 In the eTextbook of this lesson, you will find interactive links to a selection of teaching tools.

Quick recap

Find out how familiar children are with the words 'longer', 'taller' and 'shorter'.

Show children an object, such as a pencil, and ask them to find something that is longer or shorter than the pencil. This will allow you to assess whether children are familiar with these words already. Repeat for other objects.

Unit 9: Introducing length and height, Lesson 1

Discover

WAYS OF WORKING Pair work

ASK
- Question 1 a): *What is height? How can you compare height?*
- Question 1 b): *Why does Emily think she is the tallest?*

IN FOCUS Questions 1 a) and b) introduce children to comparing lengths and heights in a real-life context. Encourage them to discuss which lengths and heights they can compare from the image, noticing that as Emily is standing on the ladder, we cannot compare her height to the heights of the others.

PRACTICAL TIPS Ask children to stand back-to-back and then to compare their heights, encouraging them to say the sentences out loud in both possible ways. For example, 'Myra is taller than Anya' and 'Anya is shorter than Myra.'

ANSWERS

Question 1 a): Anya is shorter than Myra.
Myra is taller than Anya.

Question 1 b): Emily is higher, but she might not be taller. They would have to stand on the same level to check.

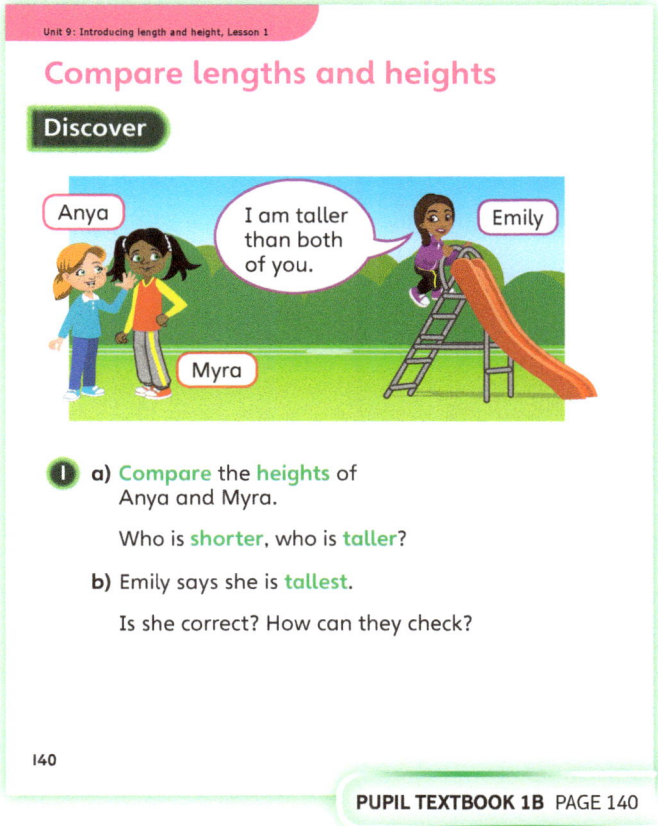

PUPIL TEXTBOOK 1B PAGE 140

Share

WAYS OF WORKING Whole class teacher led

ASK
- Question 1 a): *Why is it important that the children are lined up next to each other?*
- Question 1 a): *How can you tell that Anya is shorter than Myra? How can you tell that Myra is taller than Anya?*
- Question 1 b): *Why does Emily being higher not mean she is taller?*
- Question 1 b): *What does 'tallest' mean? What does 'shortest' mean?*

IN FOCUS In question 1 a), encourage children to discuss why, if Anya is shorter than Myra, we can also say this the other way round as Myra is taller than Anya. In question 1 b), children are introduced to the words 'shortest' and 'tallest' when comparing heights of more than two children. Repeat this activity in class with groups of three children, asking children to describe the heights out loud, ensuring they use the correct words, for example 'shortest' rather than 'shorter'.

PUPIL TEXTBOOK 1B PAGE 141

Think together

WAYS OF WORKING Whole class teacher led (I do, We do, You do)

ASK
- Question ❶: *Which skipping rope is the longest? How do you know? Which is the shortest? Why is it important that all of the skipping ropes are lined up?*
- Question ❶: *Which skipping rope is longer, A or B? How do you know?*
- Question ❷: *Is there more than one possible answer? How do you know?*

IN FOCUS In question ❶, children use their learning from the **Discover** and **Share** sections in the context of length. Encourage them to recognise that all of the skipping ropes are lined up and discuss the importance of this. In question ❷, children should recognise that any flower taller than the middle one would be a correct answer.

STRENGTHEN Provide children with objects that they can compare the lengths and heights of practically. Give children three pieces of string and ask them to put them in length order, then use their learning from this to support with question ❶. Similarly, figures or other classroom objects can be used to support question ❷.

DEEPEN In question ❸, children are stretched to also consider the words 'wider' and 'thinner'. They should recognise that just because an object is longer, it does not necessarily mean that it is also wider. Challenge them to find twigs or leaves where the taller one is also wider.

ASSESSMENT CHECKPOINT Use question ❶ to assess whether children can compare lengths using correct mathematical language. Use question ❷ to assess whether children understand how to order items based on their height.

ANSWERS

Question ❶: C is the longest.
B is the shortest.
A is longer than B.
A is shorter than C.
B is shorter than A and C.
C is longer than A and B.

Question ❷: Children should draw a flower in the third box that is taller than the flower in the middle box.

Question ❸: Children should find two real leaves to compare. They must use the comparatives longer/shorter and wider/thinner when comparing two items. The superlatives longest/shortest and widest/thinnest are only used when comparing 3 or more items.

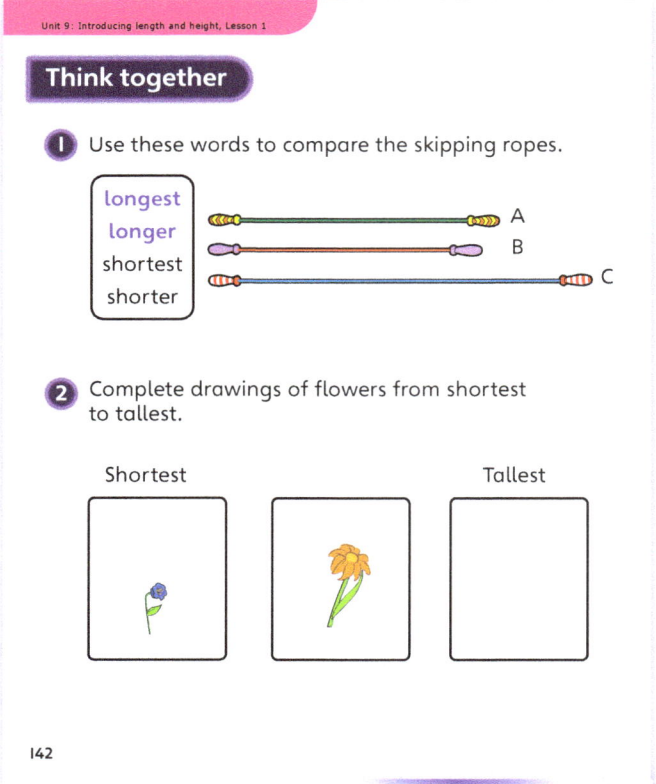

PUPIL TEXTBOOK 1B PAGE 142

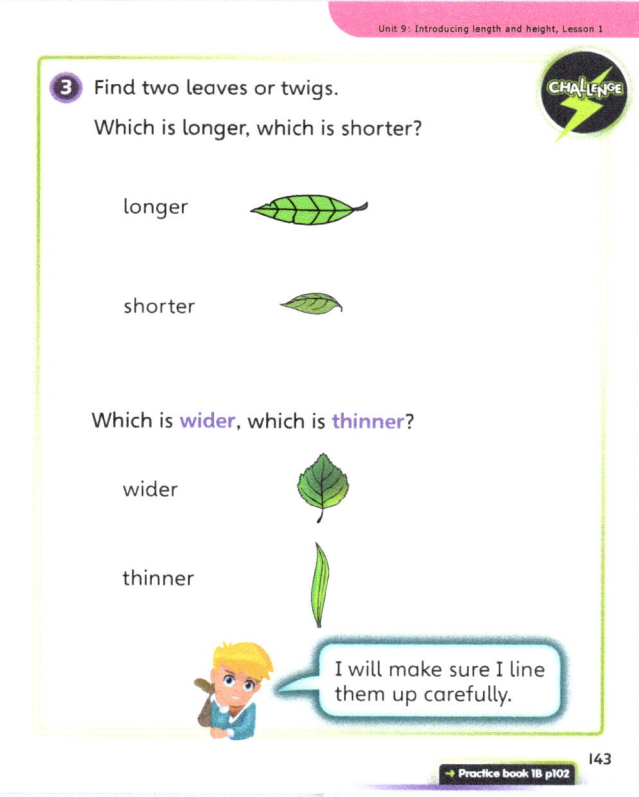

PUPIL TEXTBOOK 1B PAGE 143

Unit 9: Introducing length and height, Lesson 1

Practice

WAYS OF WORKING Independent thinking

IN FOCUS These questions provide an opportunity for children to independently practise comparing lengths and heights. Questions ❶ and ❷ focus on heights and ordering items from shortest to tallest, whilst questions ❸ and ❹ focus on lengths and comparing lengths using language such as 'longer' and 'shorter'.

STRENGTHEN Allow children to compare lengths and heights practically using objects available in the classroom. This will help to strengthen their understanding.

DEEPEN In question ❺, children compare pairs of buildings from a group. They need to identify the buildings they are comparing then choose the correct word. Encourage children to choose three objects from the classroom and compare their lengths or heights using three different sentences. Ask children to compare their answers with those of a partner. Ask: *Are your sentences the same?*

THINK DIFFERENTLY Question ❹ asks children to find some concrete objects and compare their lengths to the length of a given rectangle. As well as identifying whether their objects are longer or shorter than the rectangle, they are also asked whether any of their objects are the same length as the rectangle.

ASSESSMENT CHECKPOINT Use questions ❶ and ❷ to assess whether children can compare heights using the words 'shortest', 'shorter', 'taller' and 'tallest'. Use questions ❸ and ❹ to assess children's confidence in comparing lengths using 'shortest', 'shorter', 'longer' and 'longest'.

ANSWERS Answers for the **Practice** part of the lesson can be found in the *Power Maths* online subscription.

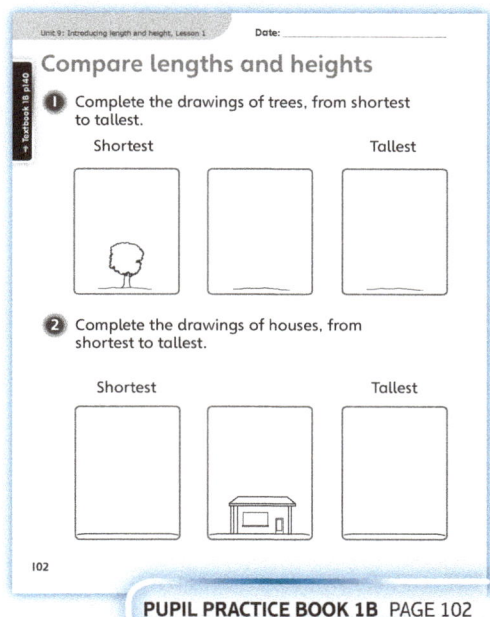

PUPIL PRACTICE BOOK 1B PAGE 102

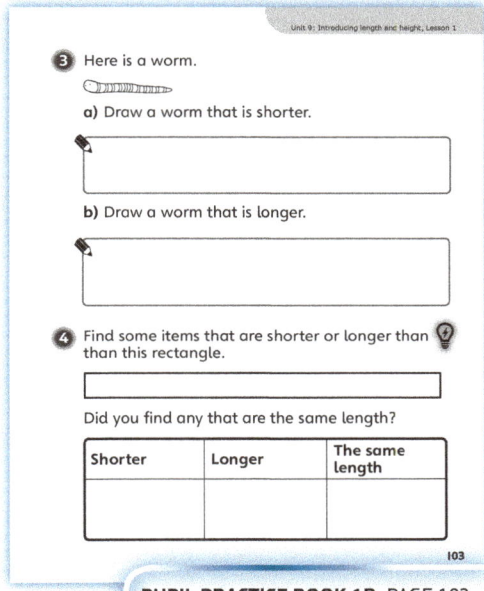

PUPIL PRACTICE BOOK 1B PAGE 103

Reflect

WAYS OF WORKING Independent thinking

IN FOCUS The **Reflect** question challenges children to demonstrate their understanding of comparing lengths by asking them to draw a longer leaf.

ASSESSMENT CHECKPOINT This question will allow you to assess whether children have understood the concept of comparing lengths. By drawing a longer leaf, children will demonstrate that they understand the term longer in the context of length.

ANSWERS Answers for the **Reflect** part of the lesson can be found in the *Power Maths* online subscription.

After the lesson ⏸

- Were children secure in how to align objects in order to compare their lengths accurately?
- Do children recognise the difference between tallest and longest?

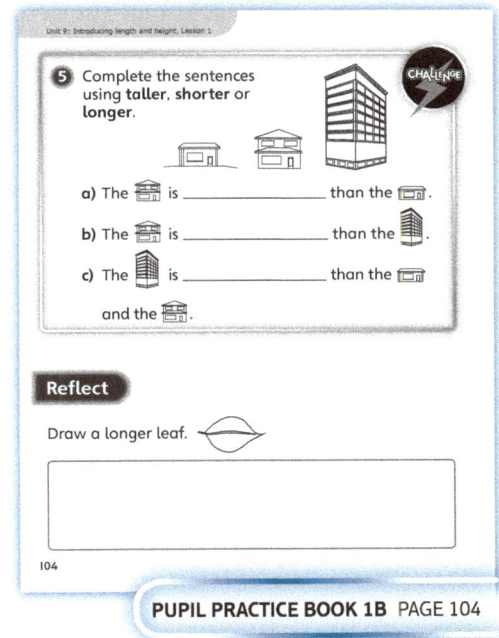

PUPIL PRACTICE BOOK 1B PAGE 104

Unit 9: Introducing length and height, Lesson 2

Measure length (non-standard units of measure)

Learning focus

In this lesson, children will measure objects using non-standard units. This links measurement directly with number and counting.

Before you teach

- Are children secure in comparing heights and lengths of objects?
- Are children secure in counting objects up to 50?
- What practical opportunities can you provide for children in this lesson?

NATIONAL CURRICULUM LINKS

Year 1 Measurement

Measure and begin to record the following: lengths and heights.

ASSESSING MASTERY

Children can use cubes as non-standard units of measurement to accurately measure length and height of objects. Children can compare the length of objects using the measurements they have made.

COMMON MISCONCEPTIONS

Children may not understand the importance of aligning the end of the cubes with the end of the object. Ask:
- *How can you use these cubes to measure the length of this object? Where do you place the first cube?*

STRENGTHENING UNDERSTANDING

Provide children with cubes and objects to measure. Prompt children to place the cubes and the objects on the table or floor to measure height or against a wall or book to measure length.

GOING DEEPER

Children may be ready to calculate and compare objects with measures of length. Ask: *How much taller/longer/shorter? What is the total length of these two objects?* Ensure that this is followed by asking children how they arrived at their answer.

KEY LANGUAGE

In lesson: length, height, **long**, **tall**, longer, shorter, taller, longest, shortest

Other language used by the teacher: compare, comparison, accurate

RESOURCES

Mandatory: multilink cubes, objects to measure

 In the eTextbook of this lesson, you will find interactive links to a selection of teaching tools.

Quick recap

Show children two objects. Ask them which is longer/taller and how they know. Then ask which is shorter and how they know. Repeat for other objects. Introduce a third object and ask: *Which is longest/tallest/shortest? How do you know?*

Unit 9: Introducing length and height, Lesson 2

Discover

WAYS OF WORKING Pair work

ASK

- Question 1 a): *What can you tell me about the cubes in the picture? Is it important that the cubes are the same colour?*
- Question 1 b): *How are the cubes being used to measure the car and the fire engine?*

IN FOCUS Questions 1 a) and b) introduce the concept of measuring the length of an object using non-standard units. It is important to highlight to children that the cubes are all the same size and this is necessary in order to use the cubes to measure and compare lengths.

ANSWERS

Question 1 a): The fire engine is longer than the car.

Question 1 b): The car is 4 cubes long.
The fire engine is 8 cubes long.

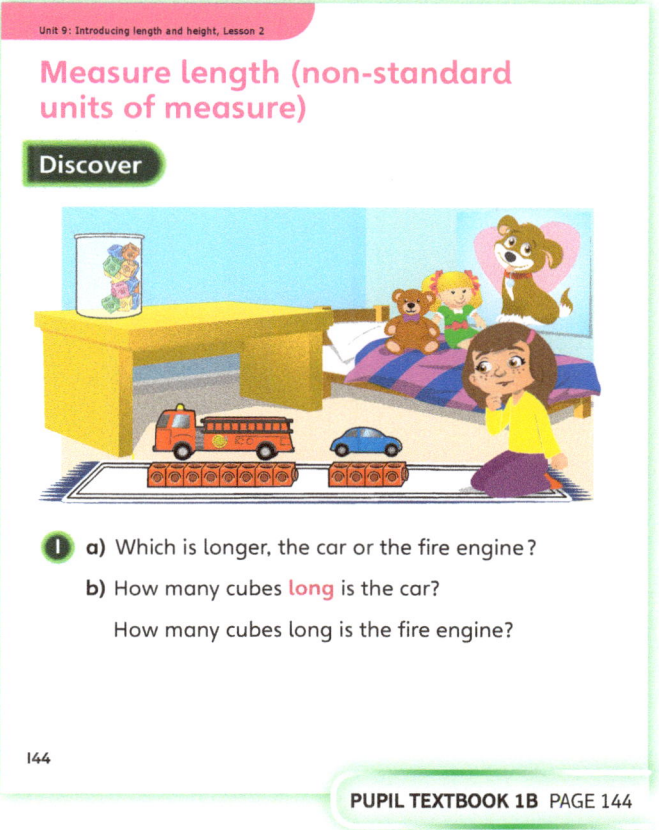

PUPIL TEXTBOOK 1B PAGE 144

Share

WAYS OF WORKING Whole class teacher led

ASK

- Question 1 a): *Can you make a stick of cubes the same length as the car and fire engine put together?*
- Question 1 a): *How do the cubes show you that the fire engine is longer than the car? What can you tell me about the length of the fire engine compared to the length of the car?*

IN FOCUS In question 1 b), Ash's comment introduces the idea that once you have measured the objects, you no longer have to place the objects side by side in order to compare them. Children can see that the fire engine is longer by either comparing the length of the stick of cubes or by using their knowledge that 8 is larger than 4.

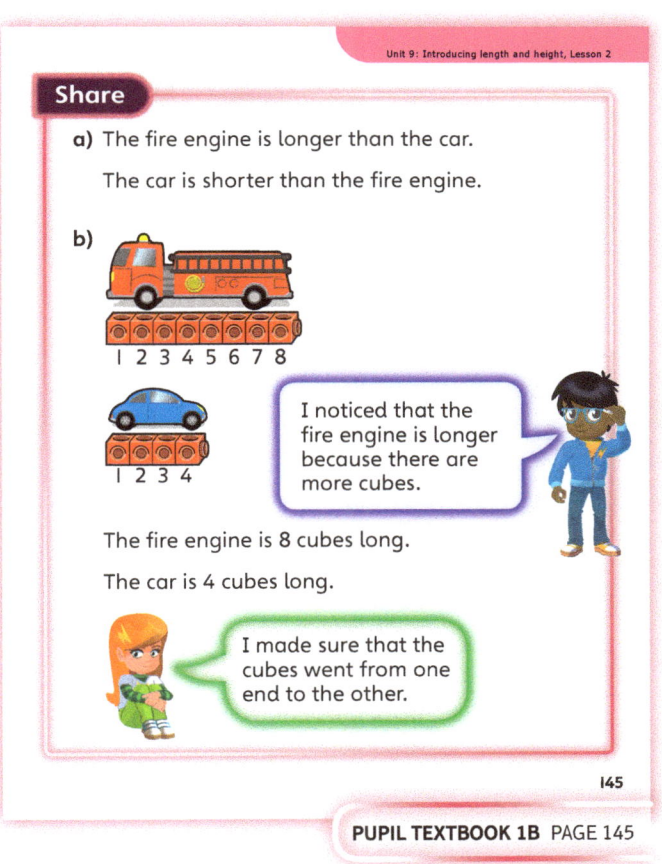

PUPIL TEXTBOOK 1B PAGE 145

Unit 9: Introducing length and height, Lesson 2

Think together

WAYS OF WORKING Whole class teacher led (I do, We do, You do)

ASK
- Question ①: *How many cubes tall is the teddy bear? How many cubes tall is the doll? Which is taller – the teddy bear or the doll?*
- Question ①: *How can you find out how much taller the doll is than the teddy bear?*
- Question ②: *Is there a quick way of counting the cubes?*
- Question ③: *How can you be sure that you are measuring accurately?*

IN FOCUS In question ①, children need to carry out a subtraction in order to work out how much shorter the teddy bear is than the doll. Ensure children are aware of the link between their previous work on subtraction by asking them to place the sticks of cubes side by side and count how many more, or to subtract the 5 from the 8 and count what is left. Question ③ requires children to make estimations of length in order to find objects that are more than, less than or exactly 10 cubes long.

STRENGTHEN Give children the opportunity to measure practically the length and height of objects using cubes. Support them in ensuring that the ends of the objects and the cubes are aligned for accurate measurement.

DEEPEN Ask children to choose an object, estimate its length and then measure the actual length using cubes. Ask: *How close was your estimate?* Ask children to find pairs of objects with the combined length of 10 or 15 cubes.

ASSESSMENT CHECKPOINT Questions ① and ② assess whether children are able to use cubes as a non-standard measure to determine the height or length of objects. Question ① will also determine whether children can identify what they need to do in order to calculate how much shorter one object is than another. In question ③, the objects children select will highlight whether they have an understanding of applying a value to height or length.

ANSWERS

Question ①: The doll is 8 cubes tall.
The teddy bear is 5 cubes tall.
The teddy bear is shorter than the doll.

Question ②: The bed is 30 cubes long.
The chair is 15 cubes tall.

Question ③: Children's answers will depend on the objects the children choose and on the size of the cubes used to measure.

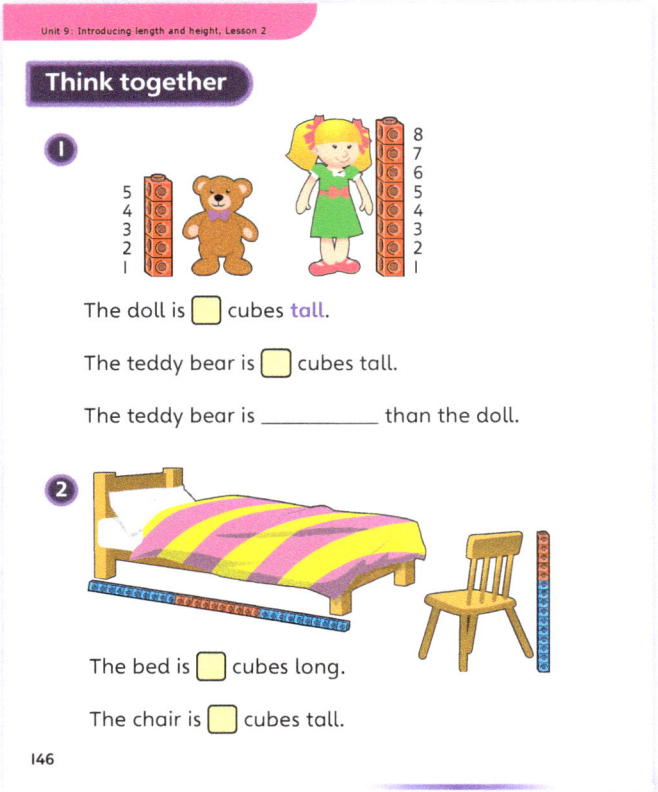

PUPIL TEXTBOOK 1B PAGE 146

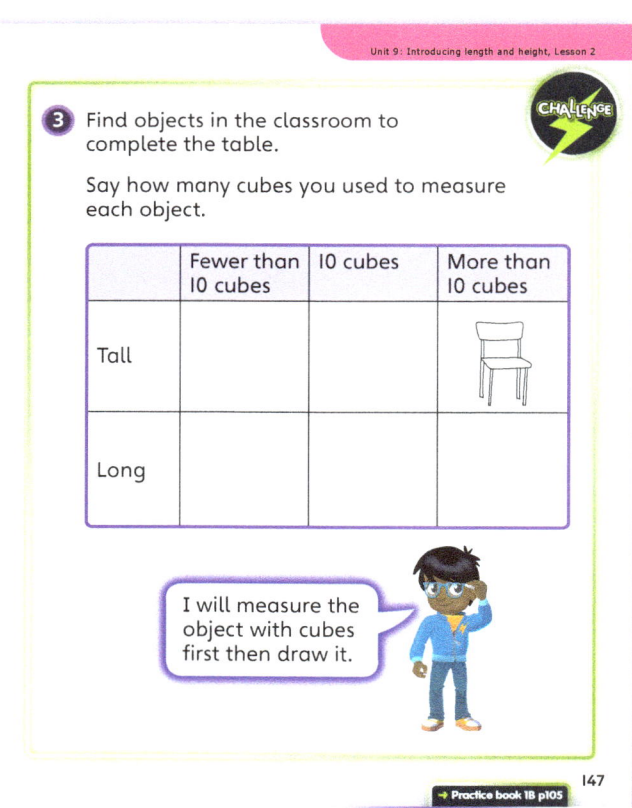

PUPIL TEXTBOOK 1B PAGE 147

Unit 9: Introducing length and height, Lesson 2

Practice

WAYS OF WORKING Pair work

IN FOCUS Question ❶ asks children to measure length and height in cubes where more cubes than are needed are displayed. Children need to ensure that they measure from the end or the bottom of the object rather than just counting all the cubes as they have done before. Question ❸ brings number bonds and measure together. You might want to help children use the part-whole model to work out how much longer the spring needs to be. 8 is the whole and 6 is one part (the current length of the spring). Children should determine that the other part should be 2, and hence extend the spring by 2 cubes. This can be repeated for the bear's hat.

STRENGTHEN Give children a rod of 10 cubes and a variety of objects that are shorter than 10 cubes. Support them in aligning the objects with the cubes and using a piece of card on the end of the object to determine the length.

For question ❶, it may help children to draw a straight line from the end or top of the object through the line of cubes to work how many cubes make up the lengths and heights. For question ❺, children should be encouraged to make sure the feet of the person line up with the bottom of the cube, and the head of the person lines up with the top of the uppermost cube. They could draw horizontal lines to help them do this.

DEEPEN Deepen understanding by asking children to choose two objects of different lengths, find their measurements in cubes and work out the difference in length. Ask children to say how much taller, longer or shorter one object is compared to another. This will strengthen the link to calculating difference. Can children find two objects where one is twice the length or height of another?

ASSESSMENT CHECKPOINT Questions ❶ and ❹ assess children's ability to use a non-standard unit to measure objects accurately. Question ❷ assesses children's ability to make reasonable estimations based on seeing one non-standard unit. Questions ❸ and ❺ assess whether children can draw accurately to a desired length.

ANSWERS Answers for the **Practice** part of the lesson can be found in the *Power Maths* online subscription.

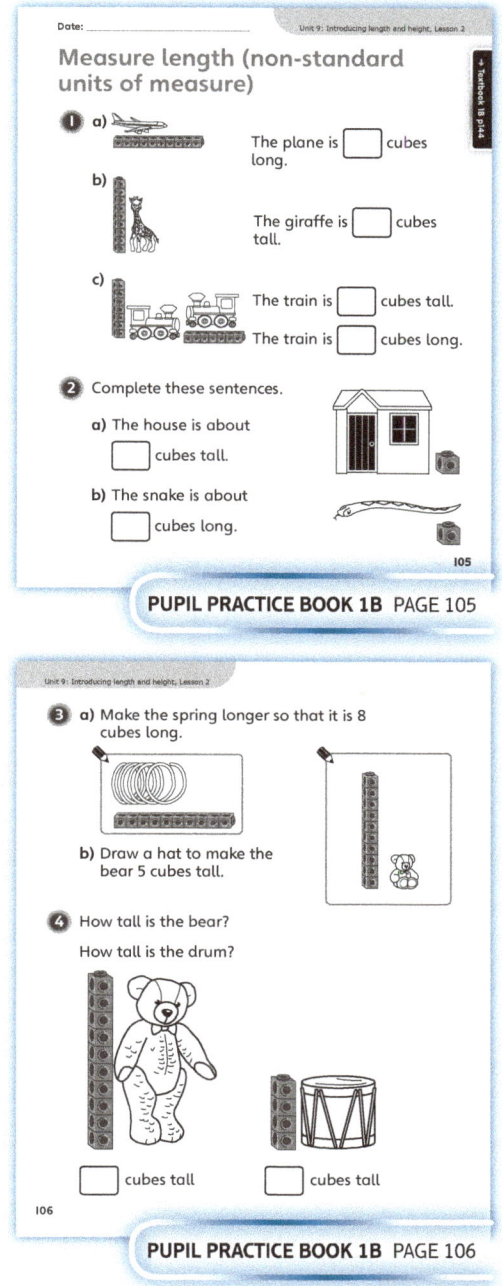

PUPIL PRACTICE BOOK 1B PAGE 105

PUPIL PRACTICE BOOK 1B PAGE 106

Reflect

WAYS OF WORKING Independent thinking

IN FOCUS This **Reflect** question requires children to justify their answer. In order to do this effectively, they need to have a secure understanding of how to measure the length of an object accurately using non-standard units of measure.

ASSESSMENT CHECKPOINT Children should be able to explain that, in order to measure the length, the object and the cubes have to start at the same point; and also that cubes are only counted to the end of the object.

ANSWERS Answers for the **Reflect** part of the lesson can be found in the *Power Maths* online subscription.

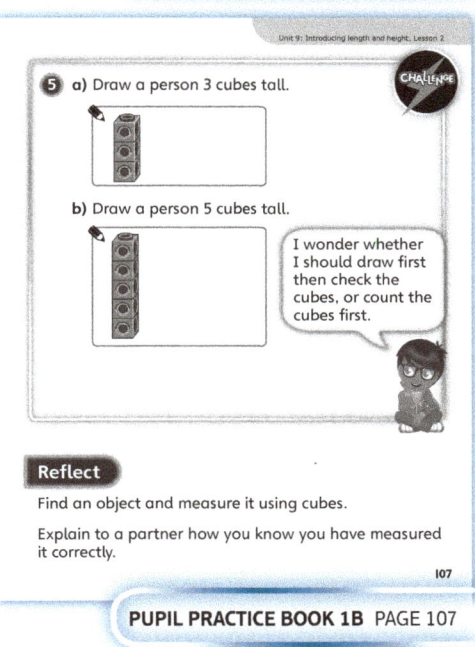

PUPIL PRACTICE BOOK 1B PAGE 107

After the lesson

- Did children have enough practical opportunities to support their understanding?
- Were children able to justify their measurements?
- Were children able to make the link between measuring length and addition and subtraction?

Unit 9: Introducing length and height, Lesson 3

Measure length (using a ruler)

Learning focus
In this lesson, children will learn that non-standardised objects can vary in size so a standard unit (centimetres) is needed. They will learn how to use a ruler correctly to measure length in centimetres accurately.

Before you teach
- Are children secure in measuring with non-standard units?
- Where can you provide practical opportunities for measuring with a ruler?

NATIONAL CURRICULUM LINKS

Year 1 Measurement

Measure and begin to record the following: lengths and heights.

ASSESSING MASTERY

Children can identify the importance of having standard units of measurement. Children can measure length accurately in centimetres using a ruler.

COMMON MISCONCEPTIONS

Children may line the end of the object with the end of the ruler or at 1 cm rather than at 0 cm. Ask:
- *How will you measure the length of this object? Where should you place the object's end on the ruler?*

STRENGTHENING UNDERSTANDING

Strengthen understanding by providing practical opportunities to measure length with a ruler. Ask children to count single jumps on a ruler as they would on a number line, so that they can see they have to start at 0 cm.

GOING DEEPER

To deepen understanding, ask children to compare the lengths of objects to work out how much shorter or longer one is from the other, drawing links to the previous lesson, but recording results in centimetres.

KEY LANGUAGE

In lesson: measure, length, ruler, centimetre, cm, compare, longer, shorter, taller

Other language used by the teacher: standard units, non-standard units

RESOURCES

Mandatory: rulers, objects to measure, multilink cubes, strips of paper

Optional: pencil cases to measure

 In the eTextbook of this lesson, you will find interactive links to a selection of teaching tools.

Quick recap

Give children some cubes and three objects. Ask them to measure the length of each object using cubes and put the objects in order, starting with the shortest. Model to children the method of measuring the length of an item using cubes. To start with, measure the object incorrectly by *not* aligning the end of the stick of cubes with the end of the object. Ensure children recognise that the object needs to line up with the cubes in order to measure the length or height.

Unit 9: Introducing length and height, Lesson 3

Discover

WAYS OF WORKING Pair work

ASK
- Question 1 b): *Where will you place the ruler?*
- Question 1 b): *Are all rulers the same? Compare with a partner.*

IN FOCUS Question 1 a) introduces the concept that non-standard units are unreliable as they can vary in size. The pieces of string both measure four cubes but are different lengths because the cubes are different sizes. This highlights the need for standard units of measure to avoid inconsistencies and confusion.

ANSWERS

Question 1 a): The string that is 4 red cubes long is longer. Both strings can be 4 cubes long because the red and yellow cubes are different sizes.

Question 1 b): The plain string is 9 cm long. The stripy string is 5 cm long.

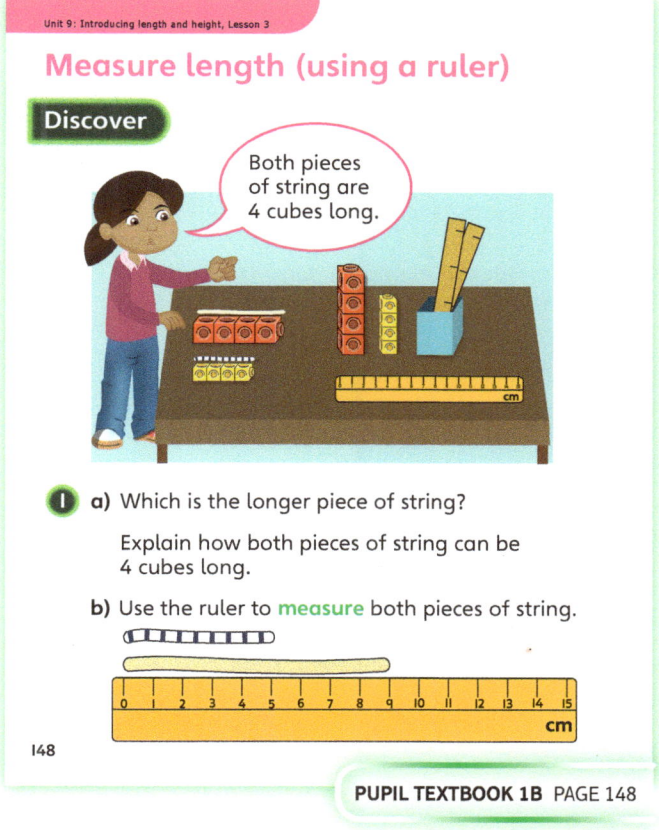

PUPIL TEXTBOOK 1B PAGE 148

Share

WAYS OF WORKING Whole class teacher led

ASK
- Question 1 b): *Where did you place the start of the string on the ruler?*
- Question 1 b): *Does using a ruler help us to compare lengths?*
- Question 1 b): *Can a centimetre ever be a different size?*

IN FOCUS In question 1 b), Sparks reinforces the importance of using the same-sized units when comparing lengths. Ensure children understand that this is the reason that units such as centimetres exist: they are the same no matter where you go.

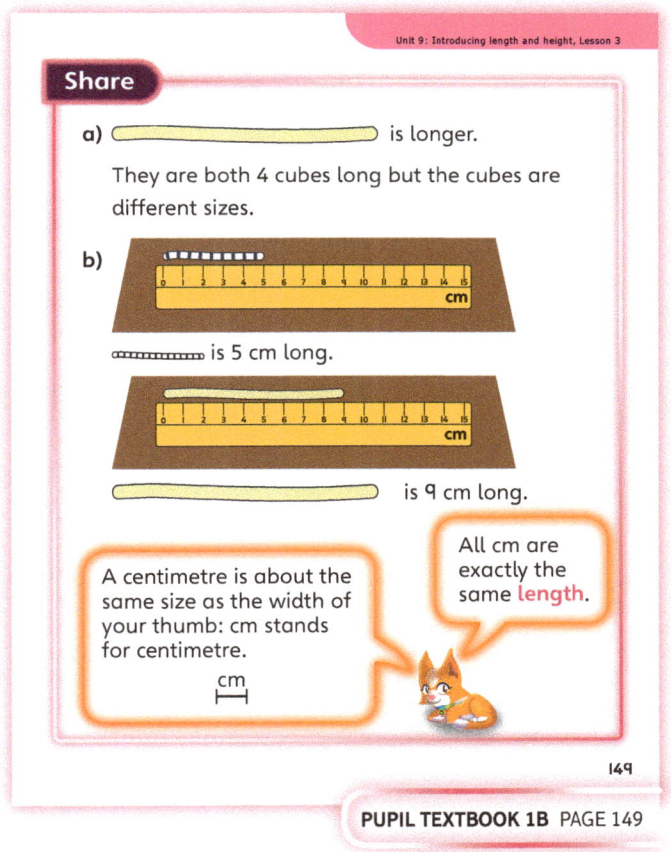

PUPIL TEXTBOOK 1B PAGE 149

187

Think together

WAYS OF WORKING Whole class teacher led (I do, We do, You do)

ASK
- Question ①: *Does the scale on the ruler remind you of anything you have used for counting and calculating?*
- Question ①: *Are your measurements the same as your partner's?*

IN FOCUS Question ③ asks children to compare objects against 10 cm, including finding objects that are 10 cm exactly. This will encourage them to use their ruler and other objects they have measured to make estimations of length. Comparing their measurements with a partner's measurements will help to identify who is using a ruler correctly, as the results should all be the same.

STRENGTHEN Provide practice measuring physical objects with a ruler. Remind children to start at 0 cm, not the end of the ruler. Children could use a straight edge on a piece of paper, aligned with the end of the object, to help read the length in centimetres on the ruler. The paper will cover up the rest of the ruler, making the scale easier to read.

DEEPEN Ask children to measure a pencil case. Challenge children to sort objects into ones that will fit and ones that will not fit inside the pencil case. Ask: *How could you check using a ruler without aligning the object alongside the pencil case?*

ASSESSMENT CHECKPOINT Assess whether children are using a ruler correctly to make accurate measurements in centimetres in questions ② and ③. Can children use what they already know to make reasonable estimations of length in question ③?

ANSWERS

Question ①: The string is 7 cm.

Question ② a): The ribbon is 5 cm.

Question ② b): The ribbon is 7 cm.

Question ② c): The ribbon is 9 cm.

Question ③: Children's answers will depend on the objects the children choose.

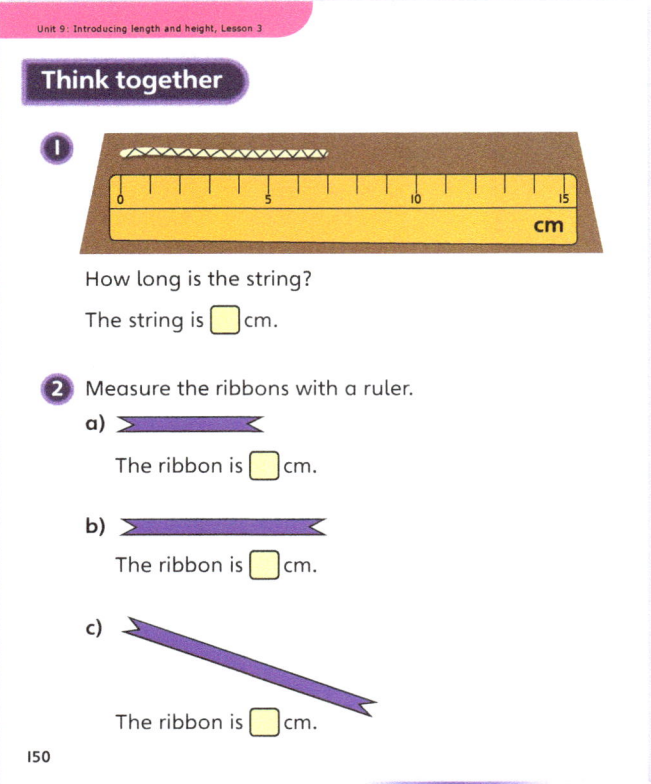

PUPIL TEXTBOOK 1B PAGE 150

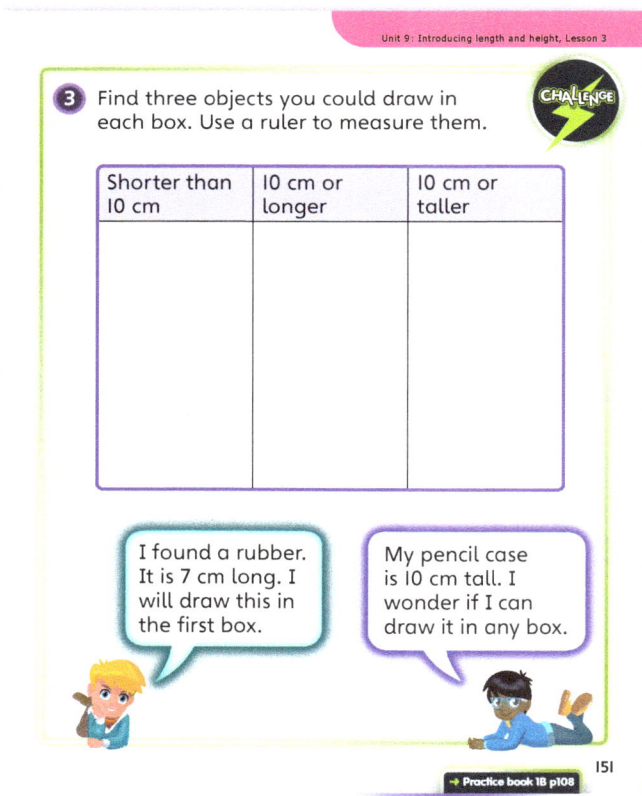

PUPIL TEXTBOOK 1B PAGE 151

Unit 9: Introducing length and height, Lesson 3

Practice

WAYS OF WORKING Pair work

IN FOCUS Question ❺ requires children to consider two dimensions, height and length, to measure and draw accurately a house of given dimensions.

STRENGTHEN When using the rulers shown in questions ❶ and ❷, ensure that children notice where the measuring starts; that it is not at the end of the ruler, but at the 0 marker. Give them practice in using a real ruler to measure items around the classroom.

DEEPEN To deepen understanding, ask children to work out how much shorter or longer one object is compared to another and to explain how they calculated it. Encourage children to use the ruler as a number line to aid their calculations. Do they realise that, if they try to measure the height of something resting on a table or the floor, the 0 on the ruler will not be aligned to the base of the object being measured?

ASSESSMENT CHECKPOINT Use question ❶ to assess whether children can correctly read the length of the object from the ruler. Assess whether children can draw an object to the desired length in question ❷.

Question ❹ will determine whether children can accurately measure lengths of more irregular shapes while question ❺ will further determine whether children can accurately draw an object to the desired dimensions and check for accuracy.

ANSWERS Answers for the **Practice** part of the lesson can be found in the *Power Maths* online subscription.

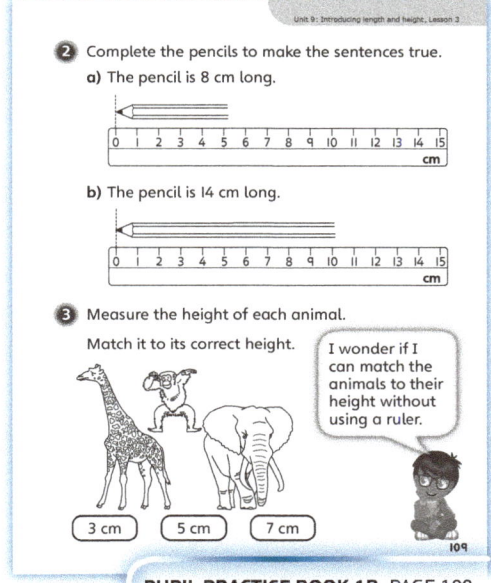

PUPIL PRACTICE BOOK 1B PAGE 108

Reflect

WAYS OF WORKING Pair work

IN FOCUS This activity requires children to explain how they use a ruler to measure length accurately. Children may randomly cut their paper into three strips resulting in lengths that cannot be measured in whole centimetres. Discuss using different units and link back to the previous lesson.

ASSESSMENT CHECKPOINT Assess whether children are making accurate measurements. Can they explain how to use a ruler correctly?

ANSWERS Answers for the **Reflect** part of the lesson can be found in the *Power Maths* online subscription.

After the lesson ⏸

- Are children confident using a ruler?
- Do children understand the necessity for standard units of measurement?
- Were you able to make links back to their work on number?

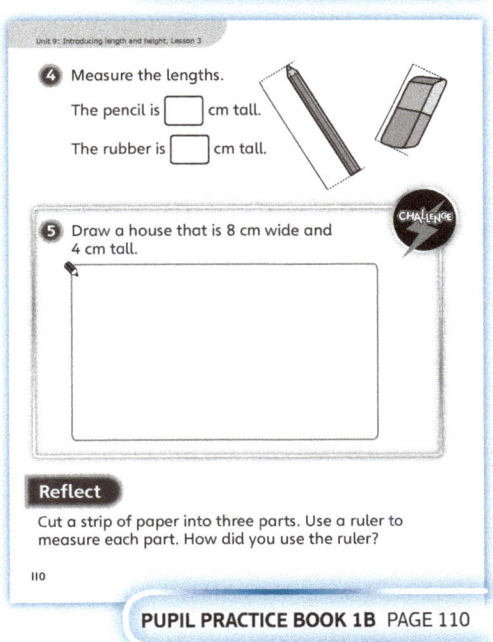

PUPIL PRACTICE BOOK 1B PAGE 110

Unit 9: Introducing length and height, Lesson 4

Solve word problems – length

Learning focus

In this lesson, children will apply what they have learned about measuring length, as well as addition and subtraction, in order to solve problems.

Before you teach

- Are children secure in making accurate measurements?
- Where can you provide practical opportunities?
- How will you model the link between measurement and addition and subtraction?

NATIONAL CURRICULUM LINKS

Year 1 Measurement

Compare, describe and solve practical problems for: lengths and heights [for example, long/short, longer/shorter, tall/short, double/half].

ASSESSING MASTERY

Children can identify which operations are required to solve a contextualised problem. Children can take the measurements they need and apply strategies and their understanding of addition and subtraction.

COMMON MISCONCEPTIONS

Children may struggle to see the link between a contextualised problem and the mathematics required in order to solve it. Ask:

- *What is this problem asking? What do you need to do to solve the problem? What do you know that can help you to solve it?*

STRENGTHENING UNDERSTANDING

Provide objects of the same length as in the questions, so that children can rearrange them to help identify what is required to solve each problem. Encourage children to use their rulers as a number line to calculate the answers. This will help them make the link between context and the number work they have done earlier in the year.

GOING DEEPER

Ask children to develop their own problems to try out on a partner. Being able to create problems will help children develop a deep and secure understanding of how the problems work.

KEY LANGUAGE

In lesson: length, shortest, longest, measure, taller, shorter, longer, centimetres, total, difference

Other language used by the teacher: zero, compare

STRUCTURES AND REPRESENTATIONS

Number line

RESOURCES

Mandatory: rulers, objects to measure

Optional: straws, items that have the same measurements as the objects in the questions

 In the eTextbook of this lesson, you will find interactive links to a selection of teaching tools.

Quick recap

Give each child an object. Ask them to find a partner whose object is shorter than theirs. Ask them to explain how they know. Ask them to check their answers by measuring the length of each object with a ruler. Repeat for finding a taller object.

Unit 9: Introducing length and height, Lesson 4

Discover

WAYS OF WORKING Pair work

ASK

- Question 1 a): *What do you know by looking at the image? What can you find out?*
- Question 1 b): *Which straw do you think is the longest? How can you check?*
- Question 1 b): *Which straw do you know the length of just by looking at the image?*

IN FOCUS This activity provides children with the opportunity to combine their learning from this unit. In question 1 b), they should recognise that they can see the length of the green striped straw from the image as it is lined up with 0 cm on the ruler. They should also recognise that the plain blue straw is the longest.

PRACTICAL TIPS Provide children with three straws or similar objects and a ruler. Ask them to order their straws or objects from shortest to longest. Now ask them to order them from longest to shortest. What do they notice?

ANSWERS

Question 1 a): The length of ▬▬▬▬▬ is 7 cm.

Question 1 b): ▬▬▬▬▬ shortest
↓
longest

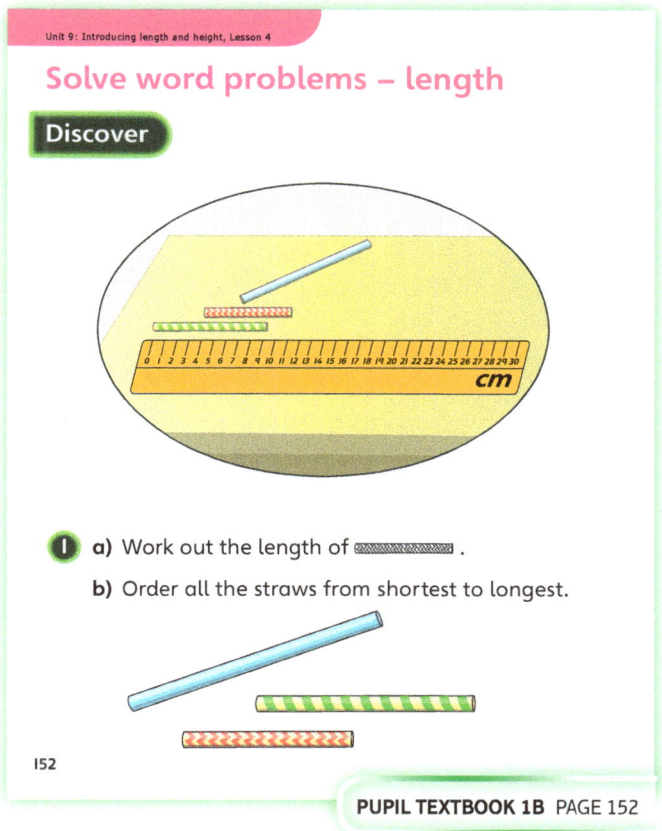

PUPIL TEXTBOOK 1B PAGE 152

Share

WAYS OF WORKING Whole class teacher led

ASK

- Question 1 a): *Why did Flo move the straw to 0? Why could you also work out 11 – 4? Why does this method also work out the length of the straw?*
- Question 1 b): *Do you need to measure the straws to order them? Why or why not?*

IN FOCUS In question 1 a), discuss with children the two different methods and why they both work. It is important that they understand that measuring from 0 will always work for finding the length, but they can also use the difference as a method of calculating length. In question 1 b), discuss with children why it was not necessary to measure the straws to put them in order, but how measuring them can support in checking their answers.

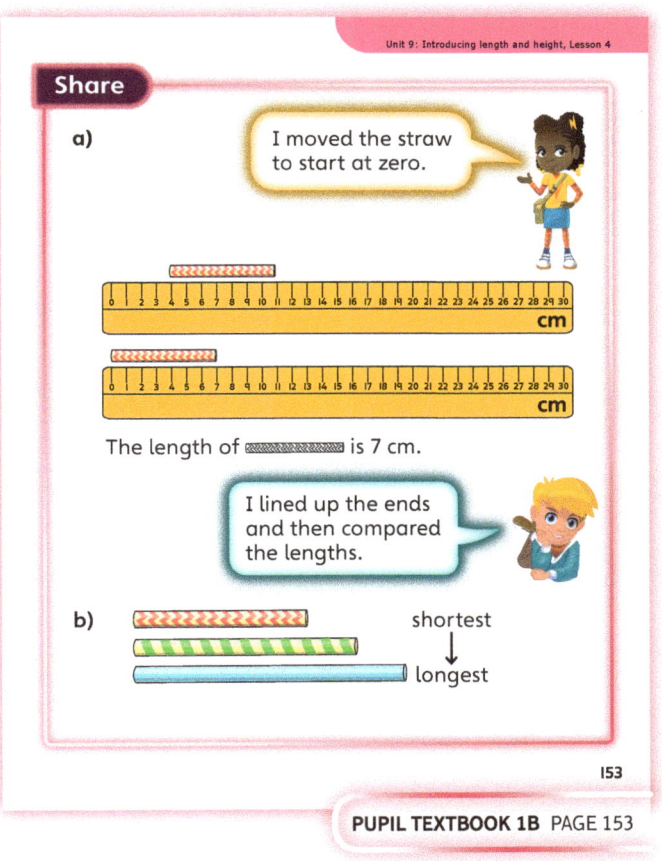

PUPIL TEXTBOOK 1B PAGE 153

Think together

WAYS OF WORKING Whole class teacher led (I do, We do, You do)

ASK
- Questions ❶ and ❸: *What calculation do you have to do? How do you know?*
- Question ❷: *What is different about how the rulers are shown? Does it change the length?*

IN FOCUS In question ❷, the scales on the rulers are represented as going from top to bottom. This is opposite to what children have been presented with before when measuring height. Question ❸ is a subtraction problem. The objects have been presented in a way that makes it easy to represent using a bar model. Show children how the number lines can be overlapped, making it easy to identify how much longer the worm is than the caterpillar.

STRENGTHEN For questions ❶ and ❸, encourage children to use their rulers as number lines to solve the calculations. This will help them to make the links between measurement and addition and subtraction. For question ❷, ask children to measure objects starting from either end so that they understand that it does not affect the length of the object, as long as the 0 marker on the ruler is aligned correctly.

DEEPEN To deepen and extend understanding, ask children to create their own problems with pictorial representations similar to those in questions ❶ and ❸ and swap them with a partner.

ASSESSMENT CHECKPOINT Questions ❶ and ❸ will determine if children are able to identify the calculation needed to solve the problem. Question ❷ will allow you to assess whether children have a secure understanding of how length can be measured and how length stays constant regardless of the orientation of the ruler.

ANSWERS
Question ❶: The total length is 16 cm.
Question ❷: The train on the left is 3 cm tall.
Question ❸: The difference in length is 5 cm.

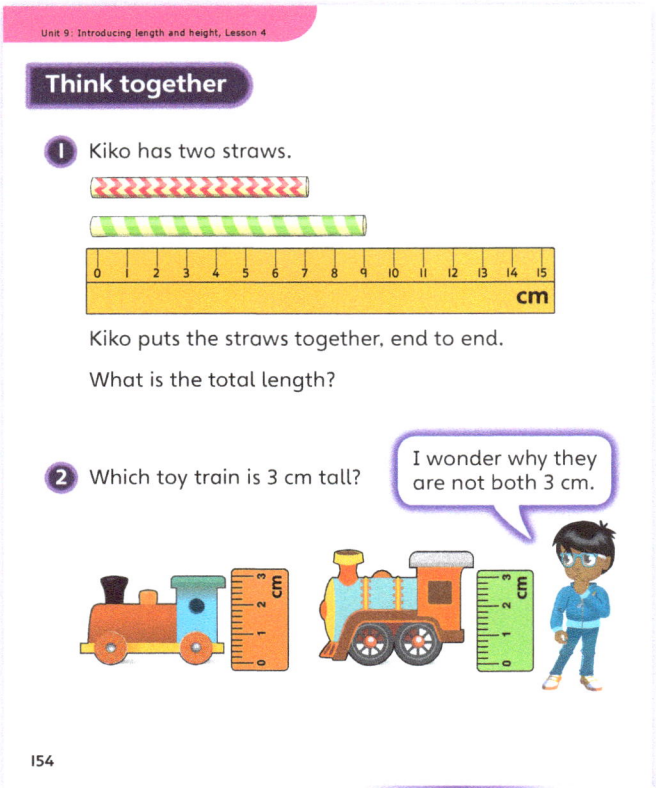

PUPIL TEXTBOOK 1B PAGE 154

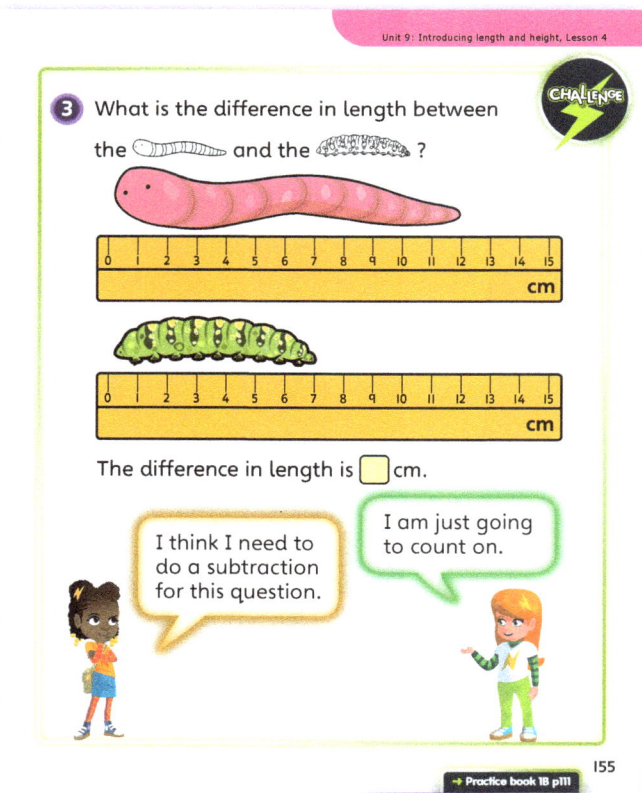

PUPIL TEXTBOOK 1B PAGE 155

Unit 9: Introducing length and height, Lesson 4

Practice

WAYS OF WORKING Pair work

IN FOCUS Question ❺ shows a worm and a caterpillar of the same length but, as the caterpillar's measurement does not begin at 0, children will not immediately be able to tell that. Discuss the importance of using a ruler correctly so that comparisons of length can be made more effectively.

STRENGTHEN Encourage children to use their rulers as number lines in order to carry out the calculations. Provide children with key vocabulary and discuss which operations the vocabulary relates to (for example, difference is subtraction and total is addition).

THINK DIFFERENTLY Question ❹ asks children to write number bonds to 10, in the context of length. They are given the total length of two objects as 10 cm and are asked to write three different possible measurements for each object.

DEEPEN Ask children to write their own problems involving three or more measurements and then test them out on a partner.

ASSESSMENT CHECKPOINT Question ❶ demonstrates if children understand that, because the ends of the objects are not at 0 cm, they cannot simply read the measurement. Can children identify the subtraction needed? As there are no picture clues, question ❷ assesses if children can identify the calculation from vocabulary alone. Use question ❸ to assess whether children can correctly justify a statement as false based on their knowledge of measurement of length.

ANSWERS Answers for the **Practice** part of the lesson can be found in the *Power Maths* online subscription.

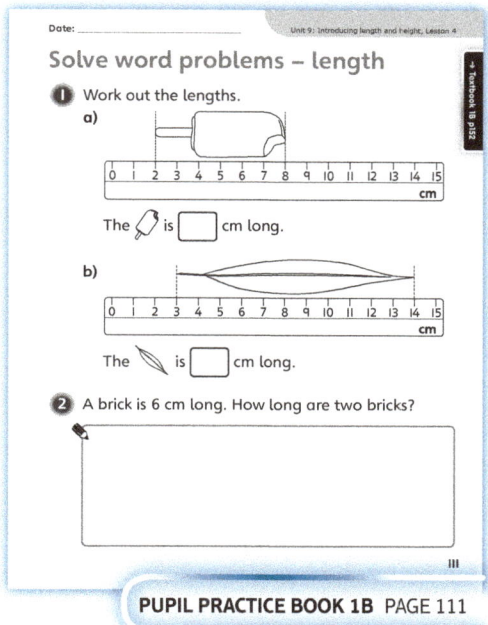

PUPIL PRACTICE BOOK 1B PAGE 111

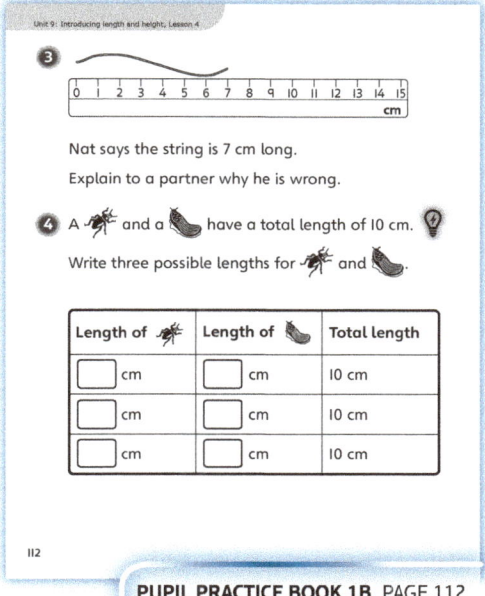

PUPIL PRACTICE BOOK 1B PAGE 112

Reflect

WAYS OF WORKING Independent thinking

IN FOCUS This question will determine if children are secure in the concept of measuring length and in how to reason and apply their skills in order to solve a problem.

ASSESSMENT CHECKPOINT Assess whether children can clearly explain how they solved the problem. Can they describe which steps and calculations are involved and why?

ANSWERS Answers for the **Reflect** part of the lesson can be found in the *Power Maths* online subscription.

After the lesson ⏸

- Were children confident in drawing upon their addition and subtraction skills in order to solve the problems?
- Were children able to reason about the choices they made in how they solved the problems?

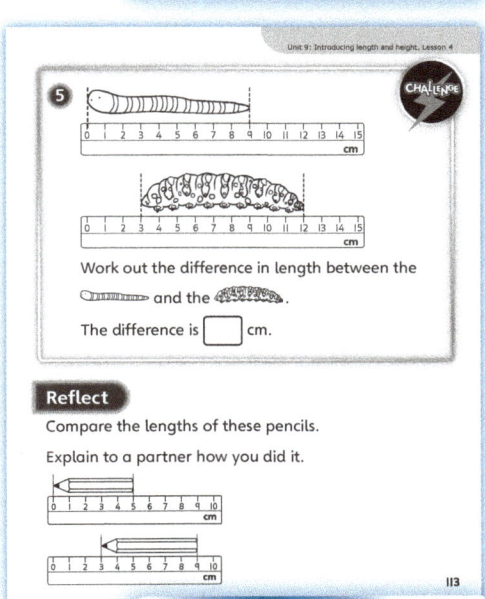

PUPIL PRACTICE BOOK 1B PAGE 113

End of unit check

> **Don't forget the unit assessment grid in your *Power Maths* online subscription.**

WAYS OF WORKING Group work adult led

IN FOCUS These questions focus on assessing children's conceptual understanding of comparing and measuring length and height.
- Question ① focuses on directly comparing the height of two or more objects.
- Question ② determines whether children understand how to measure correctly using non-standard units.
- Question ③ assesses whether children can measure accurately using a ruler.
- Question ④ determines whether children are able to use what they already know to estimate length.

Think!

WAYS OF WORKING Pair work

IN FOCUS This question focuses on how children can accurately make direct comparisons between the lengths of two objects. Although the starting points of the two pieces of string are aligned, the plain pink string is not straight, so it would be longer if straightened out. Children should reason that, although the picture shows that the start and end points of the two pieces of string are aligned, the plain pink string is longer than the spotty string as it is not laid out straight.

Key vocabulary in this question includes: long, longer, longest, short, shorter, shortest, compare.

Encourage children to think through or discuss how they know which string is longer before writing their answer in **My journal**.

ANSWERS AND COMMENTARY Children will learn how to compare and measure accurately and will understand the importance of aligning starting points. Children will draw on their knowledge of number, particularly ordering and comparing numbers. Children will also learn the relationship between number lines and scales on a ruler and use this understanding to calculate differences in length. Children will use key language such as longer, longest, shorter, shortest, taller and tallest when comparing length and height.

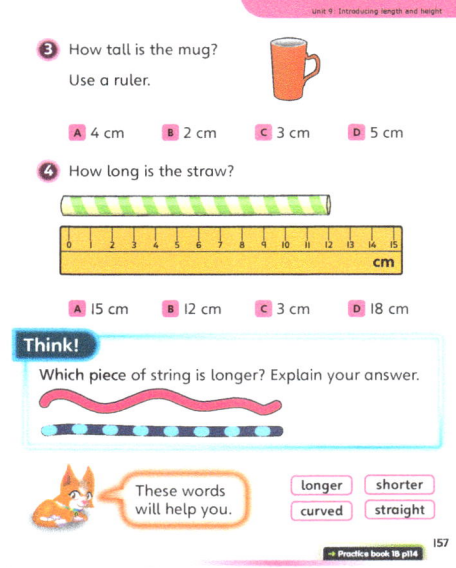

PUPIL TEXTBOOK 1B PAGE 156

PUPIL TEXTBOOK 1B PAGE 157

Q	A	WRONG ANSWERS AND MISCONCEPTIONS	STRENGTHENING UNDERSTANDING
1	C	Choosing A, B or D suggests that children do not understand the key vocabulary or that their ability to directly compare heights is not secure.	When children are asked to make comparisons between objects or to use non-standard units such as cubes, encourage them to use a flat surface such as a table, wall or book to help them align the starting points. Provide lots of practical opportunities for children to measure length using both standard and non-standard units. Discuss with children how they can ensure accuracy when measuring length.
2	B	Choosing A indicates that children do not understand that the starting points of the cubes and object have to be aligned. Children may miscount the cubes and choose C.	
3	C	Choosing A, B or D suggests that children cannot use a ruler to accurately measure length.	
4	B	Children may read the last number on the ruler and choose A.	

Unit 9: Introducing length and height

My journal

WAYS OF WORKING Independent thinking

ANSWERS AND COMMENTARY Children could answer: 'I think the top string is longer because it is not stretched out straight and it ends at the same place as the bottom string. If I stretched it out, it would be longer.'

If children struggle to understand why the top string is longer, recreate the problem using physical resources, so that they can see that the top string is longer than the bottom string once it has been stretched out.

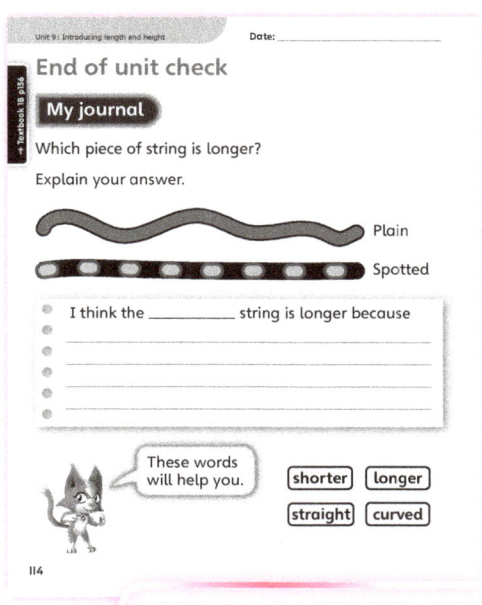

PUPIL PRACTICE BOOK 1B PAGE 114

Power check

WAYS OF WORKING Independent thinking

ASK
- Do you think you can accurately measure the length of objects using different non-standard units?
- Do you feel confident using a ruler to measure length in centimetres?
- Do you understand how you can use measurements to find the total length of an object or how much longer one object is than another?

Power play

WAYS OF WORKING Pair work or small groups

IN FOCUS This game determines whether children can make accurate measurements using standard units.

Demonstrate how to play the game so children understand how it works, then send children off to play the game in pairs. Children can play the game more than once.

ANSWERS AND COMMENTARY If children are able to make accurate measurements during the game, this suggests that they are secure in measuring length using standard units.

If children make errors with their measurements, identify where the errors occur. If children are not aligning their rulers, they may need to practise measuring from the 0 marker. If children are unsure where to take the measurement from or how to align the ruler and paper correctly, encourage them to place the ruler on top of the paper with the ends aligned and then mark off the measurement on the paper using a pencil.

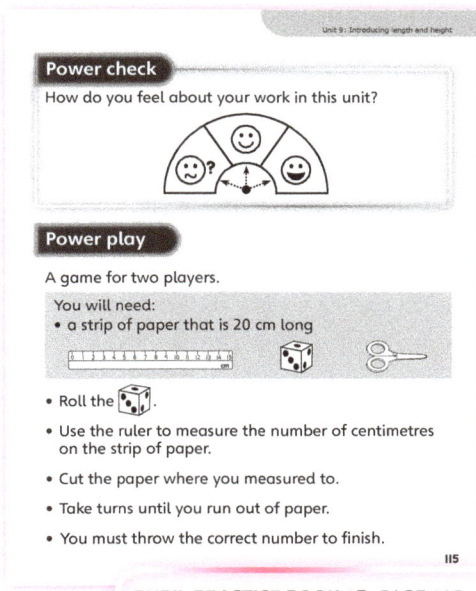

PUPIL PRACTICE BOOK 1B PAGE 115

After the unit ⏸

- Were children able to discuss how they measured the objects and how they knew why one object was longer than the other?
- Were children able to make the links between this unit and their previous work on number and calculating?

Strengthen and **Deepen** activities for this unit can be found in the *Power Maths* online subscription.

Unit 10
Introducing mass and capacity

Mastery Expert tip! 'I like to model some of the activities in this unit using real apparatus – scales and cubes, jugs of water, and so on. This is useful for all children, and it helps to keep them focused on the context of mass and capacity instead of just reducing everything to numbers.'

Don't forget to watch the Unit 10 video!

WHY THIS UNIT IS IMPORTANT

This unit establishes the use of uniform non-standard units (such as cubes and cups) to measure mass and capacity. This is an important first step towards introducing standard units and provides a simple practical context in which a range of problems involving addition and subtraction can be introduced. There is also plenty of opportunity to explore part-whole and part-part relationships: For example, if a saucepan holds 10 cups of milk and 4 cups are poured out, how much is left in the pan?

In this unit, children should also understand the difference between volume and capacity. The capacity of a container is the maximum amount of liquid that it can hold, whereas the volume of liquid in a container is the amount of liquid the container is actually holding. This can be less than the capacity.

WHERE THIS UNIT FITS

→ Unit 9: Introducing length and height
→ **Unit 10: Introducing mass and capacity**

This unit extends children's understanding of measurement by introducing the ideas of mass and capacity, both of which should already be familiar from practical, if informal, experience. The work here also reinforces understanding of comparison and ordering, including use of the inequality and equals signs.

Before they start this unit, it is expected that children:
- have had practical experience of playing with water, sand, weighing scales and a range of measuring vessels
- can carry our simple addition and subtraction calculations, supported by appropriate representations and apparatus where necessary.

ASSESSING MASTERY

Children who have mastered the work in this unit will be able to estimate, compare and order mass and capacities of a variety of familiar objects, using convenient non-standard units such as plastic cubes for mass and cups for capacity. They will be able to use simple mathematical reasoning and their understanding of addition and subtraction to solve a range of word problems involving mass and capacity.

COMMON MISCONCEPTIONS	STRENGTHENING UNDERSTANDING	GOING DEEPER
Children may fail to understand the importance of non-standard units being of a uniform size.	Use physical apparatus (weighing scales, cubes, cups, jugs) to demonstrate the importance of using uniform units.	More confident children may be able to quickly compare and order masses and capacities and solve word problems using their knowledge of number facts and operations. It is important that they realise that mass and capacity are key ideas in their own right, and not just a convenient context for calculations. Challenging children to explain their work can be a good way to keep them focused on the real-world context of mass and capacity.
Children may not be able to select the correct operation when solving problems.	Use the physical context (weight or capacity) to model the operations that are needed, for example, ask: *Do I need to add cubes to this pan, or take some away, to make the scales balance?*	

Unit 10: Introducing mass and capacity

WAYS OF WORKING

Ash's question should be used to review children's understanding of the concepts of mass and capacity, and their grasp of the associated vocabulary. Talk through the vocabulary introduced by Flo: children will use all of these words throughout the unit, and it will be very useful for them to be familiar with them from the outset.

STRUCTURES AND REPRESENTATIONS

Although there are no set mathematical structures and representations for this unit, the following concrete resources will be helpful:
- balance scales
- jugs and cups
- cubes

KEY LANGUAGE

There is some key language that children will need to know as part of the learning in this unit:
- weight, weigh, mass
- capacity, volume, contains, container
- heavier, heaviest, lighter, lightest
- more, most, fewer, less, least
- greater than (>), less than (<), equal (=)
- addition, subtraction
- balance scales, balanced
- compare, measure, estimate
- empty, full, amount, half

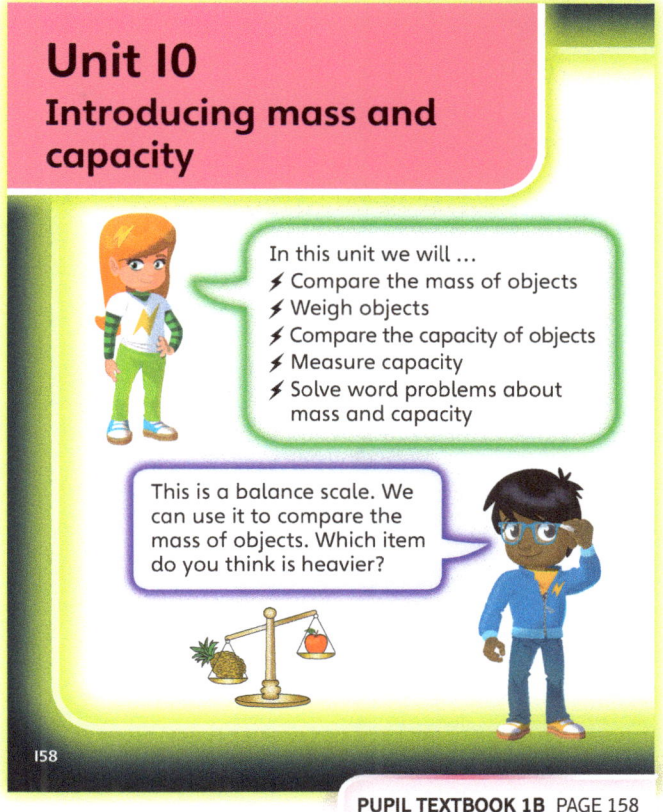

PUPIL TEXTBOOK 1B PAGE 158

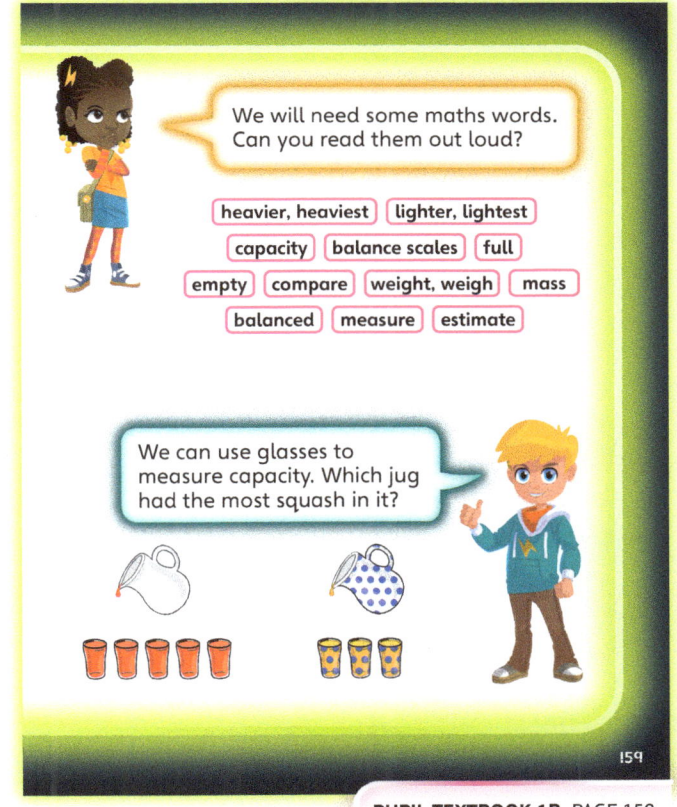

PUPIL TEXTBOOK 1B PAGE 159

Unit 10: Introducing mass and capacity, Lesson 1

Heavier and lighter

Learning focus
In this lesson, children will compare the mass of a range of familiar objects.

Before you teach
- Are children secure comparing and ordering quantities?
- Are children confident in using the inequality and equals signs to compare quantities?

NATIONAL CURRICULUM LINKS

Year 1 Measurement

Compare, describe and solve practical problems for: mass/weight (for example, heavy/light, heavier than, lighter than).

ASSESSING MASTERY

Children can understand and use appropriate language and inequality signs to express relationships between a range of objects based on their relative mass.

COMMON MISCONCEPTIONS

Children may be confused by more informal notions of 'bigger' and 'smaller', especially when these do not agree with the comparisons based purely on mass. Ask:
- Which is heavier: a balloon or a marble? And which one is bigger?

STRENGTHENING UNDERSTANDING

Ensure children understand that the comparisons being made in this lesson are based purely on mass, not on overall physical size. Use the images of sets of balance scales in the book to emphasise this idea. Ask: *Which side has gone down? Why is that?* Children may benefit from practical experience with a set of balance scales, predicting and checking what will happen when pairs of objects are compared.

GOING DEEPER

Children may be able to explore ideas of density more explicitly. To challenge children, ask: *Which would weigh more – a cup full of cubes or a cup full of sand? Why do you think that? How would you check?*

KEY LANGUAGE

In lesson: heavier, lighter, compare, greater than (>), less than (<), weigh, weight, mass, scales, balance scales, balanced

Other language to be used by the teacher: inequality signs, greater than, less than

STRUCTURES AND REPRESENTATIONS

Sets of balance scales and see-saws to compare weights

RESOURCES

Mandatory: set of balance scales

Optional: objects to weigh (teddy bear, soft toy, toy cars or lorries)

 In the eTextbook of this lesson, you will find interactive links to a selection of teaching tools.

Quick recap

Find out how familiar children are with the words 'heavier' and 'lighter'. Give children an object, such as a pencil, and ask them to find something that is heavier/lighter than that object. This will allow you to assess whether children are familiar with these words already. Repeat for other objects.

Unit 10: Introducing mass and capacity, Lesson 1

Discover

WAYS OF WORKING Pair work

ASK
- Question 1 a): *What is James putting in the bag? Which of those things is heavier? Which of those things is lighter?*

IN FOCUS Questions 1 a) and b) allow children to demonstrate that they understand both the vocabulary and the reasoning that will be used throughout the lesson. Use voice and gestures to suggest lifting heavy and light objects.

PRACTICAL TIPS Give children pairs of different objects in the classroom (or ask them to find two different objects). Ask them to compare them and say which object is heavier and which is lighter.

ANSWERS

Question 1 a): The pineapple is heavier.

Question 1 b): The toothpaste is lighter.

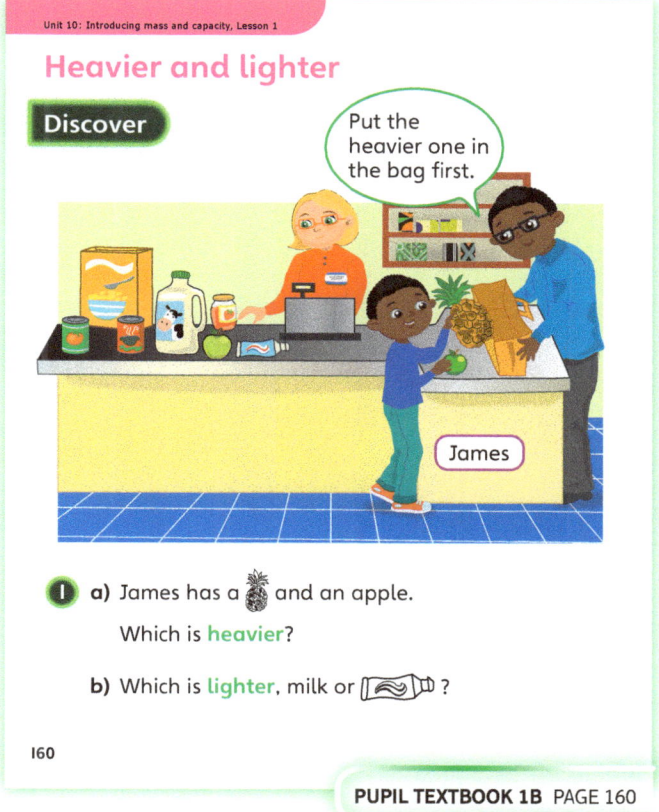

PUPIL TEXTBOOK 1B PAGE 160

Share

WAYS OF WORKING Whole class teacher led

ASK
- Question 1 a): *Who has used a set of balance scales like this?*
- Question 1 a): *How can you tell which object is heavier? When can you decide which object is heavier without having to use the balance scales?*

IN FOCUS Use this part of the lesson to reinforce the concepts of 'heavier' and 'lighter', as well as to introduce the use of a set of balance scales to compare mass (especially in cases where the result of the comparison is not immediately obvious).

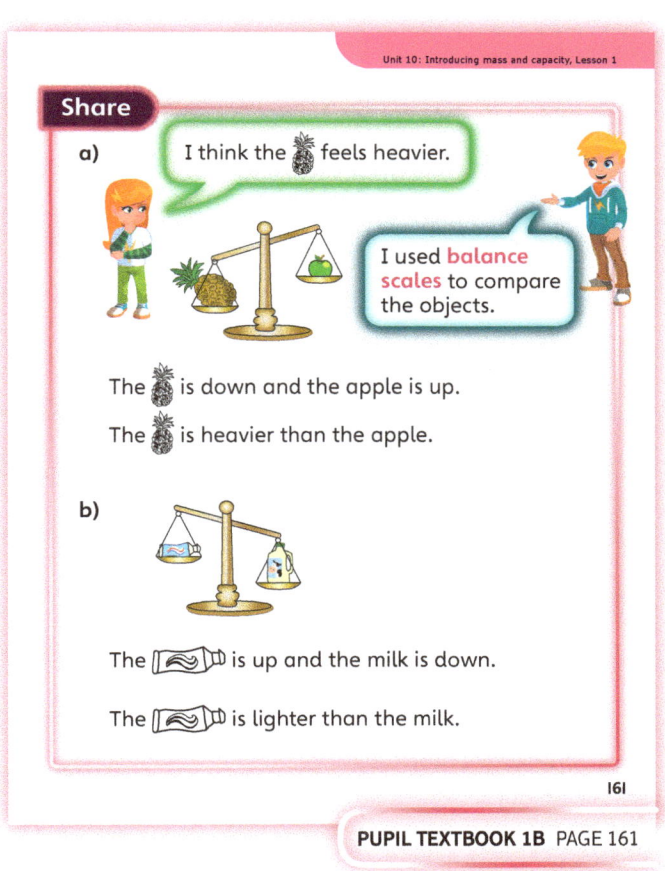

PUPIL TEXTBOOK 1B PAGE 161

Unit 10: Introducing mass and capacity, Lesson 1

Think together

WAYS OF WORKING Whole class teacher led (I do, We do, You do)

ASK
- Question ❶: *Why wouldn't you be able to guess which tin was heavier without the balance scales? How do the balance scales let you work it out?*
- Question ❷: *Which two objects weigh the same? How can you tell which items are heaviest?*

IN FOCUS Question ❸ provides a good opportunity to explore some potential areas of confusion when comparing mass. Careful discussion of the points here should identify any misconceptions.

STRENGTHEN Children who find it difficult to turn comparisons around (for example, changing 'a is lighter than b' to 'b is heavier than a') may benefit from additional practice in making this kind of change. Ask questions such as: *The dog is heavier than the cat. Now tell me something about the cat.*

DEEPEN In question ❷, encourage children to express the comparisons in their reversed forms. For example, changing 'the jam is heavier than the apple' to 'the apple is lighter than the jam'.

In question ❸, encourage children to use reasoning to explain whether each statement is correct or incorrect.

ASSESSMENT CHECKPOINT Use these questions to check whether children can compare the mass of two (or more) items using balance scales. For example, they know that the heavier one is the one that is down and that when two objects have the same mass then the scales are balanced.

ANSWERS

Question ❶: The can of soup is lighter than the can of tomatoes.
The can of tomatoes is heavier than the can of soup.

Question ❷: Children should point to the heavier item on each scale. They should notice that the heavier item is always lower.
The last scale balances so the items are the same weight.

Question ❸: Maria is correct. The balloon will go up because it is lighter.
Not all bigger items are heavier (Sam and Katie).
Not all smaller items are lighter (Hassan).
The heavier item will go down on a balance scale (Hassan).

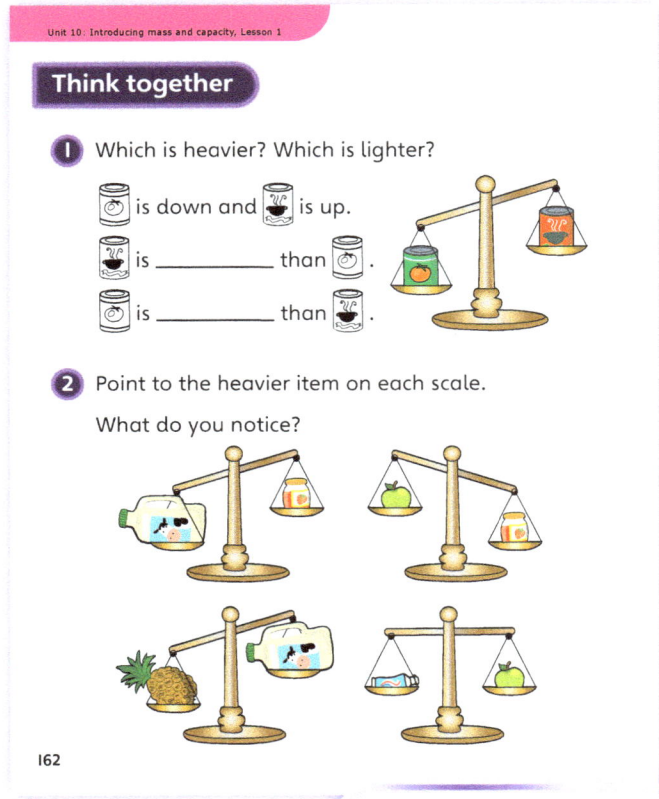

PUPIL TEXTBOOK 1B PAGE 162

PUPIL TEXTBOOK 1B PAGE 163

Unit 10: Introducing mass and capacity, Lesson 1

Practice

WAYS OF WORKING Independent thinking

IN FOCUS Question ② could involve a variety of correct answers depending on whether or not some of the items are perceived to be toys, or the real thing.

Question ③ involves working from diagrams of sets of balance scales to find corresponding inequalities. At this stage, all of the required relationships can be read from a single diagram. This will not be the case for question ⑤ – see the **Deepen** text below.

STRENGTHEN Question ② provides a further opportunity for discussion, and could be used to reinforce ideas of 'heavier' and 'lighter'. Ask: *Do you think a toy car is heavier or lighter than a teddy bear? Is a real car heavier or lighter than a teddy bear?* Use a set of balance scales to compare a teddy bear and a toy car. Ask: *Can you find a toy car that is heavier (or lighter) than this teddy bear?*

DEEPEN Question ⑤ offers a chance for some deeper mathematical thinking. Children have to combine the information shown on the separate sets of balance scales in order to find a solution. Once they solve this problem, challenge children to make up another example of the same kind.

THINK DIFFERENTLY Question ④ presents inequalities using see-saws instead of balance scales. Children should recognise that the see-saws, like balance scales, go up on the side with a lighter item and go down on the side with a heavier item.

ASSESSMENT CHECKPOINT Use questions ① and ③ to check that children have a solid understanding of the main ideas of this lesson – the concepts of 'heavier' and 'lighter', and how they relate to a set of balance scales, as well as the use of the corresponding vocabulary in this context.

ANSWERS Answers for the **Practice** part of the lesson can be found in the *Power Maths* online subscription..

Reflect

WAYS OF WORKING Independent thinking

IN FOCUS This example involves reversing a relationship: turning a 'heavier than' statement into the corresponding 'lighter than' statement.

ASSESSMENT CHECKPOINT If children find this exercise difficult, ask further questions to find out exactly where the problems lie. For example, do children understand the words 'heavier' and 'lighter'? Do they understand how the balance scales help to identify the heavier object? Do they understand how a 'heavier than' relationship can be rewritten as a 'lighter than' relationship?

ANSWERS Answers for the **Reflect** part of the lesson can be found in the *Power Maths* online subscription.

After the lesson

- Can children use the key language from this lesson in other areas of the curriculum?
- Do children understand the difference between 'bigger'/'smaller' and 'heavier'/'lighter'?
- Would children benefit from further practical experience of using balance scales, identifying heavier and lighter items?

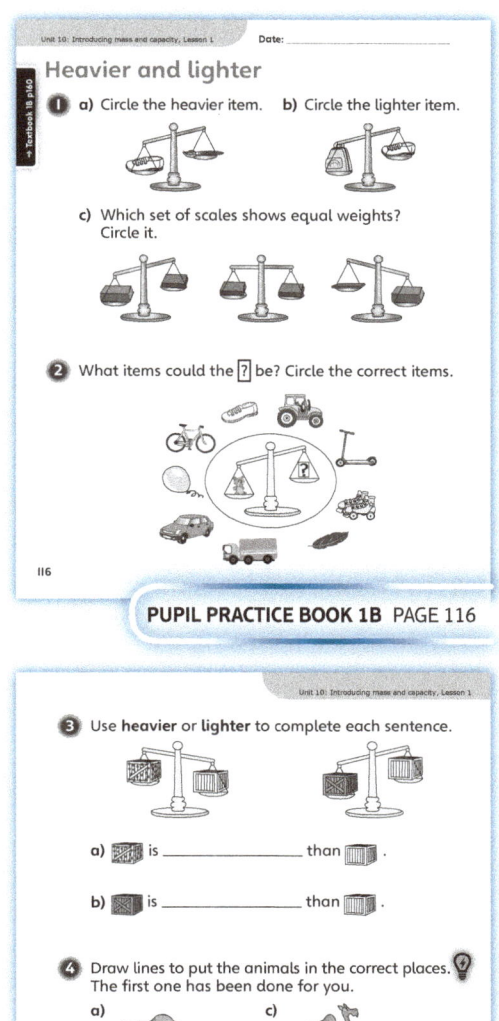

PUPIL PRACTICE BOOK 1B PAGE 116

PUPIL PRACTICE BOOK 1B PAGE 117

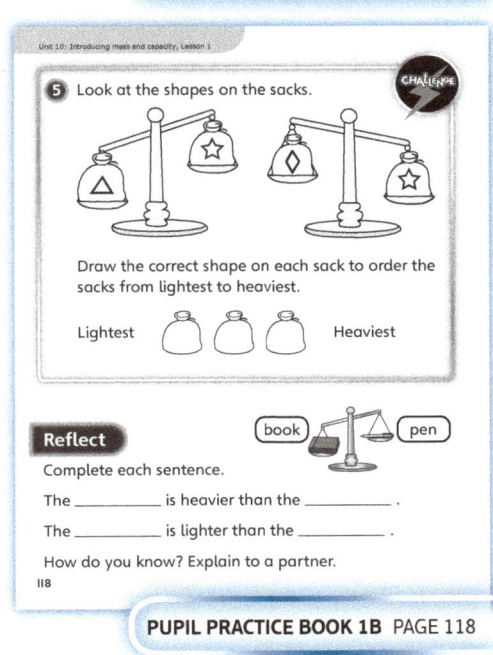

PUPIL PRACTICE BOOK 1B PAGE 118

Unit 10: Introducing mass and capacity, Lesson 2

Measure mass

Learning focus

In this lesson, children will weigh objects using a variety of non-standard units.

Before you teach

- Are children confident with the ideas of 'heavier' and 'lighter' from the previous lesson?
- Would children benefit from additional practical experience with weighing scales?
- Do children know how the balance scales show which object is heavier?

NATIONAL CURRICULUM LINKS

Year 1 Measurement

Measure and begin to record the following: mass/weight.

ASSESSING MASTERY

Children can use non-standard units to measure the mass of a range of objects.

COMMON MISCONCEPTIONS

Children may not understand that we can use any convenient object as a unit for measuring mass, provided that we have an adequate stock of the objects and that they are of uniform mass. Children may use similar objects that have non-uniform mass to measure the mass of another object. For example, they may use toy cars. Ask children why this might not be a good idea. Organise some cubes of different sizes and ask:

- *I am going to use cubes to weigh some toys. Can I just use any of these cubes?*

STRENGTHENING UNDERSTANDING

Children may benefit from practical experience of weighing objects using a set of balance scales. Ask children to weigh a selection of objects using differently-sized cubes as a non-standard unit of weight. Ask: *Will you need more or less of these smaller cubes to balance this object?*

GOING DEEPER

Challenge children to estimate and check the mass of a selection of objects using cubes and a set of balance scales; this could be a game for pairs of children and they could be asked to devise their own scoring system.

KEY LANGUAGE

In lesson: weigh, mass, balance, balances, fewer, more, measure, heavier

Other language to be used by the teacher: lighter, same size, scales, estimate, score, most, least

STRUCTURES AND REPRESENTATIONS

Balance scales, cubes

RESOURCES

Optional: sets of balance scales, cubes, a selection of objects to weigh

 In the eTextbook of this lesson, you will find interactive links to a selection of teaching tools.

Quick recap

Give children two objects each. Ask them which is heavier and how they know. Then ask them which is lighter and how they know. Repeat for other objects. Ask children to check their answers using a balance scale.

Unit 10: Introducing mass and capacity, Lesson 2

Discover

WAYS OF WORKING Pair work

ASK
- Question 1 a): *What are the children doing?*
- Question 1 a): *How is Hiro weighing his book? Would you expect Lucy to use the same number of cubes as Hiro? What is different about what Joe is doing? Do you think he is right?*

IN FOCUS Questions 1 a) and b) demonstrate a relatively complicated scenario, so children may benefit from a little help in working out what is going on. Ensure children understand that everyone in the picture is trying to do the same thing, but that they have all gone about it in slightly different ways.

PRACTICAL TIPS Recreate the problem in the classroom using balance scales, books, cubes and other items. Allow children to use the balance scales to show that the mass of an object (in this case, a book) can be measured using different non-standard units.

ANSWERS

Question 1 a): Hiro is correct. His book balances 10 cubes. Joe is wrong. One of his 5 cubes is different to the others. He should have used cubes of the same size to balance his book.

Question 1 b): Lucy needs fewer cubes than Hiro because she is using heavier cubes.

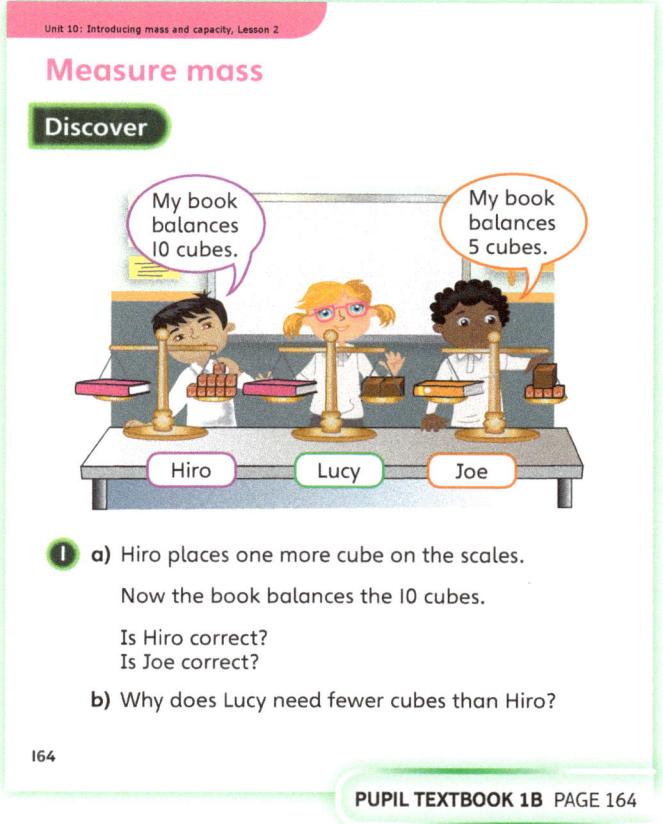

PUPIL TEXTBOOK 1B PAGE 164

Share

WAYS OF WORKING Whole class teacher led

ASK
- Question 1 a): *Why does Astrid say that the cubes need to be the same?*
- Question 1 a): *Why would Hiro need more cubes than Lucy when they are measuring the same book?*

IN FOCUS Questions 1 a) and b) establish two important learning points that will be reinforced throughout the lesson. First, it is important that the objects used as non-standard units are of uniform mass, otherwise we cannot use them to make any sensible statement about how much something weighs. Second, we need a greater number of lighter units (and a smaller number of heavier ones) to balance a given object.

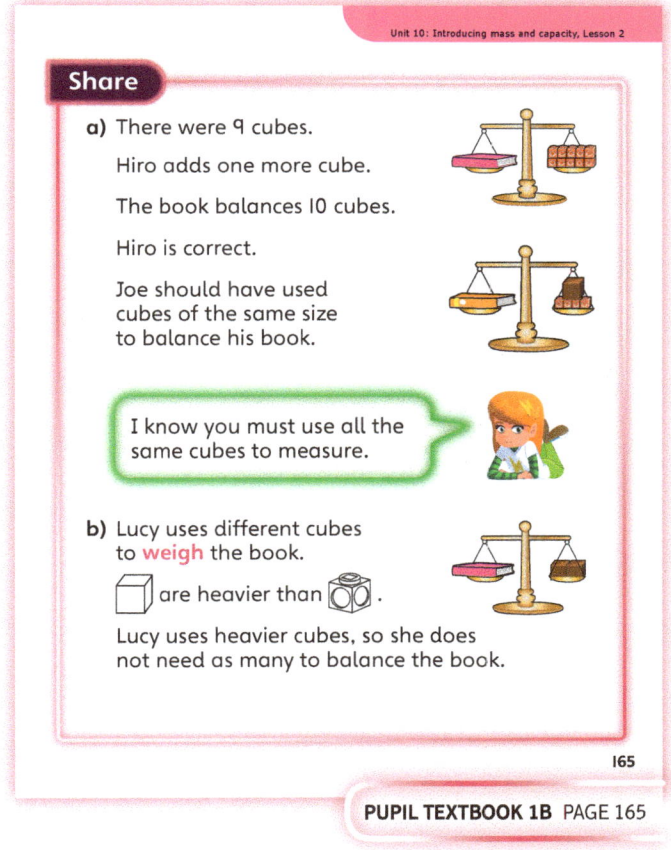

PUPIL TEXTBOOK 1B PAGE 165

Think together

WAYS OF WORKING Whole class teacher led (I do, We do, You do)

ASK
- Question ❶: *What are the different things that have been used to balance the shoe? How many cubes, bean bags or crayons are needed to weigh the same as the shoe?*
- Question ❷: *What has been used to weigh each of these objects? How can you tell which object weighs the most and which object weighs the least?*

IN FOCUS Questions ❷ and ❸ establish the idea of a uniform non-standard unit. This is simply a convenient uniform object that we can compare to other objects. It is described as 'non-standard' to distinguish it from 'standard' units of weight such as grams. Discuss how the number of cubes shows which items are heavier or lighter.

STRENGTHEN Children who are finding the idea of measuring using a 'unit' difficult should be given an opportunity to use a real set of balance scales to weigh a selection of objects against cubes or other non-standard units. Ask: *How many more cubes does this object weigh? Which object is lighter or heavier? Which object weighs more or less? How can you tell that the teddy bear weighs less than the car? Does the book weigh more or less than the car?*

DEEPEN Question ❸ involves some fairly challenging reasoning, with children having to follow a chain of reasoning to arrive at the answer. Children who answer these questions confidently could be asked to make up further examples of their own. Ask: *How many cubes do the book and teddy bear weigh altogether? How many more cubes does the book weigh than the teddy bear? How can you make the scales balance, so that the side with the teddy bear balances the side with the book?*

ASSESSMENT CHECKPOINT Check that children understand that non-standard units can be anything as long as they are of uniform size, there are enough of them to weigh heavier objects and they are not too large or heavy themselves, so that we can make the scales balance exactly.

ANSWERS

Question ❶: The shoe has a mass of 5 cubes.
The shoe has a mass of 3 bean bags.
The shoe has a mass of 9 crayons.

Question ❷:

Object	Mass in 🎲
🖊️	8
✏️ case	12
🟦 cube	10
📏 ruler	1

Question ❸: Car + book = 6 + 5 = 11 cubes
10 cubes = 4 + 6 = teddy + car
10 cubes = 5 + 5 = book + book
book + 1 cube = 6 cubes = car

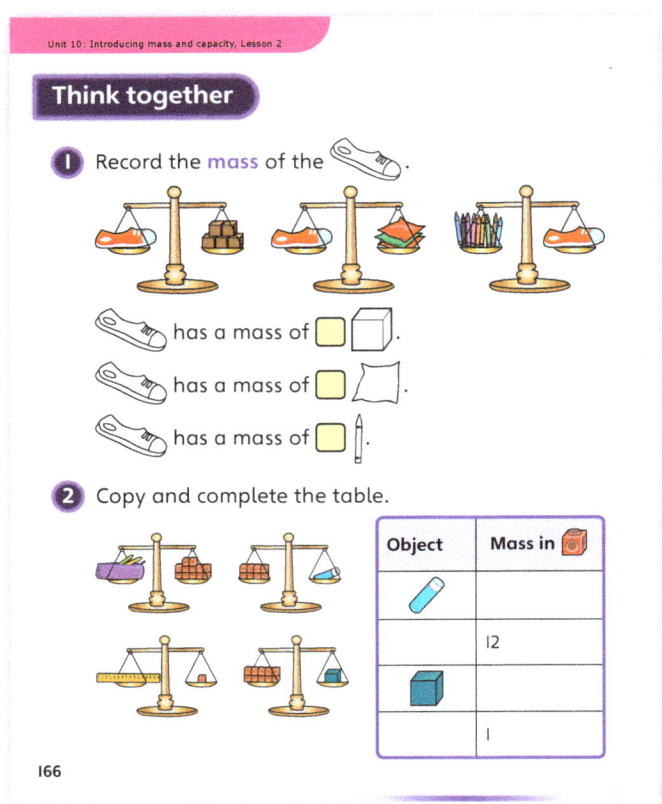

PUPIL TEXTBOOK 1B PAGE 166

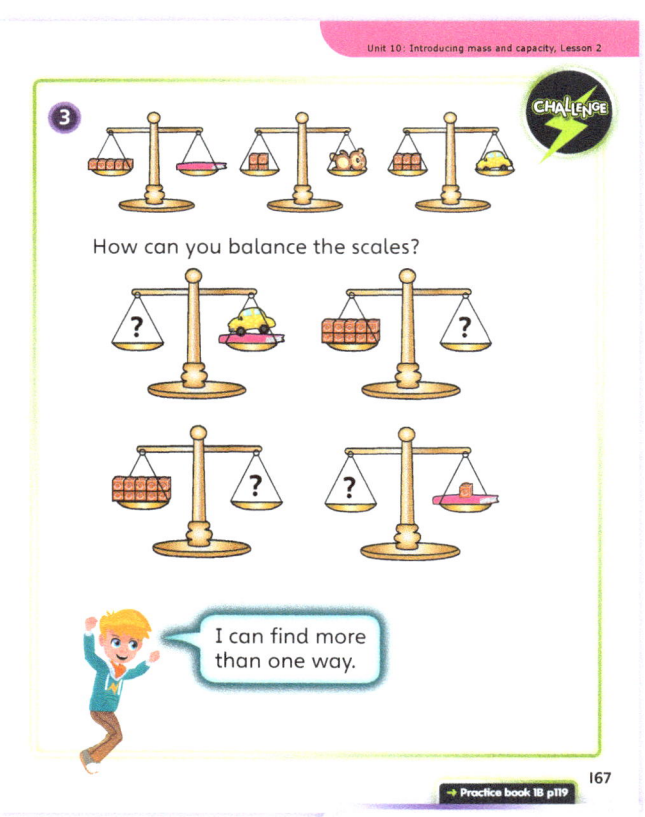

PUPIL TEXTBOOK 1B PAGE 167

Unit 10: Introducing mass and capacity, Lesson 2

Practice

WAYS OF WORKING Independent thinking

IN FOCUS Questions ❶ and ❷ encourage children to measure using non-standard units. Question ❸ involves some interesting calculations. The thinking that is needed here is essentially algebraic in nature, with children needing to balance a calculation in each case. Notice that cubes can be added to either side of the balance scales, and each part can be answered in many ways.

STRENGTHEN Depending on the accuracy of the available equipment, it may be possible to provide additional practice by modelling some calculations using real apparatus. For example, you could try weighing two toys (individually) using cubes, and then check that the total weight of both toys is the sum of the individual results.

DEEPEN Question ❺ involves some deeper mathematical reasoning – children will need to use the first diagram to work out which of the packages is heavier, and then use that information to decide which of the alternative scales follows logically. Again, this is essentially algebraic thinking, and the challenge of creating more puzzles like this would be a good way to encourage the deeper development of logical thinking.

THINK DIFFERENTLY Question ❸ uses see-saws instead of balance scales, to show balanced relationships between different sets of items. Children should recognise that, like balancing balance scales, the see-saws will be balanced if both sides have objects with equal mass. They have to use their logic to determine which objects will be needed to balance the see-saws.

ASSESSMENT CHECKPOINT Question ❷ checks the key learning points of this lesson – do children recognise that the weight of an object can be found by counting the number of units that it balances with on a set of balance scales? Can children work out that the boot must weigh 20 cubes?

ANSWERS Answers for the **Practice** part of the lesson can be found in the *Power Maths* online subscription.

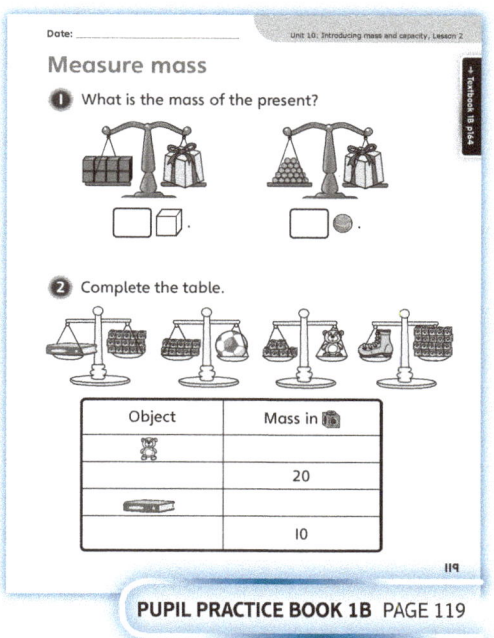

PUPIL PRACTICE BOOK 1B PAGE 119

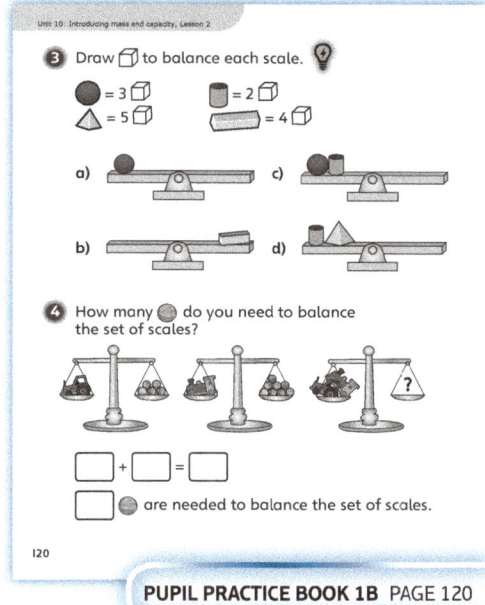

PUPIL PRACTICE BOOK 1B PAGE 120

Reflect

WAYS OF WORKING Pair work

IN FOCUS The loaf of bread in this example is weighed against two different collections of cubes. Children should notice that in the second diagram there are two different sizes of cubes, and in particular that '7 cubes' is not a good description for this diagram.

ASSESSMENT CHECKPOINT Check that children understand that the second example is different from the first. Ask: *How many of these small cubes weigh the same as one of the big cubes?*

ANSWERS Answers for the **Reflect** part of the lesson can be found in the *Power Maths* online subscription.

After the lesson

- Did children understand the idea that the 'non-standard unit' they choose needs to be of a uniform mass?
- Would children benefit from additional practical work on this topic?
- Are children able to compare mass when using the same non-standard units of measure?

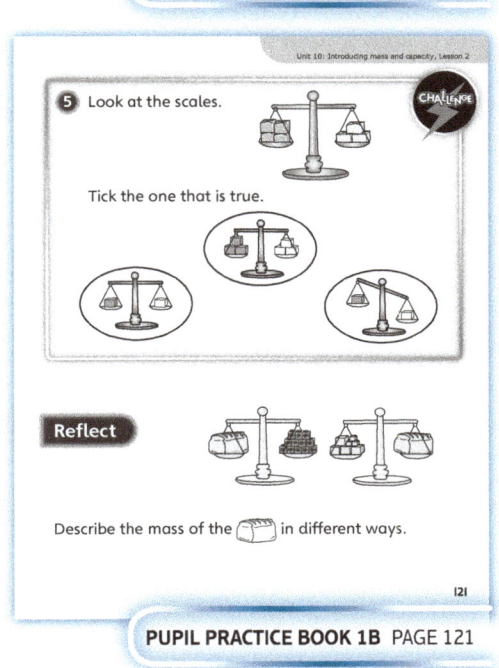

PUPIL PRACTICE BOOK 1B PAGE 121

Unit 10: Introducing mass and capacity, Lesson 3

Compare mass

Learning focus
In this lesson, children will use a variety of non-standard units to compare and order objects by their mass.

Before you teach
- Are children confident in measuring objects using non-standard units such as cubes?
- Would children benefit from using balance scales and cubes to try out some of the activities practically?
- Can children order numbers to 20?

NATIONAL CURRICULUM LINKS

Year 1 Measurement

Compare, describe and solve practical problems for: mass/weight (for example, heavy/light, heavier than, lighter than).

ASSESSING MASTERY

Children can compare and order small sets of objects, based on mass measured using non-standard units.

COMMON MISCONCEPTIONS

Children may not understand that we can compare and order a whole set of objects on the basis of their measured mass. This means that we only need to weigh (and record) each object once, rather than making repeated comparisons between different pairs. Ask:
- *Does knowing the mass of each toy make it easier to put them in order? Why not just use balance scales to compare one toy with another?*

STRENGTHENING UNDERSTANDING

It may be useful to refer back to the physical process of weighing various objects until children understand that the number of cubes is all that is really needed. Talk them through the process: *This can balances 10 cubes; this can weighs the same as 10 cubes; this can weighs 10 cubes.*

GOING DEEPER

Throughout this lesson, there are opportunities for what is essentially algebraic thinking. The process of manipulating inequalities and relationships helps to build the foundations of algebra. Encourage this type of thinking by asking questions like: *How do you know? Can you prove it? Can you find any other ways to do it?*

KEY LANGUAGE

In lesson: heaviest, lightest, heavier, lighter, greater than (>), less than (<), weights, weighs, comparing, order, equal

Other language to be used by the teacher: prove, balance, balanced

STRUCTURES AND REPRESENTATIONS

Sets of balance scales, cubes

RESOURCES

Optional: balance scales, cubes or similar to use as non-standard units of weight, small objects to weigh

 In the eTextbook of this lesson, you will find interactive links to a selection of teaching tools.

Quick recap

Ask children to measure the mass of two objects using balance scales and cubes. Ask them to compare the mass of the two objects using the balance scales. What do they notice about the mass in cubes of the heavier object? Repeat for other pairs of objects.

Unit 10: Introducing mass and capacity, Lesson 3

Discover

WAYS OF WORKING Pair work

ASK

- Question 1 a): *Can you find the shelves where the boxes have to go? How many shelves are there?*
- Question 1 a): *Why do you think the heaviest items need to go on the bottom shelf and the lightest ones on the top shelf?*
- Question 1 b): *I think someone has already started weighing the boxes. What have they found out so far?*

IN FOCUS The lesson starts with a practical situation where we might need to sort objects by mass. Before attempting question 1 a), children may need a little explanation to justify the order in which the items need to be sorted, but once that is understood the task should become clear.

PRACTICAL TIPS Give small groups of children a balance scale, cubes and three objects to weigh. Ask them to find out which object weighs the most, which object weighs the least, and to order the objects from heaviest to lightest.

ANSWERS

Question 1 a): The box of chocolate bars goes on the top shelf.

The box of cans goes on the bottom shelf.

Question 1 b): The order from heaviest to lightest is: box of cans, box of bananas, box of chocolate bars.

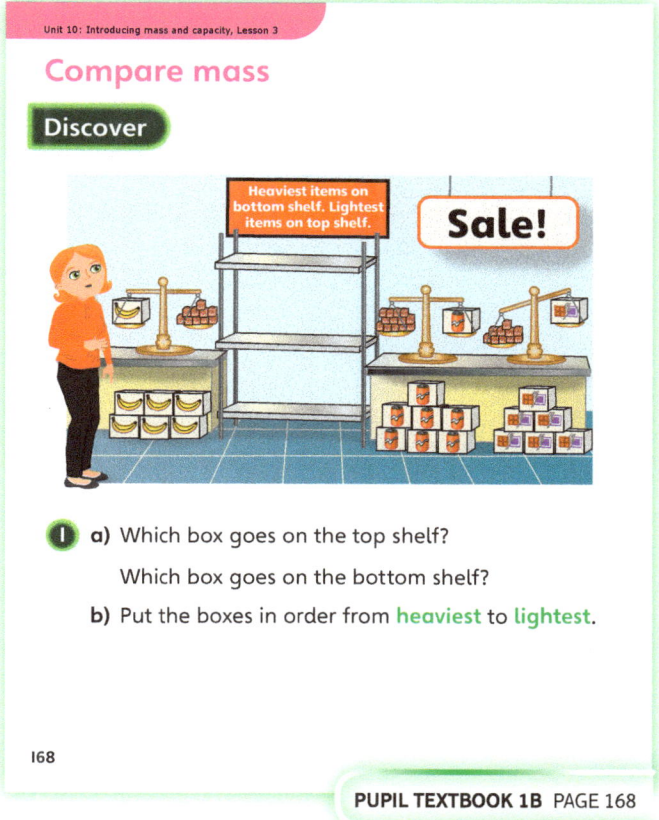

PUPIL TEXTBOOK 1B PAGE 168

Share

WAYS OF WORKING Whole class teacher led

ASK

- Question 1 a): *What does Flo mean when she says that she compared the number of cubes? How did she find out the number of cubes that balanced each box?*
- Question 1 a): *Do you remember how to put numbers in order? Why is that important here?*

IN FOCUS The key idea here is that, once we have the masses as numbers of cubes, we can effectively forget about cubes, mass and scales and just deal with the numbers themselves. This does not need to be discussed explicitly, but you could say something like: *Once we have all of the masses, we can just use what we already know about numbers to work out what order the masses must go in.*

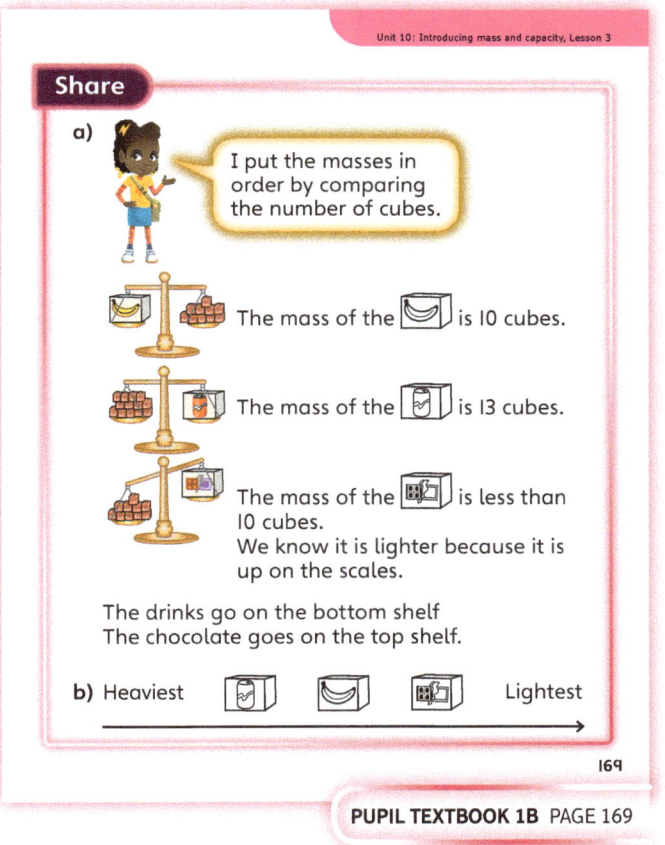

PUPIL TEXTBOOK 1B PAGE 169

Unit 10: Introducing mass and capacity, Lesson 3

Think together

WAYS OF WORKING Whole class teacher led (I do, We do, You do)

ASK
- Question ②: *Which weighs more, the teddy bear or the car?*
- Question ②: *What about the car and the horse; how would you work out which one of those two was the heavier?*
- Question ②: *What order will the three toys have to go in? Which one goes at the top?*

IN FOCUS Question ② includes most of the key learning for this lesson. Notice that the first set of balance scales is not really needed; the third set establishes that the car is heavier than the teddy bear, and the second set shows that the horse is a little heavier still. We do not need the actual masses (in cubes) to be able to order the items in this case, but it might be useful to work them out if some children find the comparison difficult.

STRENGTHEN Children who are finding the comparisons difficult could be encouraged to write the measurements down before attempting to order them. Reducing the pressure on working memory may make it easier to carry out the comparisons.

DEEPEN Challenge children to find as many answers as possible for question ③. You can make the question more challenging by removing one or more of the 'unbalanced' scales. For example, relax the conditions that the blue striped vehicle weighs more than 4 cubes, and/or the condition that the yellow spotty car weighs less than 8 cubes.

ASSESSMENT CHECKPOINT Check that children can do all three parts of question ①. There is a significant step in reasoning required here; we are comparing the masses of the bags, but the balance scales are not being used to compare the bags directly. Ensure children understand that once we have the masses (in cubes) we can find the correct order directly from the numbers.

ANSWERS

Question ① a): The bag of pears is heavier than the bag of tomatoes.

Question ① b): The bag of tomatoes is lighter than the bag of pears.

Question ① c): The bag of bananas is equal to the bag of pears.

Question ②: The teddy is the lightest (5 cubes), so it goes on the top shelf.
The horse is the heaviest (10 + 5 = 15 cubes), so it goes on the bottom shelf.
The car goes in the middle (5 + 5 = 10 cubes).

Question ③: Red tractor = blue truck + yellow car
Red tractor = 12 cubes
The blue truck is heavier than 4 cubes.
The yellow car is lighter than 8 cubes.
There are various combinations for blue truck and yellow car, but they must add up to 12:

Blue	5	6	7	8	9	10	11
Yellow	7	6	5	4	3	2	1

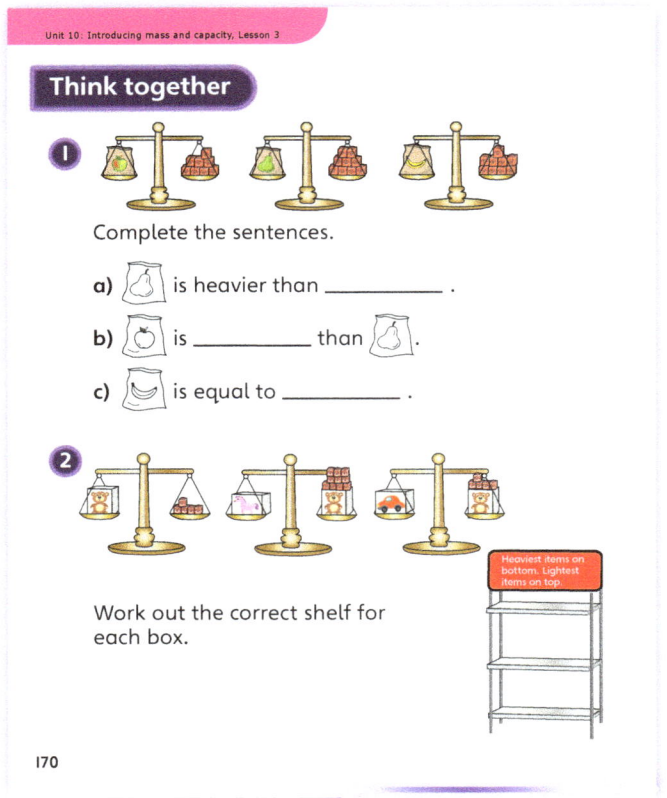

PUPIL TEXTBOOK 1B PAGE 170

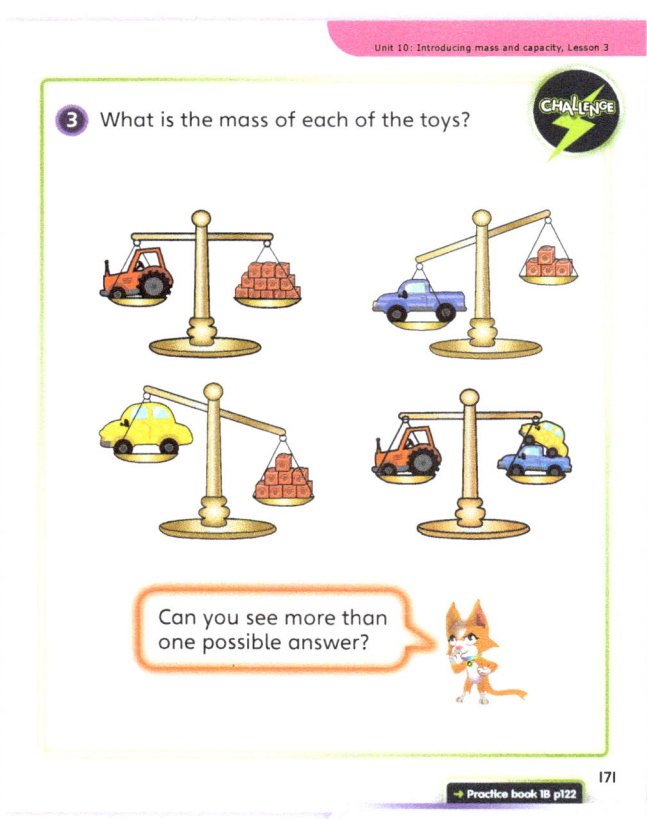

PUPIL TEXTBOOK 1B PAGE 171

Unit 10: Introducing mass and capacity, Lesson 3

Practice

WAYS OF WORKING Independent thinking

IN FOCUS Question ➋ involves a direct comparison of masses, using recorded values rather than pictures of sets of balance scales. It is worth pointing out to children that once they are happy working directly with the numbers in this way, the question actually becomes easier. Explain that they are still 'really' working with masses, but they will get the right answers if they simply use their knowledge of numbers to compare the values directly.

STRENGTHEN Where children find this work difficult, it will be important to investigate the exact nature of the difficulty. Can they successfully count and record the number of blocks for each object? Can children order the resulting list of numbers? Can they relate that order to the concepts of 'heavier' and 'lighter'? Use probing questions to investigate.

DEEPEN Question ➍ is an example of a problem where good mathematical reasoning is required. You might need to help children understand that they cannot use the balance scales themselves to determine the individual mass of each object in marbles. Instead, they should use the three balance scale images to order the masses from lightest to heaviest. From there, they can use the table to match each object with its equivalent mass in marbles. Children could be encouraged to make up more puzzles of this sort.

THINK DIFFERENTLY Question ➌ asks children to compare the masses of the fruit, without showing balanced scales for the banana or the pineapple. They will not be able to work out the exact masses in cubes for these fruits and will need to recognise what the balance scales show in relation to the apple and the orange.

ASSESSMENT CHECKPOINT Question ➋ can be used to check that children have understood the main point of this lesson – that they can use numerical values of masses to compare objects using the terms 'heavier' and 'lighter'. In question ➌, assess how children approach it. Do they understand what the scales for the banana and pineapple show in terms of whether these fruits are lighter or heavier than the apple?

ANSWERS Answers for the **Practice** part of the lesson can be found in the *Power Maths* online subscription.

Reflect

WAYS OF WORKING Independent thinking

IN FOCUS This question acts as a reminder of the importance of using uniform items as non-standard units.

ASSESSMENT CHECKPOINT Children may suggest that the third cake is the lightest, because the differently-sized cube is smaller than the others. This is not totally unreasonable, but it is still worth making the point that we cannot really tell how heavy the small cube is just by looking at it.

ANSWERS Answers for the **Reflect** part of the lesson can be found in the *Power Maths* online subscription.

After the lesson

- Are children able to compare small sets of objects using words like 'heaviest' and 'lightest'?
- Can children compare masses using the numerical values of the masses?
- Where difficulties remain, have you been able to establish what these difficulties are? For example, are children recording masses correctly but then ordering the list of masses incorrectly?

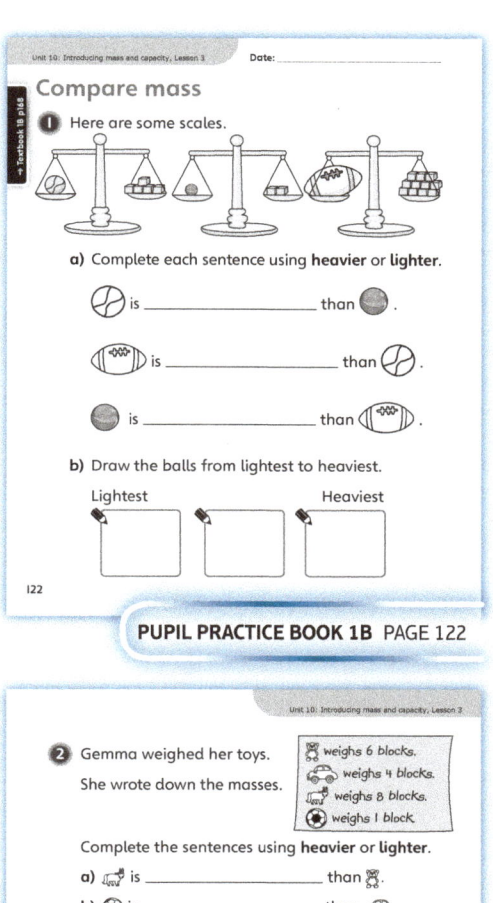

PUPIL PRACTICE BOOK 1B PAGE 122

PUPIL PRACTICE BOOK 1B PAGE 123

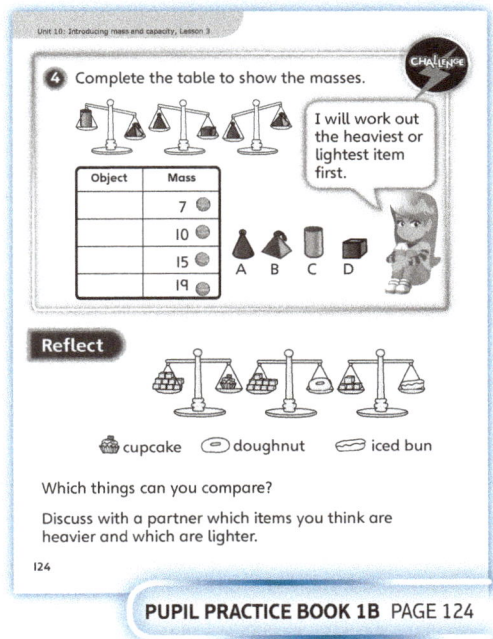

PUPIL PRACTICE BOOK 1B PAGE 124

Unit 10: Introducing mass and capacity, Lesson 4

Full and empty

Learning focus
In this lesson, children will compare a range of objects according to their capacity.

Before you teach
- Do you have suitable facilities in your classroom to manage practical work with water and containers?
- Do children understand the concept of a half?

NATIONAL CURRICULUM LINKS

Year 1 Measurement

Compare, describe and solve practical problems for: capacity and volume (for example, full/empty, more than, less than, half, half full, quarter).

Measure and begin to record the following: capacity and volume.

ASSESSING MASTERY

Children can compare and order a set of containers according to their capacity. Children can describe the volume of a container as full, empty, half full, less than half full, more than half full. Children should understand that the capacity of a container is the maximum amount that it can hold, whereas the volume of liquid in a container is the amount of liquid actually inside. This can be less than the capacity.

COMMON MISCONCEPTIONS

Children may struggle to see that containers of different shapes can have the same capacity. Display two containers of different shapes but with the same capacity. Ask:
- *Do these containers have the same capacity? How could you tell?*

STRENGTHENING UNDERSTANDING

Children will benefit from practical experience with a carefully selected range of containers. Measuring cylinders are a good choice – the shape means that the height of the liquid is directly proportional to the volume, and transparent sides mean that the level can be read easily.

GOING DEEPER

Provide children with a range of different containers and ask them to predict which one has the greatest capacity, or to order the containers from greatest to least capacity. Can they prove their predictions are correct?

KEY LANGUAGE

In lesson: full, empty, half full, less than, more than, level

Other language to be used by the teacher: capacity, container, contains, amount, glass, jug, cylinder

STRUCTURES AND REPRESENTATIONS

Glasses and jugs containing liquids

RESOURCES

Optional: plastic glasses, beakers, jugs, measuring jugs and cylinders, water or squash to fill the containers

 In the eTextbook of this lesson, you will find interactive links to a selection of teaching tools.

Quick recap
Find out how familiar children are with the words 'full' and 'empty'. Show children a number of containers with one full, one empty and others with varying amounts. Ask: *Which is full? How do you know? Which is empty? How do you know?*

Unit 10: Introducing mass and capacity, Lesson 4

Discover

WAYS OF WORKING Pair work

ASK
- Question 1 b): *What is Molly asking for?*
- Question 1 b): *How much squash do you think Fred should put into the glass?*

IN FOCUS Questions 1 a) and b) check that children can apply the ideas of 'more' and 'less' to volumes and capacities. Make sure children understand that Molly is saying she wants Fred to give her a different glass containing less squash than the one that she is pointing to.

PRACTICAL TIPS Recreate the scenario in the classroom by pouring out four glasses of squash or water, each with different amounts.

ANSWERS

Question 1 a): This glass is full:

This glass is empty:

Question 1 b): Molly wants this glass:

PUPIL TEXTBOOK 1B PAGE 172

Share

WAYS OF WORKING Whole class teacher led

ASK
- Question 1 b): *Why do you think Dexter moved the glasses so that they are side by side?*
- Question 1 b): *How does the dotted line help Dexter to compare the amount of squash in each glass?*
- Question 1 b): *The empty glass has also got less squash than the one that Molly was pointing at. Why should Fred not give her that one?*

IN FOCUS Make sure that children can 'read' the pictures in this activity properly. To answer question 1 b) in practice, you would compare the level of the liquid in each of the glasses by adjusting your eye level to that of the liquid, and the dotted line is used to indicate where the comparison needs to be made.

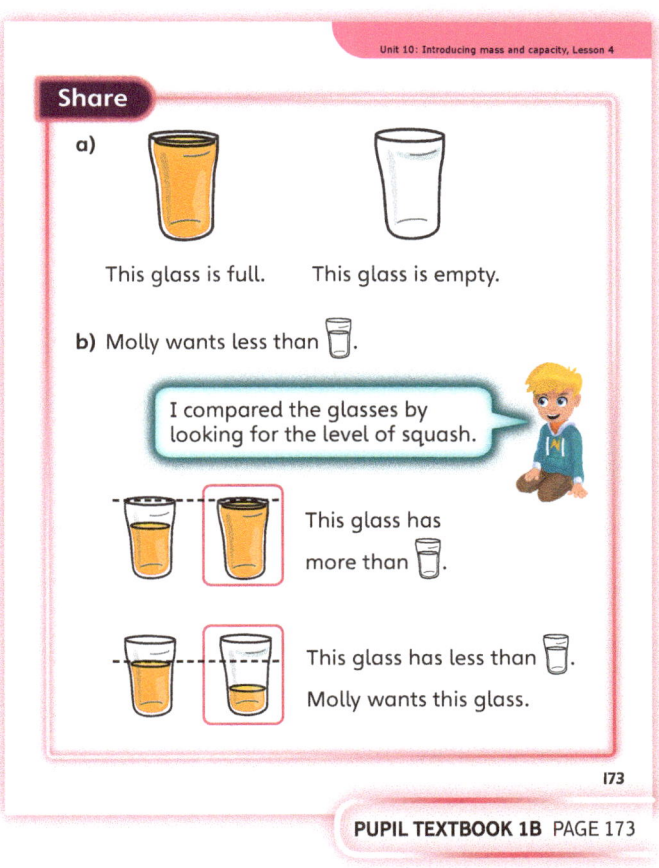

PUPIL TEXTBOOK 1B PAGE 173

Unit 10: Introducing mass and capacity, Lesson 4

Think together

WAYS OF WORKING Whole class teacher led (I do, We do, You do)

ASK
- Question ❷: *How would you find out which of the glasses holds more?*
- Question ❷: *What would happen if you filled up the jug and then tried to empty it into one of the glasses?*

IN FOCUS For question ❷, ask: *What happens when you try to fill the big jug using a smaller glass?* Look for suggestions that you would be able to fill and empty out the smaller glass several times into the jug, and that the smaller the glass, the more times you would have to pour from a smaller glass before the jug is full.

STRENGTHEN In question ❸, where children only offer one suggestion for each request, prompt them to go further. For example, ask: *Yes, the first glass is full. But is that the only one that you could give this person?*

DEEPEN Deepen question ❸ by challenging children to find all the possible combinations to satisfy the requests.

ASSESSMENT CHECKPOINT Use questions ❶ and ❸ to check children's understanding of the ideas of 'full', 'half full', and 'more than half full'. In question ❸, probe children's understanding to see whether they understand that there may be more than one drink that meets each of the requests. In particular, do they understand that a full glass is also 'more than half full'?

ANSWERS

Question ❶: E, D, A, C, B

Question ❷:  holds more.

Question ❸: Fred:

Molly:

Sam:

No one would like:

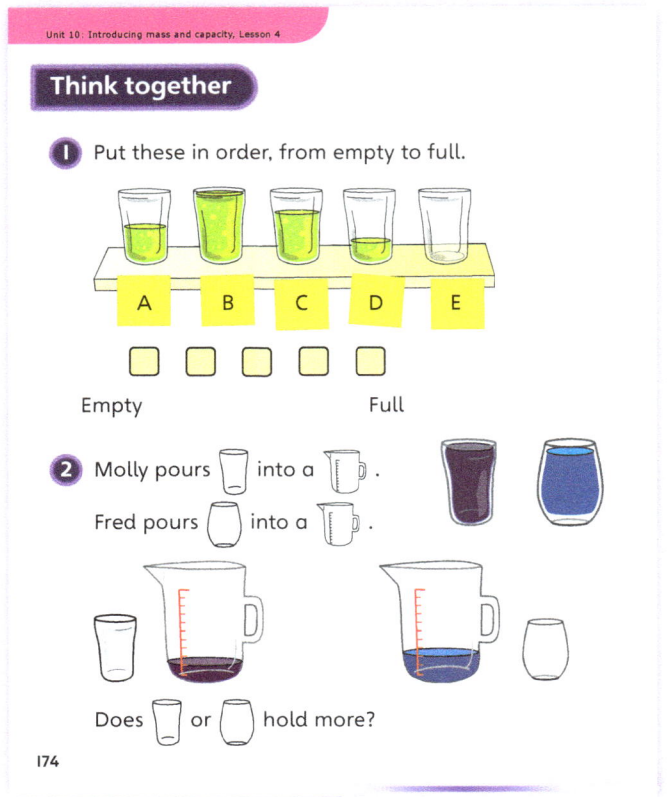

PUPIL TEXTBOOK 1B PAGE 174

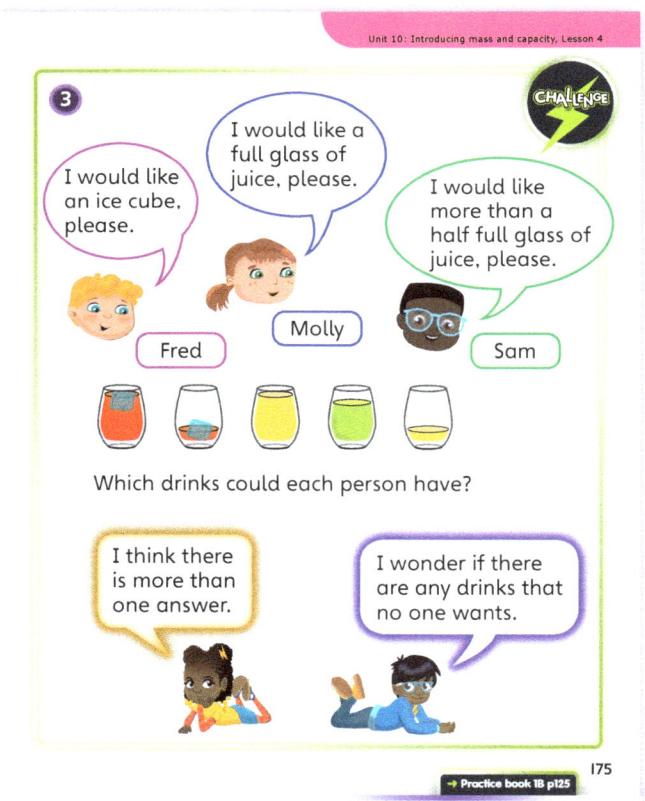

PUPIL TEXTBOOK 1B PAGE 175

Unit 10: Introducing mass and capacity, Lesson 4

Practice

WAYS OF WORKING Independent thinking

IN FOCUS Question ❶ checks children's understanding of 'empty' and 'full', and question ❷ checks that children are able to order capacity, from full to empty. Most children will do this by inspection. Question ❹ will help children to check that they understand terms such as 'more than' and 'less than' with respect to capacity.

STRENGTHEN Children who find the questions in this activity difficult could be supported by the use of practical equipment. For example, in question ❷, help children to fill a set of containers with water to the approximate levels shown, and then move them so that they are in decreasing order of 'fullness'.

DEEPEN Question ❺ could be extended by asking children to make a list of containers that they might find in a kitchen, for example, a kettle, cups, mugs, glasses, pans of various sizes, egg cups, and so on. Ask: *Can you put the list in order by capacity? How would you check that the order was correct?*

THINK DIFFERENTLY Question ❹ gives children an opportunity to use logic to work out what level of drink meets the two requests. They need to understand the terms 'more than' and 'less than' in relation to capacity to answer this question.

ASSESSMENT CHECKPOINT Question ❹ can be used to check that children have understood the ideas of 'more than' and 'less than' in relation to volume and capacity. In the first example, anything that is visibly fuller than the glass shown is correct, including a full glass. In the second example, an empty glass is not acceptable, because a minimum amount is indicated.

ANSWERS Answers for the **Practice** part of the lesson can be found in the *Power Maths* online subscription.

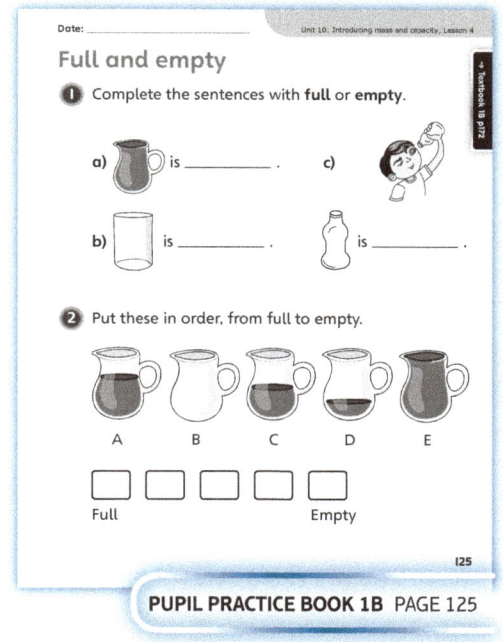

PUPIL PRACTICE BOOK 1B PAGE 125

PUPIL PRACTICE BOOK 1B PAGE 126

Reflect

WAYS OF WORKING Independent thinking

IN FOCUS This question reviews some of the key vocabulary relating to this topic.

ASSESSMENT CHECKPOINT Check that children can identify the full and empty containers, and that they understand the idea of 'more than' in relation to volumes and capacity.

ANSWERS Answers for the **Reflect** part of the lesson can be found in the *Power Maths* online subscription.

After the lesson

- Did children use practical apparatus to support their learning?
- Can children explain how to compare the differently shaped containers by pouring the volumes into identical measuring jugs and comparing the level of the liquid?
- Do children understand that a container has a maximum capacity; that any volume less than the capacity will not fill the container and any volume over the capacity will fill the container with some left over, or fill it more than once?

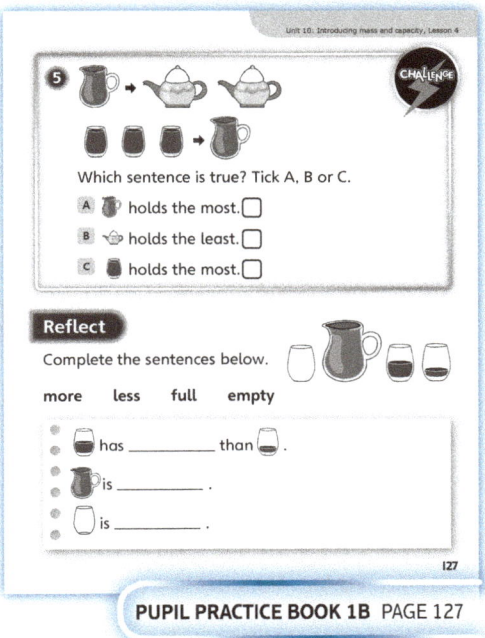

PUPIL PRACTICE BOOK 1B PAGE 127

Unit 10: Introducing mass and capacity, Lesson 5

Measure capacity

Learning focus
In this lesson, children will estimate and measure the capacity of a range of containers, using a variety of non-standard units.

Before you teach
- What practical work will you provide to reinforce the learning in this lesson?
- What equipment will you need to organise?

NATIONAL CURRICULUM LINKS

Year 1 Measurement

Measure and begin to record the following: capacity and volume.

ASSESSING MASTERY

Children can estimate and measure the capacity of a range of containers, using non-standard units such as cups and scoops. They are aware of the need to measure using uniform units and they understand the effect of altering the size of the units used.

COMMON MISCONCEPTIONS

Children may not recognise the importance of measuring using uniform units. This problem can manifest itself when there is more than one type of container of a particular type available, for example, more than one size of cup. Ask:
- *I am going to measure the capacity of this pot using cups. Can I use any cups? Can I swap cups in the middle of measuring?*

STRENGTHENING UNDERSTANDING

The relationship between the size of the unit and the number needed to fill a container may be difficult for children to grasp: the bigger the measure you use, the fewer you need to fill a container. Children may benefit from practical work with sand or water to make sense of this relationship.

GOING DEEPER

Children may be able to reason more quantitatively about the relationship between units of different sizes. For example, ask: *Suppose a container can be filled by 6 scoops or 3 cups. If a larger container could be filled by 12 scoops, how many cups would be needed to fill it?*

KEY LANGUAGE

In lesson: estimate, capacity, too few, too many

Other language to be used by the teacher: measure, container

STRUCTURES AND REPRESENTATIONS

Various familiar containers to represent different capacities, for example, buckets, spoons, cups and scoops

RESOURCES

Optional: practical apparatus and materials to model measuring including sand, rice, water, beakers, buckets, scoops, spoons

 In the eTextbook of this lesson, you will find interactive links to a selection of teaching tools.

Quick recap
Show children two different sized cups. Ask: *Which holds more? How can you check?* Ask children to investigate and find out which one holds more.

Unit 10: Introducing mass and capacity, Lesson 5

Discover

WAYS OF WORKING Pair work

ASK

- Question 1 b): *Which one do you think takes the longest to fill the bucket? Why? Which one do you think fills the bucket quickest? Why?*

IN FOCUS Questions 1 a) and b) provide an opportunity for children to think about measuring capacity in a practical context. Encourage discussion around who will take longer to fill the bucket, allowing children (at this point) to use language such as 'The spoon will take longer because it is small'. This will encourage children to think about the capacity of objects.

PRACTICAL TIPS Give children different containers, a bucket and some sand, rice, water, etc., and investigate how many times they need to fill their container in order to fill a bucket.

ANSWERS

Question 1 a): Children should estimate about 10 cups of sand will fill the bucket.

Question 1 b): Children should estimate about 3 spades of sand will fill the bucket.

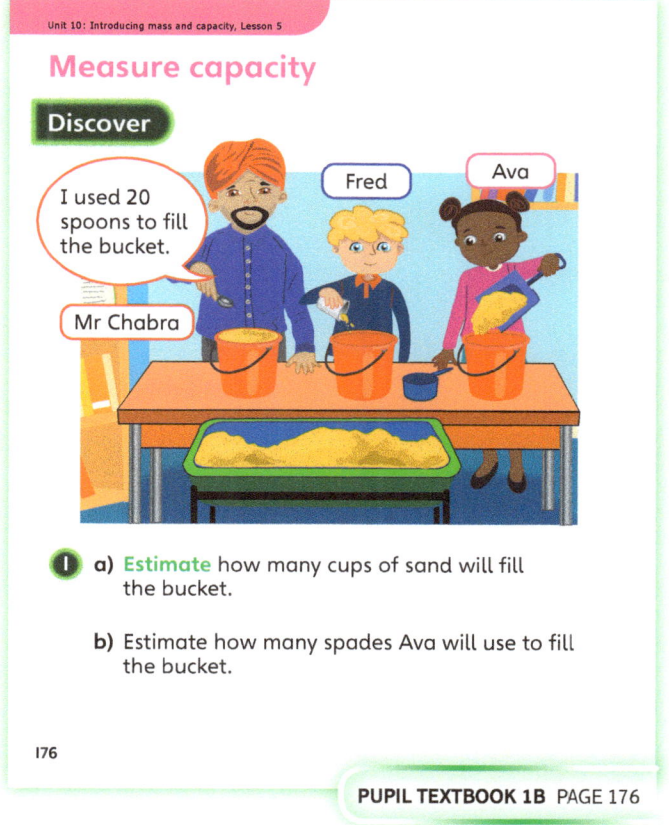

PUPIL TEXTBOOK 1B PAGE 176

Share

WAYS OF WORKING Whole class teacher led

ASK

- Question 1 a): *Why do you need fewer cups than spoons to fill the bucket?*
- Question 1 b): *Will you need more or fewer spades than cups to fill the bucket? How do you know? How many more or fewer do you think? How could you check?*

IN FOCUS In questions 1 a) and b), encourage discussion around how the size of the object changes the number of times it is needed to fill the bucket. Ensure children recognise that the capacity of the bucket does not change; it is just the units they are using to measure in that change.

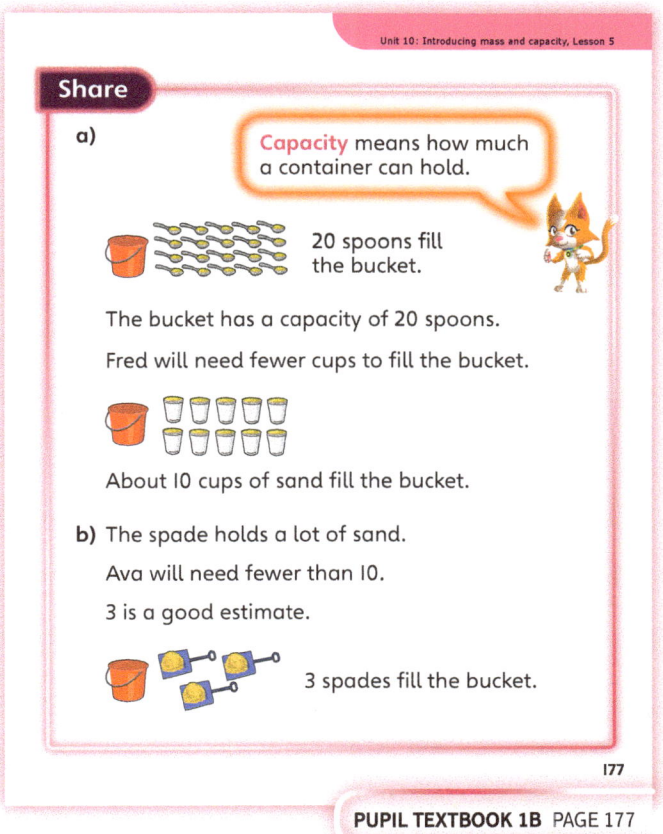

PUPIL TEXTBOOK 1B PAGE 177

215

Unit 10: Introducing mass and capacity, Lesson 5

Think together

WAYS OF WORKING Whole class teacher led (I do, We do, You do)

ASK
- Question ❶: *How much sand did each child put in the bucket? How much sand did they put in altogether?*
- Question ❷: *Which container is the biggest? Which is the smallest?*

IN FOCUS Question ❶ provides a useful context for exploring the links between addition and subtraction with capacity. You could use the same context to ask further questions. Ask: *What if Ava took three spoonfuls of sand out of the full bucket? How many spoonfuls of sand would be left in the bucket then?*

STRENGTHEN Children who find question ❸ difficult may benefit from modelling the same situation with real apparatus. Discuss what they notice about the fullness of the glasses. Ask: *Are all the glasses full?*

DEEPEN Question ❸ can be extended to provide further early experience of proportional reasoning. Ask: *Imagine you were filling glasses from a jug like this. What would happen to the number of glasses that you needed if you used bigger glasses? What would happen if you used smaller glasses? How do you know?*

ASSESSMENT CHECKPOINT Use question ❸ to check that children understand the need for uniform measurement size. In this case, the measurement size is uniform if each glass holds the same amount of liquid. Ask: *If the people in the picture are going to use glasses to measure, what do they need to check?* (Establish that all the glasses need to have the same amount of liquid in them.) Mrs Shaw's glasses do not all have the same amount amount of liquid. You might also want to elicit that Mr Chabra is incorrect as his jug holds more than 6 glasses.

ANSWERS

Question ❶: 4 + 5 = 9
9 spoons fill the bucket.

Question ❷: Children should complete the table to show:
4 bowls, 15 plain cups, 12 striped cups.

Question ❸: Mrs Hodge is right. Mr Chabra's jug holds more than 6 glasses. Mrs Shaw's jug holds less than 5 glasses because not all the glasses are full.

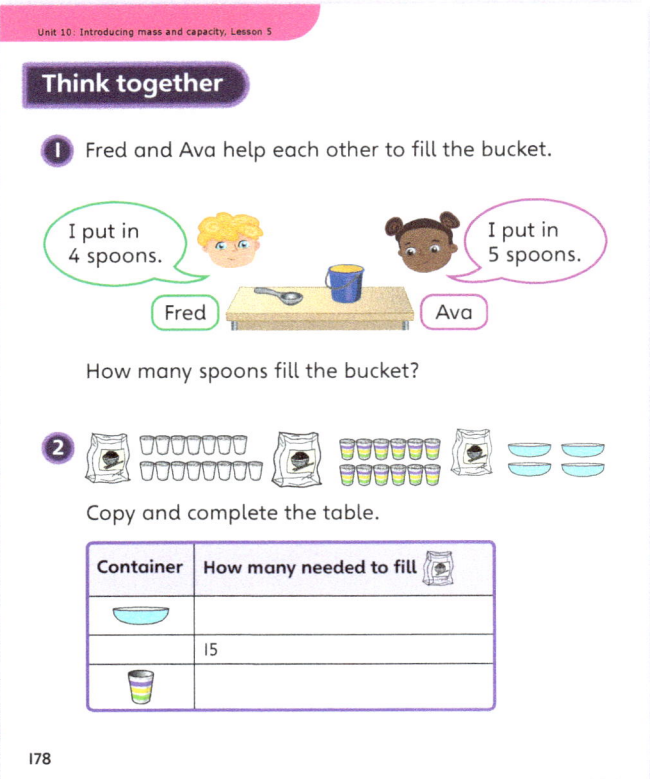

PUPIL TEXTBOOK 1B PAGE 178

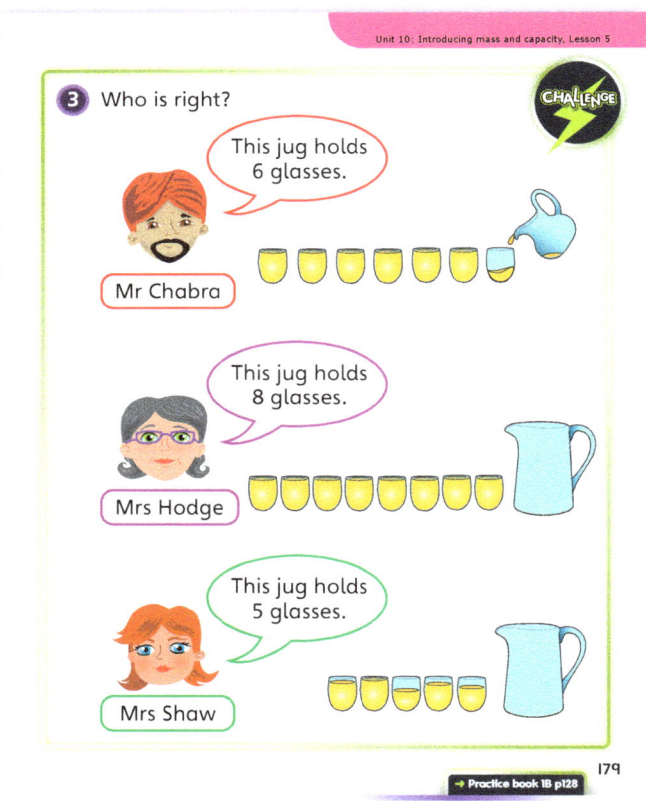

PUPIL TEXTBOOK 1B PAGE 179

216

Unit 10: Introducing mass and capacity, Lesson 5

Practice

WAYS OF WORKING Independent work

IN FOCUS Questions ❶ and ❷ use pictorial representations, so children can see non-standard units of measure being used to represent the capacity of a jug and a wheelbarrow. Question ❸ uses a glass as a non-standard unit to measure the capacity of a range of familiar containers. This scenario should make sense to most children, and provides a simple setting to explore ideas relating to addition of capacities.

STRENGTHEN It is possible to model the 'addition' in question ❸ using real apparatus, and some children will benefit from doing so. Record the number of glasses needed to fill each container, empty the containers into the saucepan, and then count how many glasses can be filled from the contents of the pan.

DEEPEN The relationships in question ❸ can form the basis for some more challenging questions, with links to both proportional and algebraic reasoning. For example, ask: *If you have 3 jugs and 2 beakers, how many bottles will you be able to fill? Which is more, 4 bottles or 3 jugs?*

In question ❹, help children to see they need to do repeated additions: 5 + 5 + 5 in a), and 5 + 5 + 5 + 5 + 5 in b). This kind of repeated addition forms an introduction to multiplication, which children will begin in the next unit (Unit 11).

ASSESSMENT CHECKPOINT Check that children recognise the need to add capacities in question ❸, and pay attention to the techniques that they use to carry out the additions. Some children may make use of the diagrams at the top of the page, counting the images of the glasses under the jug and the bottle, while others may make use of known number bonds.

ANSWERS Answers for the **Practice** part of the lesson can be found in the *Power Maths* online subscription.

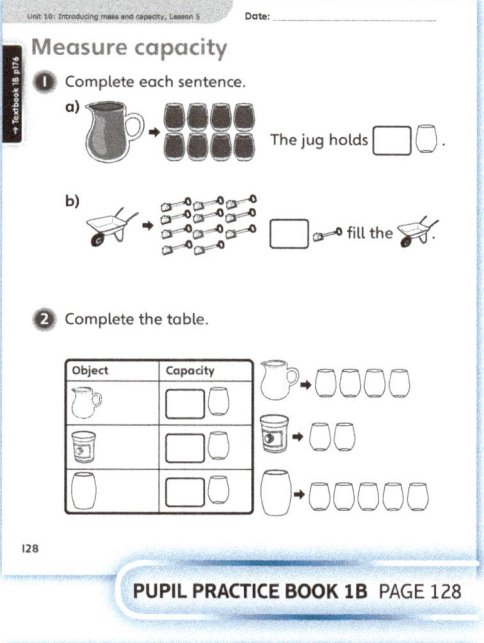

PUPIL PRACTICE BOOK 1B   PAGE 128

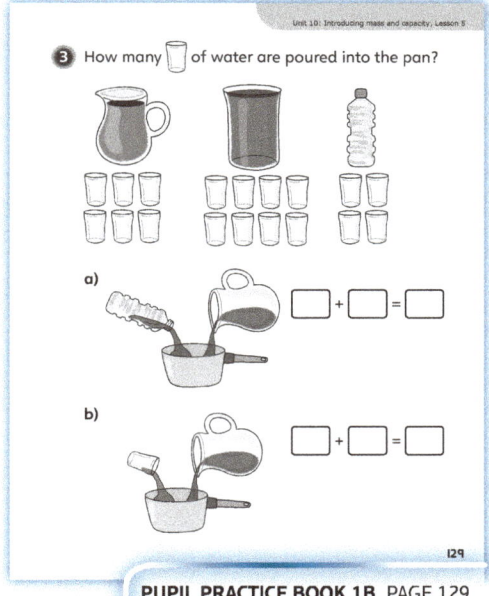

PUPIL PRACTICE BOOK 1B   PAGE 129

Reflect

WAYS OF WORKING Pair work

IN FOCUS This activity explores equivalences between larger and smaller capacities in both 'directions'. In the first example, the glass is the unit used to measure the capacity of the jug, while in the second example the jug is the unit used to measure the capacity of the pan.

ASSESSMENT CHECKPOINT Check that all children understand what is being asked here and that they understand the different roles that the jug plays in each part of the activity.

ANSWERS Answers for the **Reflect** part of the lesson can be found in the *Power Maths* online subscription.

After the lesson

- Do children understand the need for units to be of a uniform size?
- Do children understand that a container can be either the unit of measurement or the thing being measured, depending on the context?
- Do children understand the relationship between the size of the unit and the numerical value of the measured capacity?

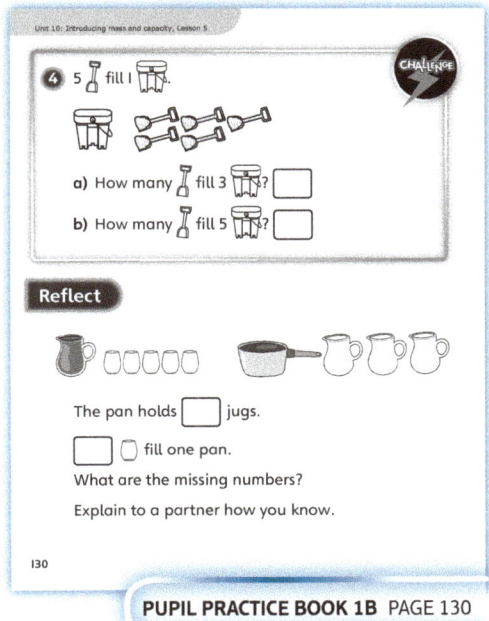

PUPIL PRACTICE BOOK 1B   PAGE 130

Unit 10: Introducing mass and capacity, Lesson 6

Compare capacity

Learning focus
In this lesson, children will use a variety of non-standard units to compare and order objects according to their capacity.

Before you teach
- Are children able to compare pairs of containers, given their respective capacities measured in convenient non-standard units?
- Would children benefit from further practical work to support their learning in this topic?
- Are children beginning to use proportional reasoning?

NATIONAL CURRICULUM LINKS

Year 1 Measurement

Compare, describe and solve practical problems for: capacity and volume (for example, full/empty, more than, less than, half, half full, quarter).

ASSESSING MASTERY

Children can compare and order sets of containers according to their capacities, measured using a variety of non-standard units.

COMMON MISCONCEPTIONS

When comparing the capacity of several containers, children may find it difficult to understand that the unit is a container like any other, but that it plays a special role and needs to fit into the containers that are being measured a convenient number of times. This means that the unit needs to be a uniform size and should be small enough, but not too small. Ask:
- I am going to measure the capacity of a pond. Should I do this using cups or buckets? Why?

STRENGTHENING UNDERSTANDING

This lesson makes use of inequality statements with three terms, such as 15 > 5 > 2. Some children may find this more difficult to understand than the more familiar two-term version, such as 5 > 2. Encourage children to read this as '15 is greater than 5 and 5 is greater than 2'; inserting 'and' in the sentence to help make the meaning clear.

GOING DEEPER

The work in this lesson provides further opportunities for informal approaches to proportional reasoning. For example, ask: *If 2 cups fill a jar and 6 cups fill a bowl, how many jars will fill a bowl?*

KEY LANGUAGE

In lesson: capacity, fill, greater than (>), less that (<), equal (=), greatest, least, container

Other language to be used by the teacher: compare, more, less, greater than, less than, biggest, inequality sign, ten frame

STRUCTURES AND REPRESENTATIONS

Ten frames to represent numbers, images of containers with a variety of capacities

RESOURCES

Optional: containers and water to model some of the comparisons, interlocking cubes

 In the eTextbook of this lesson, you will find interactive links to a selection of teaching tools.

Quick recap

Tell children that a bucket has a capacity of 10 cups. Show them pictures of other containers, such as a spoon, an egg cup, a spade, etc. Ask: *Will I need more or fewer than 10 spoons to fill the bucket? Is the capacity of the bucket in spoons greater or less than 10? How do you know?*

Unit 10: Introducing mass and capacity, Lesson 6

Discover

WAYS OF WORKING Pair work

ASK

- Question 1 a): *How many containers are there on the table?*
- Question 1 a): *The child is going to fill all three containers with water. Which one will need the most water? Which one will need the least water?*
- Question 1 a): *You are told how much water is needed, but the names of the containers are missing. Can you work out which is which?*

IN FOCUS In question 1 a), children are required to estimate the size of a range of containers from their visual appearance. They then use the resulting numbers to put the capacities in order in question 1 b).

ANSWERS

Question 1 a): C is filled by 2 cups. A filled by 15 cups. B is filled by 5 cups.

Question 1 b): The order from greatest to smallest capacity is: A, B, C.

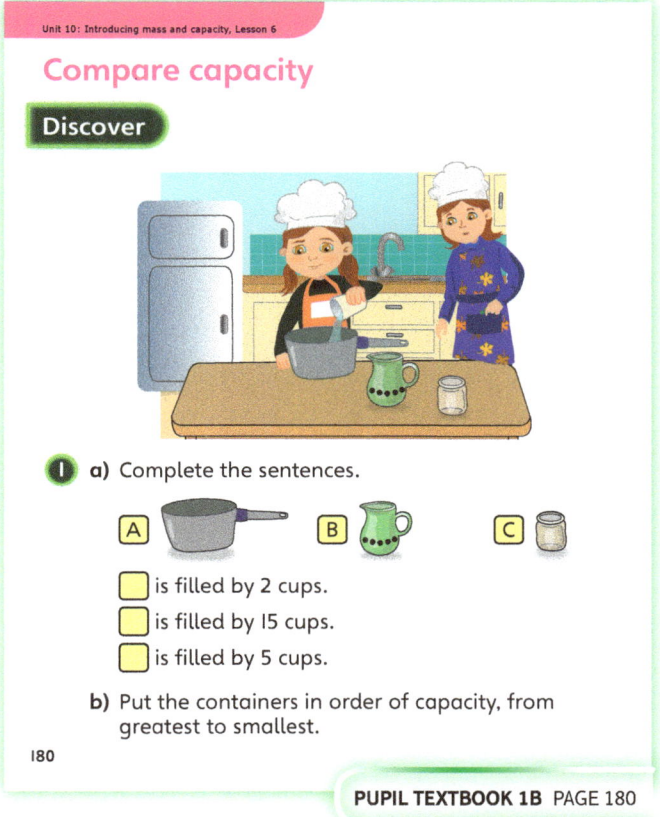

PUPIL TEXTBOOK 1B PAGE 180

Share

WAYS OF WORKING Whole class teacher led

ASK

- Question 1 a): *Do the ten frames make it easier to see which of the numbers is the biggest?*
- Question 1 a): *Look at the inequality signs that Dexter has used. How would you say this sentence?*

IN FOCUS The solution presented for question 1 a) involves first putting the numbers in order (using the ten frames as a visual aid) and then assigning the measured capacities to the appropriate containers. In question 1 b), a 'common sense' solution could have been used to put the containers in order of capacity, even without the measurements.

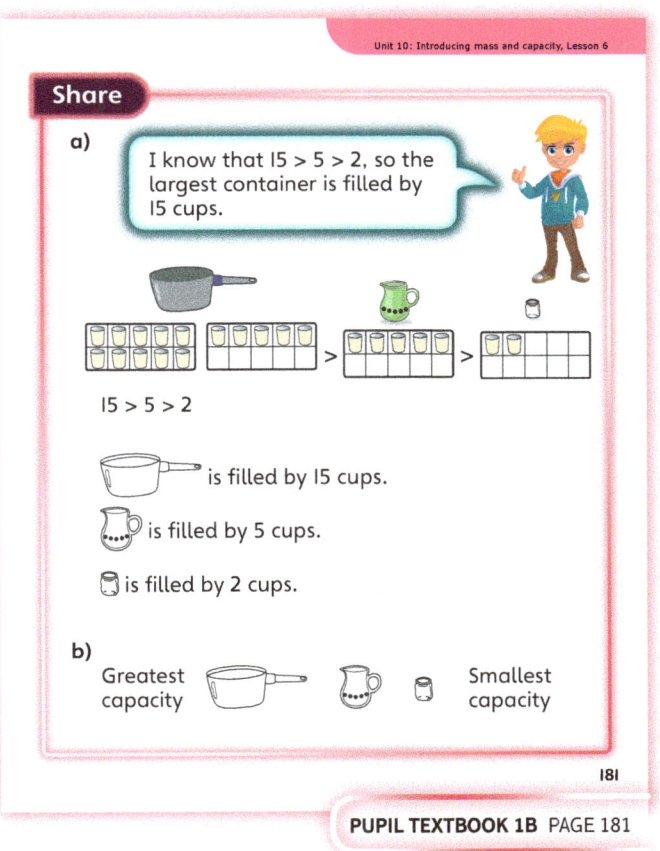

PUPIL TEXTBOOK 1B PAGE 181

Unit 10: Introducing mass and capacity, Lesson 6

Think together

WAYS OF WORKING Whole class teacher led (I do, We do, You do)

ASK
- Question ❶: *Which bucket is the biggest? Which one is the smallest? Will the biggest bucket hold more or fewer cups than the smallest bucket?*
- Question ❷: *Can you use the information in the table to put the containers in order?*

IN FOCUS Question ❸ provides an opportunity to reason about capacity in a way that goes beyond just 'counting cups'. Children need to notice that some water is left in jug B and understand that this means it has a greater capacity than jug C, despite the same number of filled glasses being shown.

STRENGTHEN If children find question ❶ difficult, encourage them to put each set of diagrams in order separately, and then to match the corresponding pictures.

DEEPEN Challenge children to think more deeply about variations on question ❸. Ask: *When does having some water left in the jug help you to answer the question?*

ASSESSMENT CHECKPOINT Use these questions to assess whether children are able to order capacities using non-standard units.

ANSWERS

Question ❶: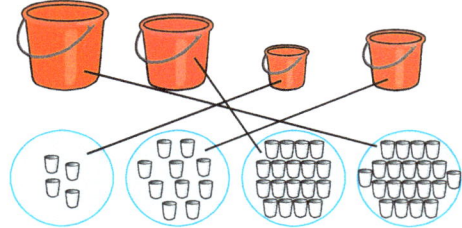

Question ❷: The order from greatest to smallest capacity is: spotted jug, plain jug, jar.

Question ❸: The order from smallest to greatest capacity is: A, C, B.

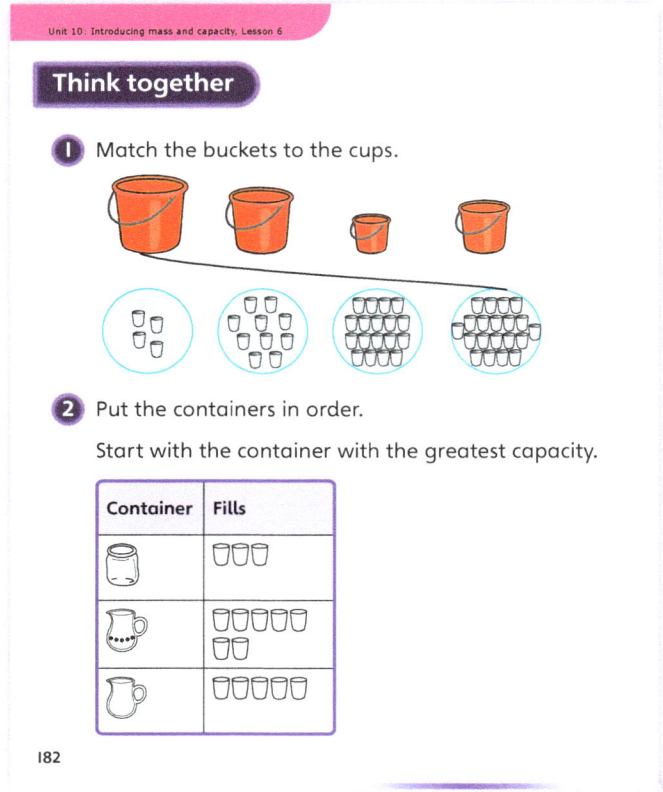

PUPIL TEXTBOOK 1B PAGE 182

PUPIL TEXTBOOK 1B PAGE 183

Unit 10: Introducing mass and capacity, Lesson 6

Practice

WAYS OF WORKING Independent thinking

IN FOCUS Question ❸ involves comparing the capacities of containers with capacities expressed in terms of multiples of a non-standard unit (in this case, a glass). Children will need to understand that the pictures are not to scale, and that they need to read the information provided in the question rather than relying on the relative sizes of the pictures.

STRENGTHEN If children find question ❸ difficult, you could try modelling the situation using interlocking cubes (use a '1' to represent each glass). Although this approach makes it easier to carry out the required comparison, the link to capacity becomes abstract; make sure that children return to a focus on capacity in question ❹.

DEEPEN Question ❹ requires some quite sophisticated reasoning about capacity. Explain: *The less water there is left in the jug, the more the container holds.* Children who cope well with this could be challenged to draw similar questions and swap them with a partner to answer. They should first write the answer secretly.

THINK DIFFERENTLY Question ❹ requires children to use reasoning to be able to order the capacities of the containers. They need to look at how much water is left in each jug to determine the capacity of the containers.

ASSESSMENT CHECKPOINT Use question ❶ to check that all children have a secure grasp of the key idea that larger containers have a greater capacity (measured in uniform but non-standard units).

ANSWERS Answers for the **Practice** part of the lesson can be found in the *Power Maths* online subscription.

Reflect

WAYS OF WORKING Pair work

IN FOCUS Make sure that children understand they are expected to describe the procedure that they would use to measure the capacity of each of the containers pictured, using suitable non-standard units of their own choice; they are not being asked to estimate or order the capacities.

ASSESSMENT CHECKPOINT Check that children specify a sensible unit (for example, a cup or small glass). Ensure the procedure they describe indicates that they are aware of the importance of the units being of uniform size and that the same unit of measure is used for each container.

ANSWERS Answers for the **Reflect** part of the lesson can be found in the *Power Maths* online subscription.

After the lesson ⏸

- Are children able to use everyday language like 'bigger' and 'smaller' in the context of capacity; for example, when placing a set of bottles in order?
- Do children understand the importance of choosing and using a suitable unit of measure to compare capacities?

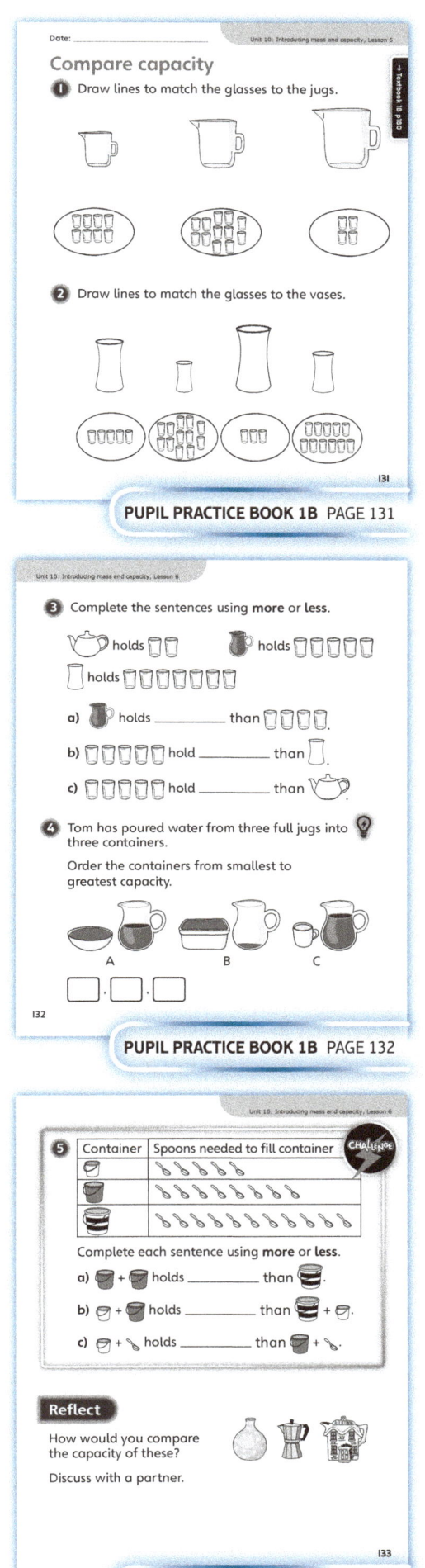

221

Unit 10: Introducing mass and capacity, Lesson 7

Solve word problems – mass and capacity

> **Learning focus**
>
> In this lesson, children use reasoning and understanding of number relationships to tackle a range of word problems involving mass and capacity.

> **Before you teach**
>
> - Do children have a basic understanding of applying addition or subtraction to word problems?
> - What resources and representations will be useful to support children's calculation strategies?

NATIONAL CURRICULUM LINKS

Year 1 Measurement

Compare, describe and solve practical problems for: capacity and volume (for example, full/empty, more than, less than, half, half full, quarter).

ASSESSING MASTERY

Children can use mathematical reasoning, addition and subtraction, and their understanding of mass and capacity to solve a variety of word problems.

COMMON MISCONCEPTIONS

Children may find it difficult to decide which operation to use to solve a word problem. The straightforward 'physical' nature of the problems in this lesson can be used to help children decide what is needed. Ask:

- *What are you doing in this question – adding something on or taking something away? So, which operation will you need – addition or subtraction?*

STRENGTHENING UNDERSTANDING

Access to a range of models, for example, counters, cubes or ten frames, can help children see what operation is required, and may help them to carry out the calculations.

GOING DEEPER

Children who cope well with these problems may enjoy making up some of their own – particularly some of the more challenging problems encountered later in the **Practice** exercise.

KEY LANGUAGE

In lesson: subtraction, weigh, weighs, weight, balance, represent, represented

Other language to be used by the teacher: addition, word problem, calculation, capacity, mass, method

STRUCTURES AND REPRESENTATIONS

Ten frames, part-whole model, cubes, sets of balance scales

RESOURCES

Optional: cubes, counters, blank ten frames, different sized containers, water

 In the eTextbook of this lesson, you will find interactive links to a selection of teaching tools.

> **Quick recap**
>
> Give each child an object. Ask them to find a partner whose object is lighter than theirs. Ask them to explain how they know. Ask them to check their answers using balance scales. Repeat with containers and finding a partner whose container has a greater capacity.

Unit 10: Introducing mass and capacity, Lesson 7

Discover

WAYS OF WORKING Pair work

ASK

- Question 1 a): *How many glasses of milk does the jug hold? How many glasses of milk does the pan hold? Does all the milk from the jug fit in the pan?*

IN FOCUS To answer question 1 a), ensure that children understand that the jug holds 10 glasses and the pan holds 6 glasses. Use the pictures to help children understand the process of first filling the jug with the milk from the glasses, and then pouring some of the milk out of the jug to fill the pan. You might want to explore how this relates to a part-whole model: the capacity of the jug is the whole (10 glasses) and the two parts are the volume of milk in the pan (6 glasses) and the remaining volume of milk in the jug (4 glasses).

PRACTICAL TIPS Use glasses of water and two different sized containers to recreate the scenario. Even if the number of glasses of water is different from the scenario, this helps children grasp the concept of a larger container having a greater capacity than a smaller container.

ANSWERS

Question 1 a) There are 4 glasses left in the jug.

Question 1 b) There will be 16 glasses in the bowl.

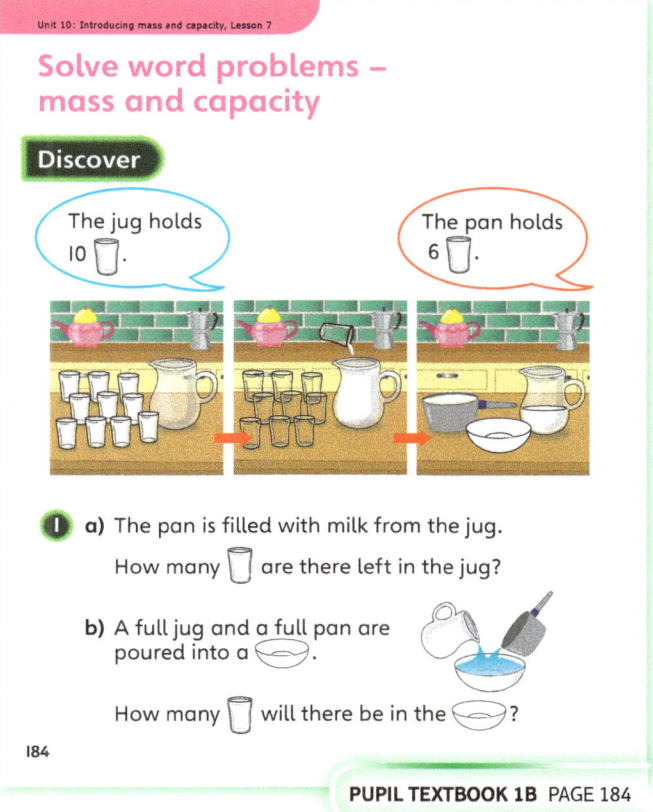

PUPIL TEXTBOOK 1B PAGE 184

Share

WAYS OF WORKING Whole class teacher led

ASK

- Question 1 a): *Astrid says that some of the milk has been poured out of the jug. How much has been poured out? How can you tell?*
- Question 1 a): *Flo says that we have to do a subtraction. Why?*
- Question 1 a): *How does this part-whole model help you to see how much is left in the pan?*

IN FOCUS Question 1 b) models the addition using two different versions of the ten frame – one using pictures of the glasses of milk that were contained in the jug and pan, and the other where the pictures are replaced with counters. Ensure children understand that the version with counters is easier to work with and will produce exactly the same answer.

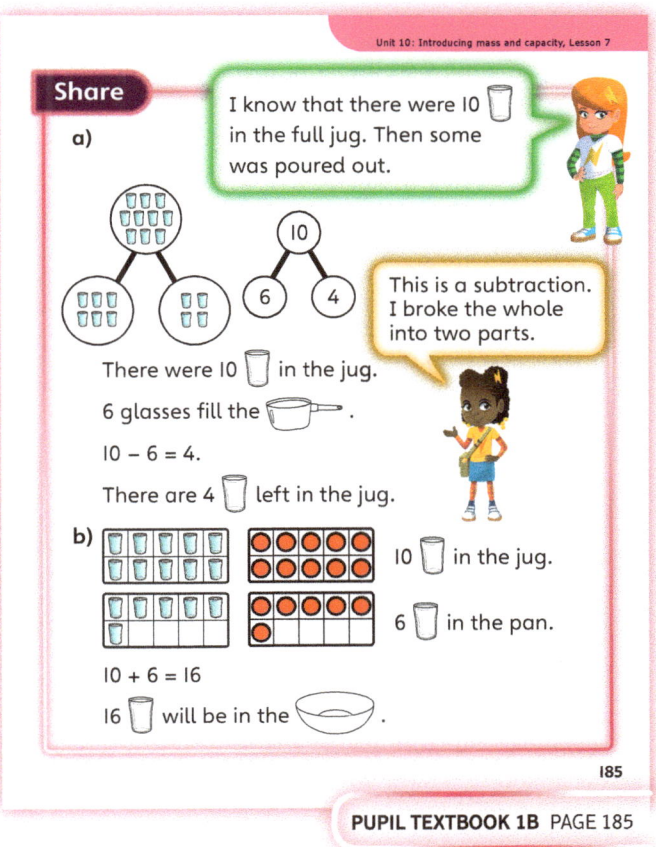

PUPIL TEXTBOOK 1B PAGE 185

Think together

WAYS OF WORKING Whole class teacher led (I do, We do, You do)

ASK
- Question ①: *Is this going to be an addition or a subtraction question? How can you tell?*
- Question ②: *What has changed between the first picture and the second one? How can that help us to find out how many cubes weigh the same as a small box?*

IN FOCUS Question ② can be tackled in a variety of ways – ask the class to explain the methods that they would use. Look out for any opportunities to build on the connections between addition and subtraction, for example, the first two diagrams can be thought of as showing 'equal subtraction' if they are read from left to right, or 'equal addition' if they are read from right to left.

STRENGTHEN If children find it difficult to understand that question ① requires an addition, try modelling the problem using cubes. Ask children to count out 8 cubes and then 6 cubes into the 'pan', and then take them out and count them (possibly using a ten frame).

DEEPEN Challenge children to make up their own variations of question ③. They could work in pairs, with each person making up a question and then passing it to their partner to solve. Being able to make up a workable problem is a good indication of children having a secure grasp of this topic.

ASSESSMENT CHECKPOINT Use these questions to assess whether children are able to solve addition and subtraction problems, in the context of mass and capacity.

ANSWERS

Question ①: 8 + 6 = 14
There are 14 cups of rice in the pan.

Question ②: 11 – 7 = 4
A small block weighs the same as 4 cubes.

Question ③ a): 3 × 2 = 6
6 sheep balance 1 horse.

Question ③ b): 3 – 1 = 2
2 cats balance 1 goose.

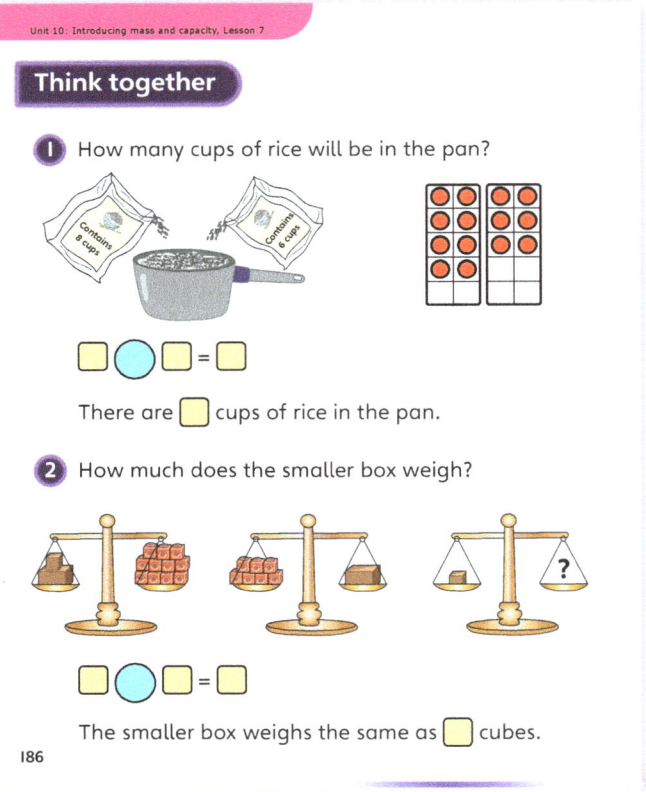

PUPIL TEXTBOOK 1B PAGE 186

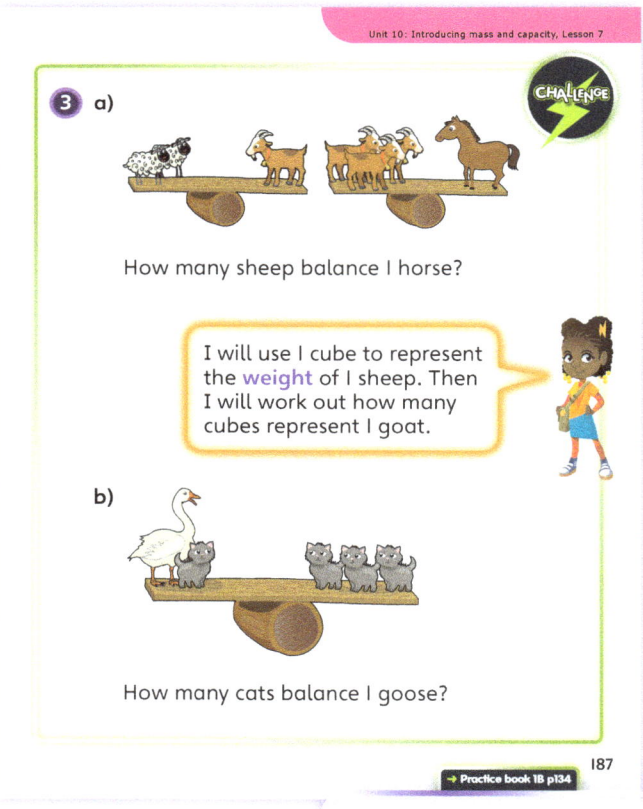

PUPIL TEXTBOOK 1B PAGE 187

Unit 10: Introducing mass and capacity, Lesson 7

Practice

WAYS OF WORKING Independent thinking

IN FOCUS Question ❷ involves a combination of addition and subtraction. Children will need to think flexibly to work out what calculations are needed in each case.

STRENGTHEN If children find the calculations in questions ❶ and ❷ difficult, suggest that they use a suitable model or apparatus to support them, for example a part-whole model, a ten frame, counters or cubes.

DEEPEN Ask children who complete the work confidently to explain their solutions to question ❸. Ask plenty of probing questions to encourage a full explanation. Ask: *Why did you choose that calculation? How would your answer be different if, instead, 4 circles balanced a triangle (for example)? How can you check your answer?*

ASSESSMENT CHECKPOINT Use question ❶ to check that children can understand simple addition and subtraction problems and carry out the required calculations accurately. It may be useful to ask some children to explain why question ❶ a) must be a subtraction problem and question ❶ b) must be an addition problem.

ANSWERS Answers for the **Practice** part of the lesson can be found in the *Power Maths* online subscription.

Reflect

WAYS OF WORKING Pair work

IN FOCUS How children feel when faced with difficult work is important. Building resilience when faced with challenges is an essential part of learning mathematics, and it is good to encourage children to keep trying, even when the work seems challenging.

ASSESSMENT CHECKPOINT Use the responses to this section to help you to analyse children's performance. Where children found the work difficult, was it because they did not understand what calculations were required or was it because they had difficulty carrying out the calculations? You will probably need to follow up children's responses with further questions if you want to form a more complete picture of the progress made by particular children.

ANSWERS Answers for the **Reflect** part of the lesson can be found in the *Power Maths* online subscription.

After the lesson ⏸

- Were children able to decide what calculations were needed in each of the word problems?
- Which calculation strategies were children using?
- How will you build children's resilience as they tackle more challenging work?

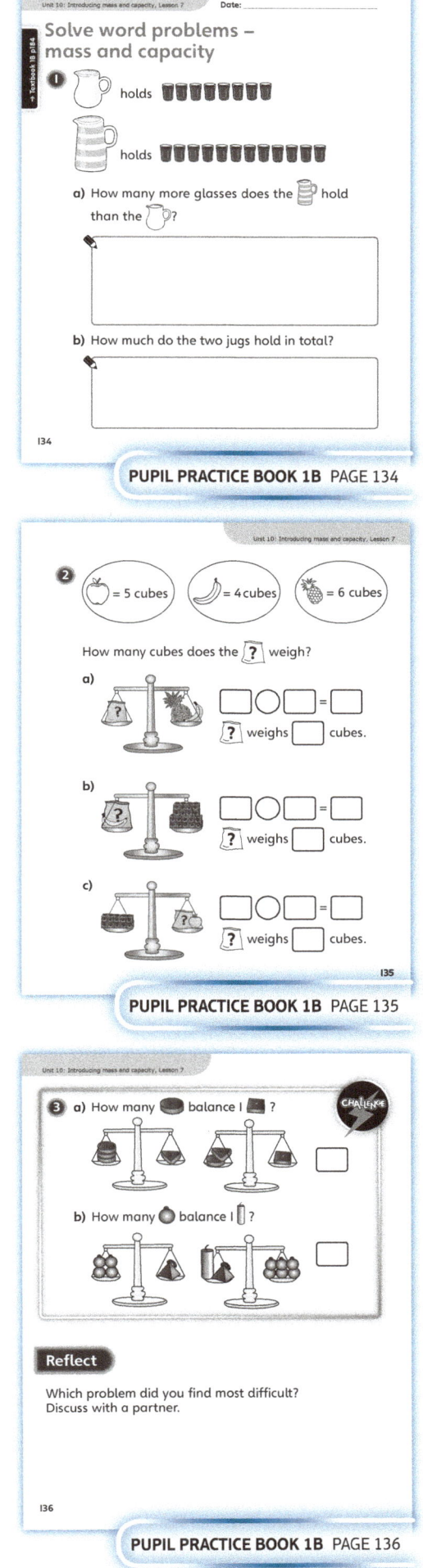

225

End of unit check

Unit 10: Introducing mass and capacity

Don't forget the unit assessment grid in your *Power Maths* online subscription.

WAYS OF WORKING Group work adult led

IN FOCUS This end of unit check focuses on using the key representations (balance scales, cups and jugs) to compare mass and capacity respectively.

- Questions 1 and 2 cover mass, testing children's understanding of how balance scales show the heavier object. Question 2 also assesses whether children recognise that the unit of measure must be the same size.
- Question 3 involves comparing capacity.
- Question 4 has a problem-solving aspect using cubes as a non-standard measure.

Think!

WAYS OF WORKING Pairs or small groups

IN FOCUS This activity provides an opportunity for children to think about the connections between the concepts of mass and capacity. A full cup of water is the uniform non-standard measure, so all the cups weigh the same.

ANSWERS AND COMMENTARY Children who demonstrate mastery of this concept will be able to use reasoning to work out mass and capacity in a problem-solving context. They will be confident in identifying and using non-standard measures, as well as demonstrating their understanding of how balance scales work.

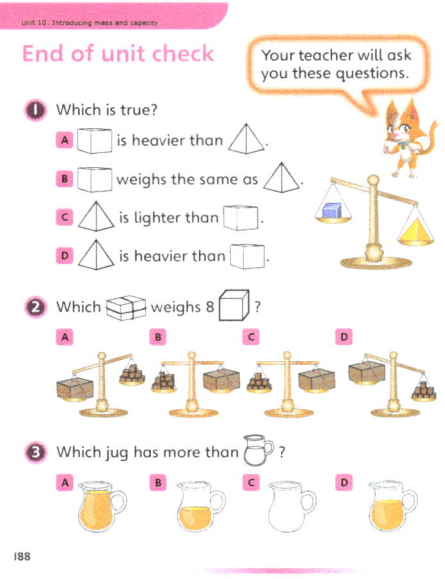

PUPIL TEXTBOOK 1B PAGE 188

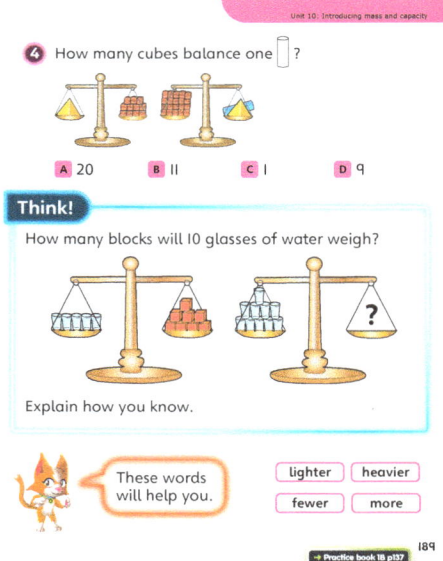

PUPIL TEXTBOOK 1B PAGE 189

Q	A	WRONG ANSWERS AND MISCONCEPTIONS	STRENGTHENING UNDERSTANDING
1	D	Other answers indicate that either children do not understand how a balance scale shows the heavier object, or that the vocabulary is confusing them.	These questions require practical experience of measurement, as well as reasoning, calculation and number skills, and an understanding of key vocabulary and language patterns. Where children find this work difficult, try to work out where the problems arise. For example: • Have they had sufficient practical experience to understand how a set of balance scales work? • Do they understand the terms 'heavier than'/'lighter than'? • Do they understand the importance of uniform units? • Can they use addition and subtraction in the context of mass and capacity?
2	C	Answer B indicates that children have not understood the importance of using uniform units of measurement. Answers A and D suggest that children have not checked that the scales are balanced.	
3	A	Answer D shows a jug with the same amount as the example shown; this answer and answer B suggests that children have not checked the level of liquid sufficiently carefully.	
4	D	Answer A suggests that children have simply counted the cubes, rather than using reasoning and calculation. B and C could indicate a miscalculation.	

Unit 10: Introducing mass and capacity

My journal

WAYS OF WORKING Independent work

ANSWERS AND COMMENTARY

10 cups of water would weigh 16 blocks.
- Proportional reasoning is needed in this activity. Children will need to reason that, as 10 cups is double 5 cups, they will also need to double the number of blocks to find the mass.
- An answer of 13 blocks indicates that children have simply added the same number of blocks as extra cups (5).

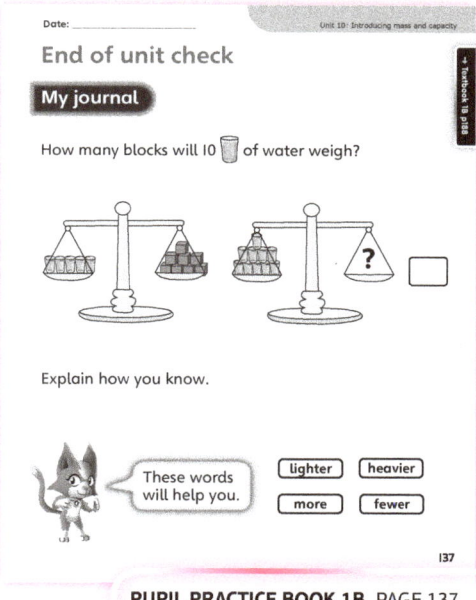

PUPIL PRACTICE BOOK 1B PAGE 137

Power check

WAYS OF WORKING Independent work

ASK
- What were the easiest parts of this unit of work? Which parts did you find most difficult?
- What made the harder parts difficult – was it the measurement, the calculating or something else?

Power play

WAYS OF WORKING Pair work or small groups

IN FOCUS This activity provides an opportunity to revisit comparing masses. You could organise this activity in pairs or small groups. Ask children to make some of the letters of their names from interlocking cubes (there is no need to make all of them as some letters are quite difficult). Ask children to write the letters in order of mass – they should be able to do this by counting cubes, but do not prompt them. Finally, ask children to check their suggested order using a set of balance scales, if they have not already done so.

ANSWERS AND COMMENTARY Allow children to carry out this task as independently as possible – their responses should then provide a good insight into some of their learning in this unit. Look carefully at how children are working. Do they use a count of the cubes to determine the mass of each letter and then check with the balance scales? Or do they start by comparing each pair of letters?

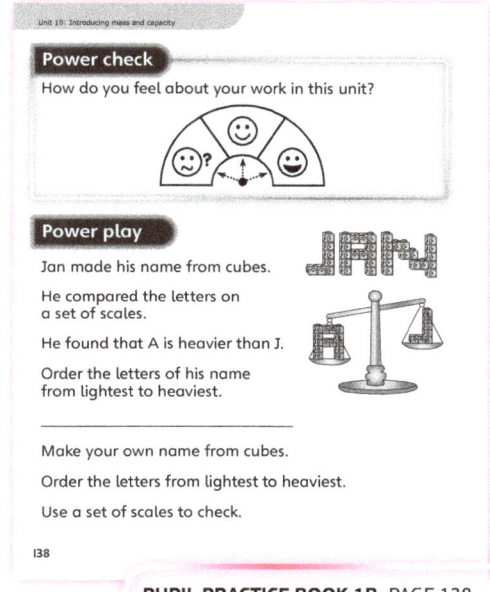

PUPIL PRACTICE BOOK 1B PAGE 138

After the unit

- Some of the comparison and matching exercises in this unit involved quite complicated reasoning – children were effectively being asked to solve simple calculations. How confident were children with this kind of reasoning?
- Which concept was it easier for children to grasp – mass or capacity?

Strengthen and **Deepen** activities for this unit can be found in the *Power Maths* online subscription.

Published by Pearson Education Limited, 80 Strand, London, WC2R 0RL.

www.pearsonschools.co.uk

Text © Pearson Education Limited 2018, 2023
Edited by Pearson and Florence Production Ltd
First edition edited by Pearson, Little Grey Cells Publishing Services and Haremi Ltd
Designed and typeset by Pearson and PDQ Digital Media Solutions Ltd
First edition designed and typeset by Kamae Design
Original illustrations © Pearson Education Limited 2018, 2023
Illustrated by Laura Aries, Phil Corbett, Nigel Dobbyn, Adam Linley and Nadene Naude at Beehive Illustration; Emily Skinner at Graham-Cameron Illustration; Kamae Design and Florence Production Ltd
Cover design by Pearson Education Ltd
Back cover illustration © Will Overton at Advocate Art and Nadene Naude at Beehive Illustration

Series editor: Tony Staneff; Lead author: Josh Lury
Authors (first edition): Tony Staneff, David Board, Julia Hayes, Tim Handley, Derek Huby, Neil Jarrett and Josh Lury
Consultants (first edition): Professor Liu Jian

The rights of Tony Staneff and Josh Lury to be identified as authors of this work have been asserted by them in accordance with the Copyright, Designs and Patents Act 1988.

This publication is protected by copyright, and permission should be obtained from the publisher prior to any prohibited reproduction, storage in a retrieval system, or transmission in any form or by any means, electronic, mechanical, photocopying, recording, or otherwise. For information regarding permissions, request forms and the appropriate contacts, please visit https://www.pearson.com/us/contact-us/permissions.html Pearson Education Limited Rights and Permissions Department.

First published 2018
This edition first published 2023

27 26 25 24 23
10 9 8 7 6 5 4 3 2 1

British Library Cataloguing in Publication Data
A catalogue record for this book is available from the British Library

ISBN 978 1 292 45048 3

Copyright notice
All rights reserved. No part of this publication may be reproduced in any form or by any means (including photocopying or storing it in any medium by electronic means and whether or not transiently or incidentally to some other use of this publication) without the written permission of the copyright owner, except in accordance with the provisions of the Copyright, Designs and Patents Act 1988 or under the terms of a licence issued by the Copyright Licensing Agency, Barnards Inn, 86 Fetter Lane, London EC4A 1EN (http://www.cla.co.uk). Applications for the copyright owner's written permission should be addressed to the publisher.

Printed in the UK by Ashford Press Ltd

For Power Maths online resources, go to:
www.activelearnprimary.co.uk

Note from the publisher
Pearson has robust editorial processes, including answer and fact checks, to ensure the accuracy of the content in this publication, and every effort is made to ensure this publication is free of errors. We are, however, only human, and occasionally errors do occur. Pearson is not liable for any misunderstandings that arise as a result of errors in this publication, but it is our priority to ensure that the content is accurate. If you spot an error, please do contact us at resourcescorrections@pearson.com so we can make sure it is corrected.